Current Topics in Microbiology 188 and Immunology

Simian Immunodeficiency Virus

Edited by N. L. Letvin and R. C. Desrosiers

With 37 Figures

Springer-Verlag
Berlin Heidelberg New York
London Paris Tokyo
Hong Kong Barcelona
Budapest

Norman L. Letvin, M.D.
Ronald C. Desrosiers, Ph.D.

New England Regional
Primate Research Center
Harvard Medical School
One Pine Hill Drive
Southborough, MA 01772-9102
USA

Cover illustration: Rhesus monkey (Macaca mulatta).
Picture by courtesy of Lyna M. Watson.

Cover design: Harald Lopka, Ilvesheim

ISSN 0070-217X
ISBN-13: 978-3-642-78538-2 e-ISBN-13: 978-3-642-78536-8
DOI: 10.1007/978-3-642-78536-8

Library of Congress Catalog Card Number 15-12910
Softcover reprint of the hardcover 1st edition 1994

Typesetting: Thomson Press (India) Ltd., New Delhi
SPIN 10096443 27/3020-5 4 3 2 1 0 – Printed on acid-free paper.

Foreword

Human immunodeficiency virus (HIV) infection and the resultant manifestation of AIDS have evolved into a national and global epidemic. The molecular anatomy of the HIV-1 genome and its replicative and transcriptive strategies are well documented, but the pathogenesis of HIV infection is poorly understood. Further, much of our data on HIV-1 replication and transcription strategies comes from tissue culture experiments. It is unlikely for it to be possible for us to understand HIV replication and its control and to successfully treat and prevent HIV without a clear understanding of the pathogenesis, i.e., how HIV causes disease. Such events will require studies in vivo to complement the in vitro observations.

Tissue injury and disease accompanying viral infection usually proceeds by one of two mechanisms. In the first, the virus directly injures cells through its ability to interfere with host protein synthesis or disrupt plasma or lysosomal membranes. An important variation on this theme is the capacity of viruses is establish a long-term presence within a cell where they can have a subtle and persistent effect such as altering the specialized function of the cell, thus interfering with its physiology. The second mechanism is that viruses, as a combination of antigens, elicit a host immune response against viral and host proteins. The components of this immune response can cause injury by combining with free-virus or viral proteins in the fluid phase. The resulting immune complexes can then deposit in arteries of selected tissues and yield immune complex mediated disease. Alternatively, or concomitantly, antibodies or cytotoxic lymphocytes generated during the antiviral immune response may react with and injure infected cells expressing viral proteins or peptides. Further, these cytotoxic lymphocytes and macrophages can release cytokines which, by themselves, can profoundly affect cells, activate the virus, and directly or indirectly cause tissue injury. Such immunopathologic responses against the virus or self-determinants (autoimmune) reflect the normal mechanism(s) by which the host recognizes an infectious agent as foreign and strives to remove it and the cells it infects in order to control and eventually exterminate the unwanted microbe. Understanding, controlling, and manipulating the immune response to both prevent viral infection from occurring and clear virally infected cells after infection are essential requirements for designing preventive vaccines or administering immunocytotherapy.

To understand the mechanisms by which HIV causes disease in an intact host, as well as the host's immune response against HIV, the use of a model system in which some or all these factors can be manipulated is important. Studies of SIV infected macaques has provided insights for identifying, in vivo, virulent genes of SIV and dissection of the immune response against this retroviral infection. Two of the major contributors in this line of research, Ron Desrosier, who established infectious clones of SIV, utilized them to map viral tropism, virulence, and attenuation, and Norman Letvin, who helped establish the cytotoxic T lymphocyte (CTL) system and began mapping the macaque's MHC glycoproteins, serve as guest editors of this volume on *Simian Immunodeficiency Virus*. They have invited several of the major contributors working in this area of research to provide not only up-to-date data as to what has occurred in evaluating the SIV model but also forecasting both what the problems and future directions will be. It is clear from the contributions in this book that important advances have been made in understanding the pathogenesis of SIV and that these may well help define some of the currently poorly understood issues with HIV. While it is also evident that a host of questions are still unanswered, it appears that the utility of the SIV animal model and others, as well as an intradisciplinary approach, will hopefully lead to knowledge that will allow a better understanding of designing effective therapy to prevent HIV.

MICHAEL B. A. OLDSTONE, M. D.
La Jolla, California

Contents

Contributors

(Their addresses can be found at the beginning of their respective chapters.)

Survey of Simian Immunodeficiency Virus Among Nonhuman Primate Populations

M. HAYAMI, E. IDO, and T. MIURA

1 Introduction

Human immunodeficiency virus type 1 (HIV-1) was found as an etiological agent of AIDS in 1983 (as early as 2 years after of the first documentation of the illness). Thereafter, another type, HIV-2, which is also capable of causing AIDS, was isolated in 1986 from West African patients (CLAVEL et al. 1986). Advanced molecular biological techniques promptly enabled us to reveal the way of replication of the viruses and their complete genomic sequences in a very short period. Nowadays, the genomical and structural features of these human exogenous retroviruses are rather well understood, and both HIVs are classified in the lentivirus subfamily.

Simian retrovirus similar to HIV was first isolated from nonhuman primates in 1985 (KANKI et al. 1985; DANIEL et al. 1985). The simian immunodeficiency virus (SIV) has many common characteristics with HIV such as Mg^{2+}-dependent reverse transcriptase activity, tropism for $CD4^+$ cells with cytopathic effects, and a virion morphology with a cylindrical nucleoid. This similarity of SIV with HIV indicates a close relationship

Research Center for Immunodeficiency Virus, Institute for Virus Research, Kyoto University, 53 Shogoinkawaharamachi, Sakyo-ku, Kyoto 606, Japan

Current Topics in Microbiology and Immunology, Vol. 188

between these two viral groups. More precisely, HIV and SIV belong to a large group of primate lentiviruses. Therefore studies on SIV should not be described without referring to those on HIV. Experimental infection of SIV using monkeys provides an animal model in AIDS research. Analysis of pathogenic and nonpathogenic SIV strains would give us a clue to understand the nature of the AIDS pathology. Furthermore, analysis of genetical relationships among the HIV/SIV groups would allow us an insight into their evolutionary origin(s). In a word, any findings and knowledges obtained from the SIV research may have more significance than we think especially when we tackle the world-threatening illness. In this chapter we describe how SIVs were recognized, together with the results of serosurveys among nonhuman primates. We also describe individual SIV isolates from a variety of monkey sources by us and other research groups. A brief introduction to the antigenic and genetic relationships among them may provide a general review of this field.

2 A Prelude to the Discovery of SIV: ATL, HTLV-1, and STLV-1

It has often been the case that a very similar but distinct micro-organism (or virus) was newly isolated from a different host after establishment of the etiological correlation between a certain disease and its pathogen. Before the story of SIV, let us begin with such an example which occurred a decade ago and probably paved a road to the discovery of SIV.

Adult T-cell leukemia (ATL), originally known to be endemic in southwestern part of Japan, was first described by Japanese scientists (UCHIYAMA et al. 1977). It was speculated that this disease was caused by a viral infection (HINUMA et al. 1981), and a retrovirus was isolated from cell lines derived from ATL patients (YOSHIDA et al. 1982) and sequenced (SEIKI et al. 1983). On the other side of the world, the first human retrovirus, human T-lymphotrophic virus (HTLV-1), was discovered from T-cell lymphoblastoid cell lines derived from a cutaneous lymphoma (mycosis fungoides); (POIESZ et al. 1980). Before long this virus was identified to be the causative agent of ATL (WATANABE et al. 1983, 1984).

A surprise then took place: Japanese monkeys (*Macaca fuscata*) were found to possess antibodies to HTLV-1 (MIYOSHI et al. 1982a, b). Subsequently the prevalence of antibody to the virus in Japanese monkeys and other nonhuman primates was investigated extensively by the author's group (HAYAMI et al. 1983a, b, 1984; ISHIKAWA et al. 1987) and others (HUNSMANN et al. 1983; MIYOSHI et al. 1983). Thus we came to recognize simian T-lymphotropic virus type 1 (STLV-1). It is now known that STLV-1 is prevalent only among Old World monkeys and apes, which are evolutionarily thought to be rather close to human, but not among New World monkeys (LOWENSTINE et al. 1986; ISHIKAWA et al. 1987). It is also known that this virus can cause ATL-like disease (NODA et al. 1986; TSUJIMOTO et al. 1987). Biological properties of STLV-1 resemble those of HTLV-1. The genetic and evolutionary relationship between these two viral groups as well as their geographical distributions are currently providing another attractive research field.

3 AIDS, HIV, and SIV

A new enigmatic disease with a breakdown of active immune system was first documented in the United States in 1981. It was soon found that the disease, now known as AIDS, was very quickly spreading not only over the country but throughout the world. This unforeseen explosion in the number of AIDS patients together with its high lethality will long be remembered in human history. Attempts were immediately made to look for a pathogen for this infectious illness, and its causative agent, HIV-1, was isolated in as early as 1983.

Soon after the appearance of AIDS, a naturally occurring immunodeficiency syndrome was reported in a captive colony of macaque monkeys (LETVIN et al. 1983). Naturally, efforts were also made to search for the causative simian retrovirus. First, a D-type virus was isolated (DANIEL et al. 1984), but its character was different from that of HIV-1 because it had a B-cell tropic nature. At last, HIV-1-like T-cell tropic virus (originally named STLV-III because HIV-1 was called HTLV-III, lymphadenopathy-associated virus (LAV) or AIDS-associated retrovirus (ARV) at that time) was isolated from rhesus macaques (KANKI et al. 1985; DANIEL et al. 1985). In addition, this new virus was soon demonstrated to be responsible for induction of the illness in the same host (LETVIN et al. 1985). Following the change in nomenclature from HTLV-III/LAV/ARV to HIV-I, STLV-III was named simian immunodeficiency virus (SIV). Surprisingly, SIV from rhesus macaques (SIV_{mac}) is antigenically and genomically very similar to HIV-2, the second human AIDS virus.

4 Seroprevalences of SIV of Various Nonhuman Primate Species

Isolation of SIV_{mac} was achieved from the rhesus macaque which showed seropositivity to HIV-1 (KANKI et al. 1985; DANIEL et al. 1985). Immunogenical analysis of SIV_{mac} revealed cross-reactivity between core proteins of

SIV_{mac} and corresponding gag proteins of HIV-1 including p24 but minimal reactivity with glycosylated env proteins of high molecular weights (KANKI et al. 1985). Since the glycoproteins of HIV-1 are the most immunogenic antigens in exposed humans, this observation indicates that the isolated SIV_{mac} is distinct from HIV-1. Likewise, new SIV isolates have been reported in most cases by finding an immunological cross-reactivity between the known virus and the tested sera. To perform serosurveys of SIV/HIV, it therefore, is important to know to what extent the viruses are related immunogenically. Nonetheless, few studies have so far been performed to clarify the comparative antigenicities. OHTA et al. (1988) reported isolation of SIV_{agm} from African green monkeys (AGM; *Cercopithecus aethiops*). The AGM sera having antibodies to SIV_{agm} reacted strongly with the glycoproteins gp32 and gp120 of SIV_{agm}, but sera of human AIDS patients showed only faint reactivity with gp120 of SIV_{agm}. In contrast, only one of nine AGM sera tested reacted with gp41 of HIV-1; the other AGM sera showed no reactivity with env proteins of HIV-1. However, when sera from various nonhuman primates were screened using SIV_{agm} as antigen, cross-reactive antibodies were detected in some of them, as described below. SCHNEIDER et al. (1987) examined the serological relationship among HIV-1 and three SIVs isolated from a rhesus monkey (*Macaca mulatta*), an African green monkey (*C. aethiops*) and a sooty mangabey (a *Cercocebus atys*). The finding was that, in addition to bidirectional cross-reaction between core polypeptides of the four viruses, all sera from three monkey species reacted with the major env glycoprotein of all SIVs but not with the analogous gp 120 of HIV-1. KRAUS et al. (1989) thereafter examined reactivities of primate sera against different lentiviruses including HIV-2 and found that the seropositive AGM sera were clearly reactive with the env proteins of HIV-2 and SIV_{mac} but not those of HIV-1. Based on subsequent studies it is now widely accepted that the antigenicity of SIV_{mac} almost coincides with that of HIV-2 (This similarity was verified by a remarkable sequence homology between two viruses; CHAKRABARTI et al. 1987.) It should be noted that SIV_{cpz} isolated from chimpanzees is antigenically very close to HIV-1 (PEETERS et al. 1989; 1992). The close relationship of SIV_{cpz} with HIV-1 was also revealed by their genomic analyses (HUET et al. 1990). For further detailed information of the antigenicities of other SIVs, one is referred to papers which describe the respective isolates (references in Table 2).

An extensive serosurvey concerning SIV among various nonhuman primates was first conducted by our group (HAYAMI et al. 1985) by immunofluorescent assay using HIV-1 as antigen. Serum samples from 1875 primates of 49 species were tested for the presence of antibodies. The result was that only 19 Old World monkeys showed seropositivity. Then one (AGM) of them was confirmed positive by western blotting assay. Thus isolation of SIV_{agm} was successful (OHTA et al. 1988). Subsequently, using the cells infected with this SIV_{agm}, more than 1700 monkey sera of 30 species were tested for reactivity (OHTA et al. 1988; TOMONAGA et al. 1993).

The seroprevalences among various nonhuman primate species, to the best of our knowledge, are summarized in Table 1. The data in Table 1 are compiled from the authors' surveys and those by LOWENSTINE et al. (1986), DANIEL et al. (1988a), and others. First, in our serosurvey specific antibodies against SIV_{agm} were found in 1/11 (9%) of de Brazza's monkeys (*Cercopithecus neglectus*), and 3/30 (10%) of mandrills (*Papio sphinx*) in addition to AGMs whose sera were positive in 90/343 (26%). The results of other species were almost all negative except in white-crowned mangabey monkeys (*Cercocebus torquatus lunulatus*) and anubis baboons (*Papio anubis*). [It should be noted that the seropositive results of mandrills and white-crowned mangabeys also led to isolation of SIV_{mnd} (TSUJIMOTO et al. 1989), another unique virus, and SIV_{wcm} (TOMONAGA et al. 1993), a close relative of SIV_{agm}, as described below.] Our result of seropositivity to SIV_{agm} in Sykes' monkeys (*C. mitis*) was rather high 23/73 (32%). Yet another group has recently isolated a new virus SIV_{syk} from this species (EMAU et al. 1991). A high seroprevalence (over 30%) of SIV_{agm} among AGMs was also reported by KRAUS et al. (1989). SIV_{agm} is thought to be non-pathogenic because neither AIDS-like signs nor immunosuppression were observed in naturally or experimentally infected monkeys (OHTA et al. 1988; MIURA et al. 1989; HONJO et al. 1990). Probably the virus has been modestly circulating among this species and has yielded a relatively high seroprevalence as a result.

It is noteworthy that, in contrast to the high seroprevalence of SIV among monkeys of African origin, detection of SIV antibodies in a captive colony of macaque monkeys in the United States was very low (3/848, 0.35%; DANIEL et al. 1988a). Of three positive macaques two were rhesus monkeys (*M. mulatta*), and one was a cynomolgus monkey (*M. fascicularis*). Although this research group examined the serum specimens taken from 160 Caribbean AGMs all were SIV seronegative whereas approximately 30% of the 50 sera from eastern AGMs were SIV seropositive. A similar result of seroprevalence that antibodies to SIV_{agm} were detected in AGMs but not in Caribbean AGMs was reported by HENDRY et al. (1986). The latter AGMs were imported from Africa to the New World as early as the sixteenth century. This provides information that may give a clue to the evolution of SIV.

On the other hand, evident seropositivity of SIV in sooty mangabeys (*Cercocebus atys*) housed in regional primate centers in the United States (79/138: FULTZ et al. 1990; 13/44: MURPHEY-CORB et al. 1986) is particularly of interest. Sooty mangabeys are naturally inhabiting in West Africa. At present, no other reports have described the presence of SIVs in nonhuman primate species inhabiting Asia or the New World in nature. Thus, detection of SIV in macaques in primate centers in United States [rhesus macaque and cynomolgus monkey; pig-tailed macaque (*M. nemestrina*) by BENVENISTE et al. (1986); stump-tailed macaque (*M. arctoides*) by GARDNER and LUCIW (1988)] is considered as a result of artificial transmission in the facilities, probably from sooty mangabeys (NOVEMBRE et al. 1992).

Table 1. Seroprevalences of SIV in various nonhuman primates

Species	Number positive/number tested			
	Hayami[a]	Lowenstine[b]	Daniel[c]	Others
New World monkeys				
Saimiri sciurea	0/4			0/15[e]
Ateles geoffroyi	0/1			
Callithrix jacchus	0/10			
Callithrix argentata	0/1			
Saguinus oedipus	0/2			
Saguinus mystax	0/11			
Saguinus labiatus	0/10			
Old World monkeys				
Cercopithecus aethiops	90/343	61(0)/116		91/242[e]
Cercopithecus aethiops	11/31*			
Cercopithecus sabaeus				4/10[f] 131/325[g]
Cercopithecus tantalus				30/80[g]
Cercopithecus cephus	0/31	4(0)/5		
Cercopithecus neglectus	1/2	2(0)/4		
Cercopithecus neglectus	0/9*			
Cercopithecus alboglaris kolbi		1(2)/11		
Cercopithecus mitis[d]	13/24*			
Cercopithecus mitis[d]	9/35*			
Cercopithecus mitis[d]	1/14*			
Cercopithecus mitis stuhlmanni		0/2		
Cercopithecus diana		2(0)/8		
Cercopithecus hamlyni		0/3		
Cercopithecus nictitans		1(0)/7		
Cercopithecus mona		0/1		
Cercopithecus campbelli lowei		0/5		
Allenopithecus nigroviridis		1(0)/8		
Miopithecus talapoin		9(11)/15		
Papio anubis	0/109			
Papio anubis	8/83*			
Papio hamadryas	0/107			
Papio hamadryas cynocephalus	0/27*			2/124[h]
Papio mandrillus sphinx	3/30	11(4)/23		
Papio mandrillus leucophaeus		0/3		
Papio papio		1(0)/10		
Papio ursinus				0/96[e]
Theropithecus gelada	0/2	0/1		
Macaca assamensis	0/2			
Macaca mulatta	0/99	0/1	2/439	0/24[e]
Macaca cyclopis	0/16		0/48	
Macaca arctoides	0/13		0/28	0/12[e]
Macaca nemestrina	0/14	0/6		0/8[e]
Macaca radiata	0/30	0/1		
Macaca sinica	0/65			
Macaca fascicularis	0/427	4(0)/23	1/333	33/112[e]
Macaca maura	0/3	0(0)/5		
Macaca fuscata	0/108			
Macaca silensus		17(0)/26		
Macaca nigra		0(0)/12		
Macaca tonkeana		1(0)/21		
Macaca sylvanus		0/12		

Table 1. (*Continued*)

Species	Number positive/number tested			
	Hayami[a]	Lowenstine[b]	Daniel[c]	Others
Cercocebus atys		2(4)/7		79/138[i] 13/44[j] 2/25[k]
Cercocebus torquatus torquatus		0/10		
Cercocebus torquatus lunulatus	6/9*	0/1		
Cercocebus albigena		0/2		
Erythrocebus patas		0/2		
Colobus angolensis	0/4*			
Colobus guereza		1(0)/14		
Colobus guereza kukuyensis		1(2)/24		
Presbytis entelles		0(0)/6		
Presbytis obscurus		0/15		
Presbytis cristatus		0/12		
Presbytis françoisi		0/7		
Presbytis senex		0/2		
Pygathrix namaeus		0/5		
Nasalis larvatus		0/3		
Rhinopithecus roxellanae		0/2		
Apes				
Hylobates lar	0/16	0/7		
Hylobates concolor		1(0)/11		
Hylobates agilis		0/1		
Hylobates agilis x hoolock		0/2		
Hylobates sp.		0/1		
Symphalangus syndactylus		0/10		
Pongo pygmaeus	0/5	2(0)/19		
Pan troglodytes troglodytes	0/64	0(0)/2		2/50[l] 1/44[m]
Pan paniscus		0(0)/6		
Gorilla gorilla	0/11	0/26		

* Data from Tamonaga et al. 1993 (see footnote a)

[a] The results of Hayami's laboratory were compiled from two independent surveys: OHTA et al. (1988) and TOMONAGA et al. (1993) (marked by asterisks); judged by IFA using SIV_{agm} as antigen.

[b] LOWENSTINE et al. (1986); ELISA (western blot) for HIV-1.

[c] DANIEL et al. (1988a); ELISA for SIV_{mac} confirmed by radioimmunoprecipitation assay.

[d] *Cercopithecus mitis* was separated into three subspecies based on the wild habitat.

[e] KRAUS et al. (1989); judged by presence of *env* bands on western blotting using SIV_{agm} as antigen.

[f] ALLAN et al. (1991); western blotting using SIV_{agm} as antigen.

[g] MÜLLER et al. (1993); ELISA for HIV-1 and HIV-2.

[h] KODAMA et al. (1989); ELISA using SIV_{agm} as antigen.

[i] FULTZ et al. (1990); IFA using SIV_{smm} as antigen confirmed by radioimmuno-precipitation assay.

[j] MURPHEY-CORB et al. (1986); ELISA using HIV-1 as antigen.

[k] MARX et al. (1991); western blotting using SIV_{mac} as antigen.

[l] PEETERS et al. (1989); ELISA for HIV-1 and HIV-2.

[m] PEETERS et al. (1992); ELISA and western blotting for HIV-1 and HIV-2.

Although our survey in Table 1 showed no seropositivity in chimpanzees, PEETERS et al. (1989, 1992) have reported positive cases (3/94) and isolated SIV_{cpz}. At this moment, whether chimpanzees naturally possess this virus or not is an unsettled question because of low positivity. In any case, a unique distribution of SIVs in nonhuman primates contrasts with that of STLV-1, which is spreading over various species in Africa and Asia (but not in the New World). This difference may have something to do with the evolutionary origin of SIV and awaits further investigation.

5 Isolation of SIV from Various Nonhuman Primates

As shown in Table 1, it is now known that at least ten additional primate species harbor related lentiviruses on the basis of serological cross reactivity with either SIV or HIV. In fact, a barrage of reports describing isolation of new SIVs have appeared in recent years from various nonhuman primate sources. Representative SIV isolates and common and species names of their sources are summarized in Table 2.

Among the many SIV isolates it should be noted that SIV_{cpz} isolated from chimpanzees (PEETERS et al. 1989; 1992) is closely related to HIV-1 antigenically and genomically (HUET et al. 1990). In a similar manner, some of the other SIVs/HIVs belong to the same group. It is especially clear that SIV_{mac}, SIV_{mne}, SIV_{smm}, SIV_{stm}, and various HIV-2 strains are closely related to each other and constitute one large group. Several reports have described the isolation of a different SIV_{agm}, constituting another divergent group. In addition, SIV_{mnd} isolated from mandrills (Tsujimoto et al. 1988) and SIV_{syk} from Sykes' monkeys (EMAU et al. 1991) were found to be positioned very distantly from any other primate lentiviruses (TSUJIMOTO et al. 1989; HIRSCH et al. 1993). Characterization of these five SIV/HIV groups and genetic relationships among them are described below.

6 Genomic Structures of SIV in Comparison with HIV

A body of sequence data on the SIV full or partial genomes has been reported to date: SIV_{mac}, CHAKRABARTI et al. (1987), KESTLER et al. (1988), REGIER and DESROSIERS (1990); SIV_{smm}, HIRSCH et al. (1989), DEWHURST et al. (1990), MARX et al. (1991); SIV_{mne}, HENDERSON et al. (1988); SIV_{stm}, NOVEMBRE et al. (1992); SIV_{agm}, FUKASAWA et al. (1988), DANIEL et al. (1988b), BAIER et al. (1990), ALLAN et al. (1990, 1991), FORMSGAARD et al.

Table 2. Representative SIV isolates and their sources

Isolate	Source		Reference
	Common name	Species name	
SIV_{mac}	Rhesus monkey	*Macaca mulatta*	KANKI et al. (1985); DANIEL et al. (1985)
SIV_{cyn}	Cynomolgus monkey	*Macaca fascicularis*	DANIEL et al. (1988a); KESTLER et al. (1988)
SIV_{mne}	Pig-tailed macaque	*Macaca nemestrina*	BENVENISTE et al. (1986)
SIV_{stm}	Stump-tailed macaque	*Macaca arctoides*	GARDNER and LUCIW (1988); NOVEMBRE et al. (1992)
SIV_{smm}	Sooty mangabey monkey	*Cercocebus atys*	FULTZ et al. (1986); LOWENSTINE et al. (1986) MURPHEY–CORB et al. (1986); HIRSCH et al. (1989)
SIV_{wcm}	African white-crowned mangabey monkey	*Cercocebus torquatus lunulatus*	TOMONAGA et al. (1993)
SIV_{agm}	African green monkey	*Cercopithecus aethiops*	OHTA et al. (1988); FUKASAWA et al. (1988) DANIEL et al. (1988b); KRAUS et al. (1989)
SIV_{agm}	African green monkey	*Cercopithecus pygerythrus*	ALLAN et al. (1990)
SIV_{agm}	African green monkey	*Cercopithecus sabaeus*	ALLAN et al. (1991)
SIV_{agm}	African green monkey	*Cercopithecus tantalus*	MÜLLER et al. (1993)
SIV_{syk}	African Sykes' monkey	*Cercopithecus mitis*	EMAU et al. (1991); HIRSCH et al. (1993)
SIV_{mnd}	Mandrill	*Papio (Mandrillus) sphinx*	TSUJIMOTO et al. (1988, 1989)
SIV_{cpz}	Chimpanzee	*Pan troglodytes troglodytes*	PEETERS et al. (1989, 1992); HUET et al. (1990)

(1991), MÜLLER et al. (1993); SIV_{wcm}, TOMONAGA et al. (1993); SIV_{mnd}, TSUJIMOTO et al. (1989); SIV_{cpz}, HUET et al. (1990); SIV_{syk}, HIRSCH et al. (1993). It was previously recognized that primate lentiviruses, including HIV-1 and HIV-2, could be classified into four groups based upon the sequence and functional similarity of the viral genes (DESROSIERS 1990; SAKAI et al. 1991; LUCIW et al. 1992). However, the genomic analysis of SIV_{syk} has revealed that the virus is genetically equidistant from other primate lentiviruses, thus suggesting that the primate lentiviruses can be classified into five major groups.

Figure 1 illustrates the genomic organizations of the five groups: the HIV-1 group (HIV-1/SIV_{cpz}), HIV-2 group (HIV-2/SIV_{smm}/SIV_{nme}/SIV_{mac}), SIV_{agm} group (SIV_{agm}/SIV_{wcm}), SIV_{mnd}, and SIV_{syk}. Three major open reading frames (ORFs), *gag*, *pol*, and *env* that encode viral structural proteins are commonly present in all the retroviruses, and much more so the five lentivirus groups. Other two in-frame genes, *tat* and *rev*, are unique for the lentivirus subfamily and located within the *env* ORF. Processing of these two gene products by splicing and their regulatory mechanisms on the viral gene expression are well characterized and described elsewhere in this volume. While these five genes are essential for replicative growth, whether two other commonly existing genes, *nef* and *vif*, are essential or not is unclear (SHIBATA and ADACHI 1992). There have been conflicting reports on the function of *nef* for HIV-1, but it seems dispensable as far as in vitro tissue culture experiment is concerned. Likewise, the physiological significance of *vif* is also not yet fully clarified. However, we should acknowledge a paper describing the importance of *nef* gene of SIV_{mac} for development of AIDS in vivo (KESTLER et al. 1991).

It is the combination of three other nonessential accessory genes that distinguishes each group from the other, namely, *vpu* is unique for the HIV-1 group, *vpr* is seen in the HIV-1, HIV-2, SIV_{mnd} and SIV_{syk} groups, and *vpx* is present in only the HIV-2 and SIV_{agm} groups. At present, information about the physiological functions of these accessory genes is limited. At least mutational analyses of the three accessory genes of SIV_{agm}, HIV-1, and HIV-2 have indicated that they are dispensable in the in vitro tissue culture level (SHIBATA et al. 1990a, b). However, another recent paper reporting the importance of *vpr* of SIV_{mac} for in vivo infection to monkeys (LANG et al. 1993) is important. The *vpx* and *vpr* genes are similar to one another, and it is speculated that the *vpx* of the HIV-2 subgroup might have arisen by the duplication of an ancestral *vpr* gene (TRISTEM et al. 1992). TRISTEM et al. (1992) also suggested that *vpx* of SIV_{agm} be reclassified as *vpr* because of insufficient backgrounds from sequence data although the SIV_{agm} *vpx* protein is actually rather homologous to the HIV-2 group *vpx* (approximately 36%–39%). Interestingly, the *vpr* of SIV_{syk} is most similar to the HIV-1 *vpr* product, according to HIRSCH et al. (1993) although the homology (30%) is still not sufficiently high. Furthermore, the genomic organization of SIV_{syk} is very similar to that of SIV_{mnd} in spite of being genetically distant from one

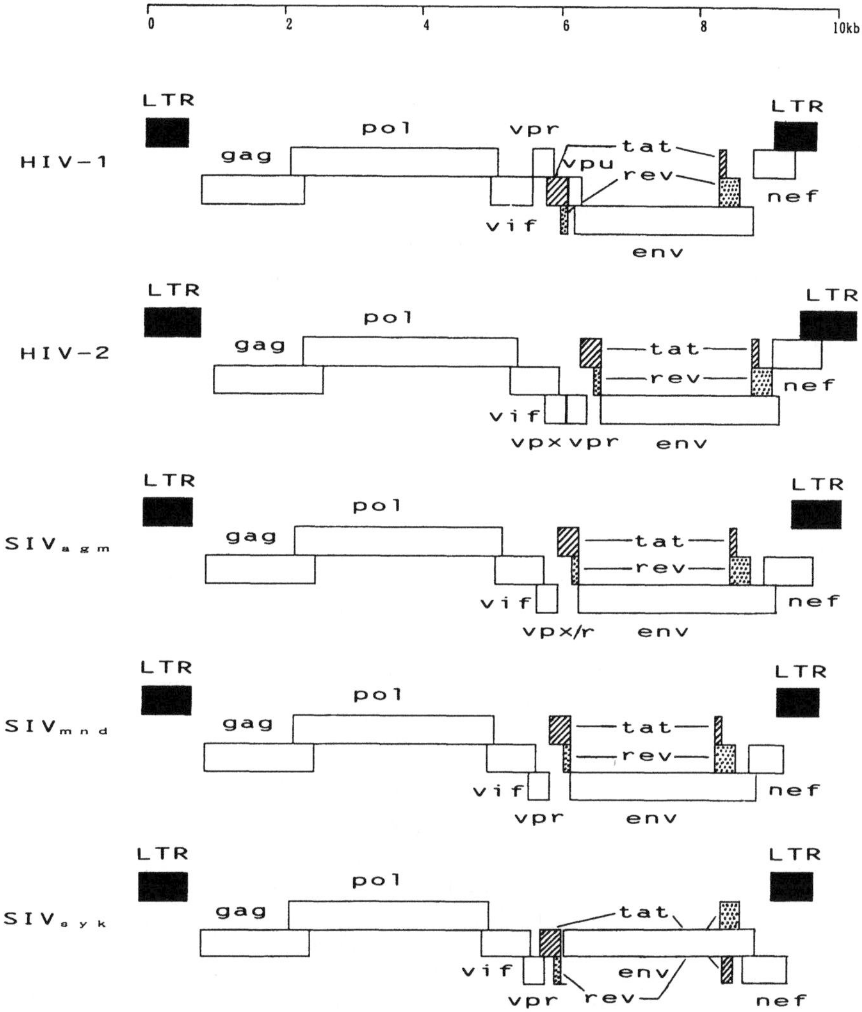

Fig. 1. Genome organizations of the five major groups of primate lentiviruses. The HIV-1 group includes HIV-1 and SIV_{cpz}; HIV-2 group includes HIV-2, SIV_{smm}, SIV_{mne} and SIV_{mac} subgroups. The genome organizations of SIV_{agm}, SIV_{mnd} and SIV_{syk}, which are genetically equidistant from other primate letiviruses and thus considered to constitute distinct groups, are fundamentally very similar

another. Provided that the *vpx* of SIV_{agm}, the *vpr* of SIV_{mnd}, and the *vpr* of SIV_{syk} are all the "*vpr*" homolog, these three groups may be very similar and constitute a so-called prototype SIV group with a single copy of "*vpr*" as far as the genomic organization is concerned. Since papers describing the in vivo functions of these accessory genes have recently appeared, the adequacy of such a classification would have been long since confirmed by their clarified biological roles.

7 The Phylogenetic Relationship of the Primate Lentiviruses

Phylogenetic analysis is one of the strong weapons in obtaining insight into the genetic relationship. Several groups have reported phylogenetic trees (SMITH et al. 1988; SHARP and LI 1988; GOJOBORI et al. 1990; MIURA et al. 1990; MYERS et al. 1992) using the available sequence data from various HIV/SIV isolates. Although there were some differences with respect to the estimated dates of divergence, the general appearance of the trees are very similar.

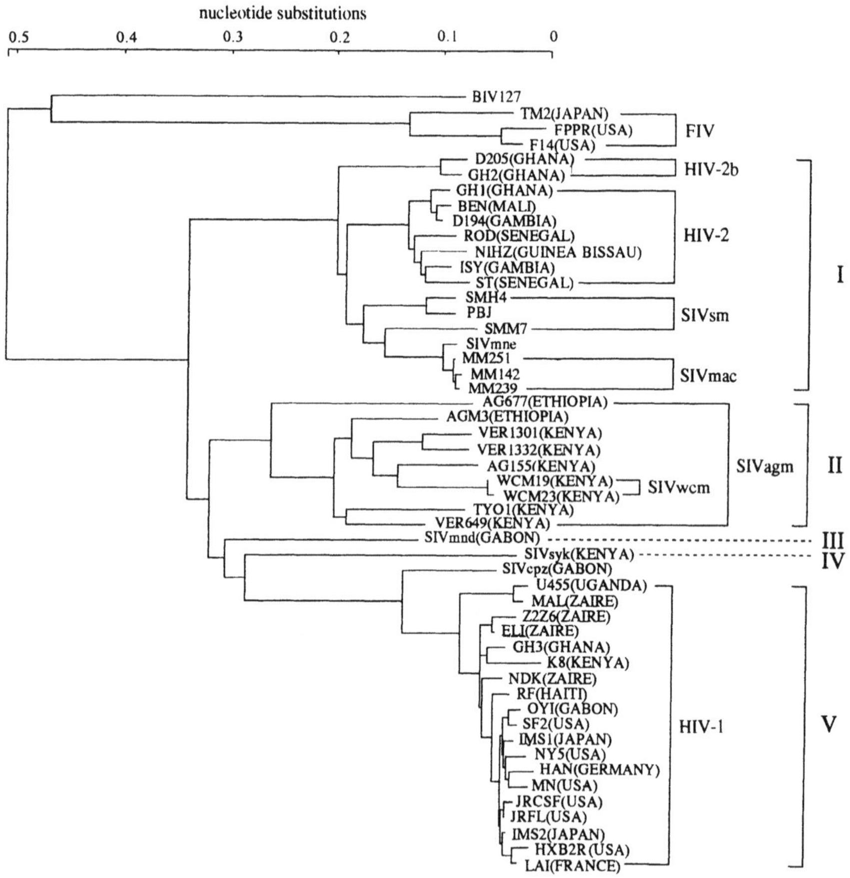

Fig. 2. Phylogenetic tree of AIDS-related viruses. The tree was constructed by the neighbor-joining method (SAITO and NEI 1987) after alignment of the nucleotide sequence of a part of *pol* region (integrase) (MIURA et al. 1990). SCALE, the number of nucleotide substitutions per site estimated by the six-parameter method. (GOJOBORI et al. 1982); horizontal branch length indicates the genetic distance

Our phylogenetic tree of AIDS-related viruses (including bovine and feline immunodeficiency viruses) based on the up-to-date information is presented in Fig. 2. The tree was constructed by the neighbor-joining method (SAITO and NEI 1987) after alignment of the nucleotide sequences of part of *pol* (integrase) region (MIURA et al. 1990). As shown in Fig. 2, the tree clearly indicates the presence of the five groups of the primate lentiviruses. From top to bottom the HIV-2 group consists of HIV-2/SIV_{smm}/SIV_{mne}/SIV_{mac} subgroups. This group is very diverse, yet HIV-2 and SIV_{smm}, etc. evidently belong to the same group, giving a hint to the origin of HIV-2, as described below. SIV_{agm} group, the majority of which are from AGMS, also show fairly large extent of diversity and constitute the second large group together with SIV_{wcm} from African white-crowned mangabey monkeys (TOMONAGA et al. 1993). The SIV_{wcm} subgroup is completely included in the SIV_{agm} group instead of SIV_{smm} subgroup. The white-crowned mangabey monkey belongs to the genus *Cercocebus*, to which the sooty mangabey monkey also does. Therefore, it is most probable that an interspecies transmission might have occurred in the past. SIV_{mnd} and SIV_{syk} diverged almost at the same time from other groups, currently being equidistantly far from each other. The last large group consists of HIV-1/SIV_{cpz}. Although many HIV-1 strains have been isolated from AIDS patients in the American and European continents and other parts of the world, they are fairly similar. However, it should be noted that some of the isolates from Africa appear to have more sequence variations.

Where did HIVs and SIVs come from? And When? It is now known that HIV-2 is genetically very close to SIV_{smm}/SIV_{mne}/SIV_{mac} (CHAKRABARTI et al. 1987; HIRSCH et al. 1989; DIETRICH et al. 1989). Moreover, human infecton by genetically diverse SIV_{smm} related HIV-2 was reported in West Africa (GAO et al. 1992), an epidemiological locus of HIV-2. The natural habitat of sooty mangabeys is also in West Africa. Therefore, one can postulate that HIV-2 may have originated from SIV_{smm}. The phylogenetic tree tells more detailed stories. SIV_{mac} and SIV_{mne} (originally isolated from macaques in primate centers in the Unite States) diverged rather lately from SIV_{smm}. Nonetheless, the serosurvey to date indicates puzzlingly that no antibody to SIV_{mac} has been detected in the natural hosts. This virus probably spread recently among new hosts through artificial transmission in the American primate centers, as mentioned above. According to the tree, however, it appears that there is a distinct subgroup termed HIV-2b (Fig. 2; DIETRICH et al. 1989; MIURA et al. 1990). This subgroup diverged even before the diversion of HIV-2 and SIV_{smm}/SIV_{mne}/SIV_{mac}, suggesting at least some HIV-2 varieties have been possessed by man for a very long time. For further delineation about the origin of HIV-2, more information is needed.

The divergence between $SIV_{cpz-gab}$ from Gabon (PEETERS et al. 1989; HUET et al. 1990) and the HIV-1 group occurred rather recently. Another isolate $SIV_{cpz-ant}$ (PEETERS et al. 1992) from Belgium (originally from Zaire), which is also cross-reactive to HIV-1, is said to possess glycoproteins with different molecular sizes and to be antigenically distinguishable from $SIV_{cpz-gab}$. In addition, the time of branching of most African isolates seemed a little earlier than the appearance of American/European strains. Especially a Cameroonian HIV-1 isolate known as ANT-70 (DE LEYS et al. 1990) diverged earlier than $SIV_{cpz-gab}$ did from HIV-1s (MYERS et al. 1992), and $SIV_{cpz-ant}$ diverged at the earliest (PEETERS et al. 1993). These observations suggest that ancestral HIV-1 may have existed in either man or chimpanzee in central Africa (the HIV-1 African origin theory).

Estimated dates of radiation of all primate lentiviruses vary from millions of years to several decades (depending on what molecular clock is used as a standard). But the average and perhaps consensual time span lies around 150–400 years (SHARP and LI 1988; QUERAT et al. 1990) except for one report which estimated HIV-1 might have evolved from its common ancestor with HIV-2 as recently as 40 years ago (SMITH et al. 1988). However, these earlier studies were undertaken before the recent findings of highly divergent strains. More detailed consideration may be needed to determine the time of divergence (MYERS et al. 1992). Further extensive research, especially characterization of the lentiviruses in the African continent, will be very useful to elucidate the true origins of HIV/SIV.

8 Current Topics of Virus Isolation

Attempts at virus isolation from new host primates are still continuing. For example, a new SIV_{agm} has lately been isolated from tantalus monkeys (*Cercopithecus tantalus*; MÜLLER et al. 1993). AGMs are a diversified superspecies and have often been sources of various SIV_{agm} strains. It has been noted that several nucleotide sequences of these SIV_{agm} isolates are distinct from others (FUKASAWA et al. 1988; BAIER et al. 1990; FORMSGAARD et al. 1991). AGMs can be divided into four subspecies according to their phenotypic differences and geographic distributions: commonly termed vervets (*C. pygerythrus*) in southern Ethiopia to South Africa, grivets (*C. aethiops*) in Ethiopia and Sudan, tantalus monkeys (*C. tantalus*) in Central Aftica, and sabaeus monkeys (*C. sabaeus*) in West Africa. Up to the present SIV_{agm} varieties have been obtained from each of these four (OHTA et al. 1988; DANIEL et al. 1988b; KRAUS et al. 1989; ALLAN et al. 1990; 1991; MÜLLER et al. 1993).

A novel SIV of particular interest is SIV_{syk} from Sykes' monkeys (EMAU et al. 1991; HIRSCH et al. 1993). Phylogenetic analysis has indicated a unique position of this virus, thus proposing that the primate lentiviruses consist of five major groups, from a genetical viewpoint. The genomic organization of SIV_{syk}, however, is similar to that of SIV_{mnd}, as mentioned above. Thus, too, that of SIV_{agm}, cosidering that its *vpx* is a "*vpr*" homolog. Interestingly,

Sykes' monkey belongs to the genus *Cercopithecus*, as does the AGM. Therefore, it is still possible that a new virus which demonstrates the presence of such a prototype SIV or links it to SIV_{cpz}/HIV-1 group will be isolated from AGM or other monkeys of the *Cercopithecus* genus.

9 Discussion

Who could have predicted the current flourishing of the SIV research 10 years ago when an immune deficiency syndrome similar to AIDS was found in captive macaques? Since then the progress in this field has been remarkable. At this moment the pathogenic mechanism of HIV is still at work despite enormous efforts of a number of scientists in the world. The need of experimental animal model of AIDS using monkeys may have accelerated studies on SIV. HIV-1 is infectious only to the human race and to apes such as chimpanzees (and possibly pig-tailed macaques). Unfortunately, experimental infection of HIV-1 did not induce any AIDS-like symptoms in chimpanzees (JOHNSON et al. 1993). In addition, the latter species is difficult to use for experimental use. Thus, attempts were made to establish more appropriate animal models of AIDS using various SIVs and monkeys. SIV_{mac} and SIV_{smm} were able to cause the disease in rhesus macaques, pig-tailed macques, and cynomolgus monkeys (LETVIN et al. 1985; FULTZ et al. 1989; PUTKONEN et al. 1992). Molecularly cloned SIV_{mac239} also succeeded in causing AIDS through the persistent infection state in infected monkeys (NAIDU et al. 1988; KESTLER et al. 1990). A particular strain, $SIV_{smm-PBj14}$, isolated from sooty mangabey monkeys was always capable of causing fatal disease within 1–2 weeks after inoculation in pig-tailed macaques (FULTZ et al. 1989; DEWHURST et al. 1990) although it was not in SIV_{smm}-seropositive sooty mangabeys. Experimental infection of SIV_{agm} to AGMs and cynomolgus monkeys (HONJO et al. 1990) and of SIV_{mnd} to cynomolgus monkeys, cottontop tamarins (*Saguinus oedipus*), and a common marmoset (*Callithrix jacchus*; MIURA et al. 1989; and unpublished data) resulted in no observation of AIDS-like clinical signs. The pathogenic properties of various SIVs are described in a review (ALLAN 1991) and somewhere else in this volume. In order to fully understand the pathogenic nature of these immunodeficiency viruses, further comparative studies on the pathological mechanisms of HIVs/SIVs are necessary.

A recent serosurvey of HIV and related viruses carried out in West Africa has revealed that some AIDS cases in the region are not associated with either HIV-1 or HIV-2, suggesting a possibility of presence of new HIV (HIV-3?; HISHIDA et al. 1992). It is also probable that in Africa more striking HIV/SIV strains may exist which can explain the divergence of HIV-1 and HIV-2 groups. In conclusion, further research for such novel viruses through serosurveys followed by virus isolation and sequencing should be continued. We believe such efforts in concert with the data from known isolates will surely deepen our understanding about the primate lentiviruses, and the combined knowledge will be useful in finding a cure for AIDS.

Acknowledgments. Our work has relied upon the cooperation of many physicians and postgraduate students at the Tsukuba Primate Center, NIH of Japan, Institute of Primates, Kyoto University, Institute of Medical Science, University of Tokyo, and the authors' present laboratory.

References

Allan JS (1991) Pathogenic properties of simian immunodeficiency viruses in nonhuman primates. In: Koff WC, Wong-Staal F, Kennedy RC (eds) AIDS research review, vol 1. Dekker, New York, pp 191–206

Allan JS, Kanda P, Kennedy RC, Cobb EK, Anthony M, Eichberg JW (1990) Isolation and characterization of simian immunodeficiency viruses from two subspecies of African green monkeys. AIDS Res Hum Retroviruses 6: 275–285

Allan JS, Short M, Taylor ME, Su S, Hirsch VM, Johnson PR, Shaw GM, Hahn BH (1991) Species-specific diversity among simian immunodeficiency viruses from African green monkeys. J Virol 65: 2816–2828

Baier M, Garber C, Müller C, Cichutek K, Kurth R (1990) Complete nucleotide sequence of a simian immunodeficiency virus from African green monkeys: a novel type of intragroup divergence. Virology 176: 216–221

Benveniste RE, Arthur LO, Tsai C, Sowder R, Copeland TD, Henderson LE, Oroszlan S (1986) Isolation of a lentivirus from a mecaque with lymphoma: comparison with HTLV-III/LAV and other lentiviruses. J Virol 60: 483–490

Chakrabarti L, Guyader M, Alizon M, Daniel MD, Desrosiers RC, Tiollais P, Sonigo P (1987) Sequence of simian immunodeficiency virus from macaque and its relationships to other human and simian retroviruses. Nature 328: 543–547

Clavel F, Guétard D, Brun-Vézinet F, Chamaret S, Rey M-A, Santos-Ferreira MO, Laurent AG, Dauguet C, Katlama C, Rouzioux C, Klatzmann D, Champalimaud JL, Montagnier L (1986) Isolation of a new human retrovirus from West African patients with AIDS. Science 233: 343–346

Daniel MD, King NW, Letvin NL, Hunt RD, Sehgal PK, Desrosiers RC (1984) A new type D retrovirus isolated from macaques with an immunodeficiency syndrome. Science 223: 602–605

Daniel MD, Letvin NL, King NW, Kannagi M, Sehgal PK, Hunt RD, Kanki PJ, Essex M, Desrosiers RC (1985) Isolation of T-cell tropic HTLV-III-like retovirus from macaques. Science 228: 1201–1204

Daniel MD, Letvin NL, Sehgal PK, Schmidt DK, Silva DP, Solomon KR, Hodi FS Jr, Ringler DJ, Hunt RD, King NW, Desrosiers RC (1988a) Prevalence of antibodies to three retroviruses in a captive colony of macaque monkeys. Int J Cancer 41: 601–608

Daniel MD, Li Y, Naidu YM, Durda PJ, Schmidt DK, Troup CD, Silva DP, MacKey JJ, Kestler HW III, Sehgal PK, King NW, Ohta Y, Hayami M, Desrosiers RC (1988b) Simian immunodeficiency virus from African green monkeys. J Virol 62: 4123–4128

De Leys R, Vanderborght B, Haeseveldt MV, Heyndrickx L, van Geel A, Wauters C, Bernaerts R, Saman E, Nijs P, Willems B, Taelman H, van der Groen G, Piot P, Tersmette T, Huisman JG, van Heuverswyn (1990) Isolation and partial characterization of an unusual human immunodeficiency retrovirus from two persons of West-Central African origin. J Virol 64: 1207–1216

Desrosiers RC (1990) A finger of the missing link. Nature 345: 288–289

Dewhurst S, Embretson JE, Anderson DC, Mullins JI, Fultz PN (1990) Sequence analysis and acute pathogenicity of molecularly cloned $SIV_{smm-PBj14}$. Nature 345: 636–640

Dietrich U, Adamski M, Kreutz R, Seipp A, Kühnel H, Rübsamen-Waigmann H (1989) A highly divergent HIV-2-related isolate. Nature 342: 948–950

Emau P, McClure HM, Isahakia M, Else JG, Fultz PN (1991) Isolation from African Sykes' monkeys (Cercopithecus mitis) of a lentivirus related to human and simian immunodeficiency viruses. J Virol 65: 2135–2140

Formsgaard A, Hirsch VM, Allan JS, Johnson PR (1991) A highly divergent proviral DNA clones of SIV from a distinct species of African green monkey. Virolgy 182: 397–402

Fukasawa M, Miura T, Hasegawa A, Morikawa S, Tsujimoto H, Miki K, Kitamura T, Hayami M (1988) Sequence of simian immunodeficiency virus from African green monkey, a new member of the HIV/SIV group. Nature 333: 457–461

Fultz PN, McClure HM, Anderson DC, Swenson RB, Anando R, Srinivasan A (1986) Isolation of a T-lymphotropic retrovirus from naturally infected sooty mangabey monkeys (Cercocebus atys). Proc Natl Acad Sci USA 83: 5286–5290

Fultz PN, McClure HM, Anderson DC, Switzer WM (1989) Identification and biologic characterization of an acutely lethal variants of simian immunodeficiency virus from sooty mangabeys (SIV/SMM). AIDS Res Hum Retroviruses 5: 397–409

Fultz PN, Gordon TP, Anderson DC, McClure HM (1990) Prevalence of natural infection with simian immunodeficiency virus and simian T-cell leukemia virus type I in a breeding colony of sooty mangabey monkeys. AIDS 4: 619–625

Gao F, Yue L, White AT, Pappas PG, Barchue J, Hanson AP, Greene BM, Sharp PM, Shaw GM, Hahn BH (1992) Human infection by genetically diverse SIV_{sm}-related HIV-2 in West Africa. Nature 358: 495–499

Gardner MB, Luciw P (1988) Simian immunodeficiency viruses and their relationship to the human immunodeficiency viruses. AIDS 2: S3–S10

Gojobori T, Ishii K, Nei M (1982) Estimation of average number of nucleotide substitutions when the rate of substitution varies with nucleotide. J Mol Evol 18:414–423

Gojobori T, Moriyama EN, Ina Y, Ikeo K, Miura T, Tsujimoto H, Hayami M, Yokoyama S (1990) Evolutionary origin of human and simian immunodeficiency viruses. Proc Natl Acad Sci USA 87: 4108–4111

Hayami M, Ishikawa K, Komuro A, Kawamoto Y, Nozawa K, Yamamoto K, Ishida T, Hinuma Y (1983a) ATLV antibody in cynomolgus monkeys in the wild. Lancet 2: 620

Hayami M, Komuro A, Ishikawa K (1983b) Natural existence of antibody to adult T-cell leukemia virus in non-human primates. Leuk Rev Int 1: 76–77

Hayami M, Komuro A, Nozawa K, Shotake T, Ishikawa K, Yamamoto K, Ishida T, Honjo S, Hinuma Y (1984) Prevalence of antibody to adult T-cell leukemia virus-associated antigens (ATLA) in Japanese monkeys and other non-human primates. Int J Cancer 33: 179–183

Hayami M, Ohta Y, Hattori T, Nakamura H, Takatsuki K, Kashiwa A, Nozawa K, Miyoshi I, Ishida T, Tanioka Y, Fujiwara T, Honjo S (1985) Detection of antibodies to human T-lymphotropic virus type III in various non-human primates. Jpn J Exp Med 55: 251–255

Henderson LE, Benveniste RE, Sowder R, Copeland TD, Schultz AM, Oroszlan S (1988) Molecular characterization of gag proteins from simian immunodeficiency virus (SIV_{mne}). J Virol 62: 2587–2595

Hendry RM, Wells MA, Phelan MA, Scheider AL, Epstein JS, Quinnan GV (1986) Antibodies to simian immunodeficiency virus in African green monkeys in Africa in 1957–62. Lancet 2: 455

Hinuma Y, Nagata K, Hanaoka M, Nakai M, Matsumoto T, Kinoshita K, Shirakawa S, Miyoshi I (1981) Adult T-cell leukemia: antigen in an ATL cell line and detection of antibodies to the antigen in human sera. Proc Natl Acad Sci USA 78: 6476–6480

Hirsch VM, Olmsted RA, Murphey-Corb M, Purcell RH, Johnson PR (1989) An African primate lentivirus (SIV_{sm}) closely related to HIV-2. Nature 339: 389–392

Hirsch VM, Dapolito GA, Goldstein S, McClure H, Emau P, Fultz PN, Isahakia M, Lenroot R, Myers G, Johnson PR (1993) A distinct African lentivirus from Sykes' monkeys. J Virol 67: 1517–1528

Hishida O, Ido E, Igarashi T, Hayami M, Miyazaki A, Ayisi NK, Osei-Kwasi M (1992) Clinically diagnosed AIDS cases without evident association with HIV type 1 and 2 infections in Ghana. Lancet 340: 971–972

Honjo S, Narita T, Kobayashi R, Hiyaoka A, Fujimoto K, Takasaka M, Mukai R, Ishikawa K, Ohta Y, Hayami M (1990) Experimental infection of African green monkeys and cynomolgus monkeys with a SIV_{agm} strain isolated from a healthy African green monkey. J Med Primatol 19: 9–20

Huet T, Cheynier R, Meyerhans A, Roelants G, Wain-Hobson S (1990) Genetic organization of a chimpanzee lentivirus related to HIV-1. Nature 345: 356–359

Hunsmann G, Schneider J, Schmitt J, Yamamoto N (1983) Detection of serum antibodies to adult T-cell leukemia virus in non-human primates and in people from Africa. Int J Cancer 32: 329–332

Ishikawa K, Fukasawa M, Tsujimoto H, Else JG, Isahakia M, Ubhi NK, Ishida T, Takenaka O, Kawamoto Y, Shotake T, Ohsawa H, Ivanoff B, Cooper RW, Frost E, Grant FC, Spriatna Y, Sutarman, Abe K, Yamamoto K, Hayami M (1987) Serological surveys and virus isolation of simian T-cell leukemia/T-lymphotropic virus type 1 (STLV-1) in non-human primates in their native countries. Int J Cancer 40: 233–239

Johnson BK, Stone GA, Godec MS, Asher DM, Gajdusek DC, Gibbs CJ Jr (1993) Long-term observations of human immunodeficiency virus-infected chimpanzees. AIDS Res Hum Retroviruses 9: 375–378

Kanki PJ, McLane MF, King NW Jr, Letvin NL, Hunt RD, Sehgal P, Daniel MD, Desrosiers RC, Essex M (1985) Serological identification and characterization of a macaque T-lymphotropic retrovirus closely related to HTLV-III. Science 228: 1199–1201

Kestler HW III, Li Y, Naidu YM, Butler CV, Ochs MF, Jaenel G, King NW, Daniel MD, Desrosiers RC (1988) Comparison of simian immunodeficiency virus isolates. Nature 331: 619–621

Kestler HW III, Kodama T, Ringler D, Marthas M, Pedersen N, Lackner A, Regier D, Sehgal P, Daniel M, King N, Desrosiers R (1990) Induction of AIDS in rhesus monkeys by molecularly cloned simian immunodeficiency virus. Science 248: 1109–1112

Kestler HW III, Ringler DJ, Mori K, Panicali DL, Sehgal PK, Daniel MD, Desrosiers RC (1991) Importance of the nef gene for maintenance of high virus loads and for development of AIDS. Cell 65: 651–662

Kodama T, Silva DP, Daniel MD, Phillips-Conroy JE, Jolly CJ, Rogers J, Desrosiers RC (1989) Prevalence of antibodies to SIV in baboons in their native habitat. AIDS Res Hum Retroviruses 3: 337–343

Kraus G, Werner A, Baier M, Binniger D, Ferdinand FJ, Norley S, Kurth R (1989) Isolation of human immunodeficiency virus-related simian immunodeficiency viruses from African green monkeys. Proc Natl Acad Sci USA 86: 2898–2896

Lang SM, Weeger M, Stahl-Hennig C, Coulibaly C, Hunsmann G, Müller J, Müller-Hermelink H, Fuchs D, Wachter H, Daniel MH, Desrosiers RC, Fleckenstein B (1993) Importance of vpr for infection of rhesus monkeys with simian immunodeficiency virus. J Virol 67: 902–912

Letvin NL, Eaton KA, Aldrich WR, Sehgal PK, Blake BJ, Schlossman SF, King NW, Hunt RD (1983) Acquired immunodeficiency syndrome in a colony of macaque monkeys. Proc Natl Acad Sci USA 80: 2718–2722

Letvin NL, Daniel MD, Sehgal PK, Desrosiers RC, Hunt RD, Waldron LM, Mackey JJ, Schmidt DK, Chalifoux LV, King NW (1985) Induction of AIDS-like disease in macaque monkeys with T-cell tropic retrovirus STLV-III. Science 230: 71–73

Lowenstine LJ, Pederson NC, Higgins J, Pallis KC, Uyeda A, Marx P, Lerche NW, Munn RJ, Gardner MB (1986) Seroepidemiologic survey of çaptive Old-World primates for antibodies to human and simian retroviruses, and isolation of a lentivirus from sooty mangabeys (Cercocebus atys). Int J Cancer 38: 563–574

Luciw PA, Shaw KS, Unger RE, Planelles V, Stout MW, Lackner JE, Pratt-Lowe E, Leung NJ, Banapour B, Marthas ML (1992) Genetic and biological comparisons of pathogenic and nonpathogenic molecular clones of simian immunodeficiency virus (SIV_{mac}). AIDS Res Hum Retroviruses 8: 395–402

Marx PA, Li Y, Lerche NW, Sutjipto S, Gettie A, Yee JA, Brotman BH, Prince AM, Hanson A, Webster RG, Desrosiers RC (1991) Isolation of a simian immunodeficiency virus related to human immunodeficiency virus type 2 from a West African pet sooty mangabey. J Virol 65: 4480–4485

Miura T, Tsujimoto H, Fukasawa M, Ohta Y, Honjo S, Hayami M (1989) Genetic analysis and infection of SIV_{AGM} and SIV_{MND}. J Med Primatol 18: 255–259

Miura T, Sakuragi J, Kawamura M, Fukasawa M, Moriyama EN, Gojobori T, Ishikawa K, Mingle JAA, Nettey VBA, Akari H, Enami M, Tsujimoto H, Hayami M (1990) Establishment of a phylogenetic survey system for AIDS-related lentiviruses and demonstration of a new HIV-2 subgroup. AIDS 4: 1257–1261

Miyoshi I, Ohtsuki Y, Fujishita M, Yoshimoto S, Kubonishi I, Minezawa M (1982a) Detection of type-C virus paticles in Japanese monkeys seropositive to adult T-cell leukemia-associated antigen. Gann 73: 848–849

Miyoshi I, Yoshimoto S, Fujishita M, Taguchi H, Kubonishi I, Niiya K, Minezawa M (1982b) Natural adult T-cell leukemia virus infection in Japanese monkeys. Lancet 2: 658

Miyoshi I, Fujishita M, Taguchi H, Matsubayashi K, Miwa N, Tanioka Y (1983) Natural infection in non-human primates with adult T-cell leukemia virus or a closely related agent. Int J Cancer 32: 333–336
Müller MC, Sakusena NK, Nerrienet E, Chappey C, Hervé VMA, Durand J-P, legal-Champodonico P, Lang M-C, Digoutte J-P, Georges AJ, Georges-Courbot M-C, Sonigo P, Barré-Sinoussi F (1993) Simian immunodeficiency viruses from central and western Africa: evidence for a new species-specific lentivirus in tantalus monkeys. J Virol 67: 1227–1235
Murphey-Corb M, Martin LN, Rangan SRS, Baskin GB, Gormus BJ, Wolf RH, Andes WA, West M, Montelaro RC (1986) Isolation of an HTLV-III-related retrovirus from macaques with simian AIDS and its possible origin in asymptomatic mangabeys. Nature 321: 435–437
Myers G, MacInnes K, Korber B (1992) The emergence of simian/human immunodeficiency viruses. AIDS Res Hum Retroviruses 8: 373–386
Naidu YM, Kestler HW III, Li Y, Butler CV, Silva DP, Schmidt DK, Troup CD, Sehgal PK, Sonigo P, Daniel MD, Desrosiers RC (1988) Characterization of infectious molecular clone of simian immunodeficiency virus (SIV_{mac}) and human immunodeficiency virus type 2: persistent infection of rhesus monkeys with molecularly cloned SIV_{mac}. J Virol 62: 4691–4696
Noda Y, Ishikawa K, Sasagawa A, Honjo S, Mori S, Tsujimoto H, Hayami M (1986) Hematologic abnormalities similar to the preleukemic state of adult T-cell leukemia in African green monkeys naturally infected with simian T-cell leukemia virus. Jpn J Cancer Res 77: 1227–1234
Novembre FJ, Hirsch VM, McClure HM, Fultz PN, Johnson PR (1992) SIV from Stump-tailed macaques: molecular characterization of a highly transmissible primate lentivirus. Virology 186: 783–787
Ohta Y, Masuda T, Tsujimoto H, Ishikawa K, Kodama T, Morikawa S, Nakai M, Honjo S, Hayami M (1988) Isolation of simian immunodeficiency virus from African green monkeys and seroepidemiological survey of the virus in various non-human primates. Int J Cancer 41: 115–122
Peeters M, Honoré C, Huet T, Bedjabaga L, Ossari S, Bussi P, Cooper RW, Delaporte E (1989) Isolation and partial characterization of an HIV-related virus occurring naturally in chimpanzees in Gabon. AIDS 3: 625–630
Peeters M, Fransen K, Delaporte E, van den Haesevelde M, Gershy- Damet G-M, Kestens L, van der Groen G, Piot P (1992) Isolation and characterization of a new chimpanzee lentivirus (simian immunodeficiency virus isolate cpz-ant) from a wild-captured chimpanzee. AIDS 6: 447–451
Peeters M et al. (1993) 9th International conference on AIDS, Berlin
Poiesz BJ, Ruscetti FW, Gazdar AF, Bunn PA, Minna JD, Gallo RC (1980) Detection and isolation of type-C retrovirus from fresh and cultured lymphocytes of patients with cutaneous T-cell lymphoma. Proc Natl Acad Sci USA 77: 7415–7419
Putkonen P, Kaaya EE, Böttiger D, Li S-L, Nisson C, Biberfeld P, Biberfeld G (1992) Clinical features and predictive markers of disease progression in cynomolgus monkeys experimentally infected with simian immunodeficiency virus. AIDS 6: 257–263
Querat G, Audoly G, Sonigo P, Vigne R (1990) Nucleotide Sequence analysis of SA-OMVV, a visna-related ovine lentivirus: phylogenetic history of lentiviruses. Virology 175: 434–447
Regier DA, Desrosiers RC (1990) The complete nucleotide sequence of a pathogenic molecular clone of simian immunodeficiency virus. AIDS Res Hum Retroviruses 6: 1221–1231
Saito N, Nei M (1987) The neighbor-joining method: a new method for reconstructing phylogenetic trees. Mol Biol Evol 4: 406–425
Sakai H, Shibata R, Sakuragi J, Kiyomasu T, Kawamura M, Hayami M, Ishimoto A, Adachi A (1991) Compatibility of rev gene activity in the four groups of primate lentiviruses. Virology 184: 513–520
Schneider J, Jurkiewicz E, Hayami M, Desrosiers R, Marx P, Hunsmann G (1987) Serological and structural comparison of HIV, SIV_{mac}, SIV_{agm}, and SIV_{sm}, four primate lentiviruses. Ann Inst Pasteur Virol 138: 93–99
Seiki M, Hattori S, Hirayama Y, Yoshida M (1983) Human adult T- cell leukemia virus: complete nucleiotide sequence of the provirus genome integrated in leukemia cell DNA. Proc Natl Acad Sci USA 80: 3618–3622
Sharp PM, Li W-H (1988) Understanding the origin of AIDS viruses. Nature 336: 315
Shibata R, Adachi A (1992) SIV/HIV recombinants and their use in studying biological properties. AIDS Res Human Retroviruses 8: 403–409

Shibata R, Miura T, Hayami M, Ogawa K, Sakai H, Kiyomasu T, Ishimoto A, Adachi A (1990a) Mutational analysis of the human immunodeficiency virus type 2 (HIV-2) genome in relation to HIV-1 and simian immunodeficiency virus SIV_{agm}. J Virol 64: 742–747

Shibata R, Miura T, Hayami M, Sakai H, Ogawa K, Kiyomasu T, Ishimoto A, Adachi A (1990b) Construction and characterization of an infectious DNA clone and of mutants of simian immunodeficiency virus isolated from the African green monkey. J Virol 64: 307–312

Smith TF, Srinivasan A, Schochetman G, Marcus M, Myers G (1988) The phylogenetic history of immunodeficiency viruses. Nature 333: 573–575

Tomonaga K, Katahira J, Fukasawa M, Hassan MA, Kawamura M, Akari H, Miura T, Goto T, Nakai M, Suleman M, Isahakia M, Hayami M (1993) Isolation and characterization of simian immunodeficiency virus from African white-crowned mangabey monkeys (Cercocebus torquatus lunulatus). Arch Virol 129: 77–92

Tristem M, Marshall C, Karpas A, Hill F (1992) Evolution of the primate lentiviruses: evidence from vpx and vpr. EMBO J 11: 3405–3412

Tsujimoto H, Noda Y, Ishikawa K, Nakamura H, Fukasawa M, Sakakibara I, Sasagawa A, Honjo S, Hayami M (1987) Development of adult T-cell leukemia-like disease in African green monkey associated with clonal integration of simian T-cell virus type-1. Cancer Res 47: 269–274

Tsujimoto H, Cooper RW, Kodama T, Fukasawa M, Miura T, Ohta Y, Ishikawa K, Nakai M, Frost E, Roelants GE, Roffi J, Hayami M (1988) Isolation and characterization of simian immunodeficiency virus from mandrills in Africa and its relationship to other human and simian immunodeficiency viruses. J Virol 62: 4044–4050

Tsujimoto H, Hasegawa A, Maki N, Fukasawa M, Miura T, Speidel S, Cooper RW, Moriyama EN, Gojobori T, Hayami M (1989) Sequence of a novel simian immunodeficiency virus from a wild-caught African mandrill. Nature 341: 539–541

Uchiyama T, Yodoi J, Sagawa K, Takatsuki K, Uchino H (1977) Adult T-cell leukemia: clinical and hematologic features of 16 cases. Blood 50: 481–492

Watanabe T, Seiki M, Yoshida M (1983) Retrovirus terminology. Science 222: 1178

Watanabe T, Seiki M, Yoshida M (1984) HTLV type 1 (U.S. isolate) and ATLV (Japanese isolate) are the same species of human retrovirus. Virology 133: 238–241

Yoshida M, Miyoshi I, Hinuma Y (1982) Isolation and chracterization of retrovirus from cell lines of human adult T-cell leukemia and its implication in the disease. Proc Natl Acad Sci USA 79: 2031–2035

Simian Immunodeficiency Viruses of African Green Monkeys

R. KURTH and S. NORLEY

1 The Host

When virologists talk about African green monkeys (AGM), they usually mean species belonging to the subgenus *Cercopithecus aethiops* of the genus *Cercopithecus*. This genus can be subdivided into eight subgenera (*C. mona, C. cephus, C. mitis, C. l'hoesti, C. hamlyni, C. neglectus, C. diana, C. aethiops*). A subgenus comprises several species and is defined by the common ecological specialization ("life-style") of its member species. Therefore, species belonging to the same subgenus are geographically separated, as they would otherwise compete for resources. Some of the

Paul-Ehrlich-Institute, Paul-Ehrlich-Str. 51–59, 63225 Langen, Germany

Current Topics in Microbiology and Immunology, Vol. 188

species can be further subdivided into altogether over 70 subspecies (for the entire genus) distinguished by subtle morphological differences. Members of similar subspecies usually do not mate in the wild. There are a number of excellent books delineating monkey species (GRZIMEK 1988; HILL 1972; NAPIER and NAPIER 1985).

Four Old World monkey species belong to the *C. aethiops* subgenus (Table 1). They live in different habitats in sub-Saharan Africa (Fig. 1). As we now know (see below), these species harbor different SIV_{agm} groups, and future descriptions of virus strains should include their precise origin (see Table 1). If not designated otherwise, all early investigations of the SIV_{agm}/AGM animal model were carried out with $SIV_{agm/ver}$ and vervets.

Table 1. Genus Cercopithecus, subgenus *C. aethiops* (African green monkey)

Species	Popular name	SIV designation
C. ae. aethiops	Grivet or savannah monkey	$SIV_{agm/gri}$
C. ae. tantalus	Tantalus monkey	$SIV_{agm/tan}$
C. ae. sabaeus	Sabaeus or green monkey	$SIV_{agm/sab}$
C. ae. pygerythrus	Vervet monkey	$SIV_{agm/ver}$

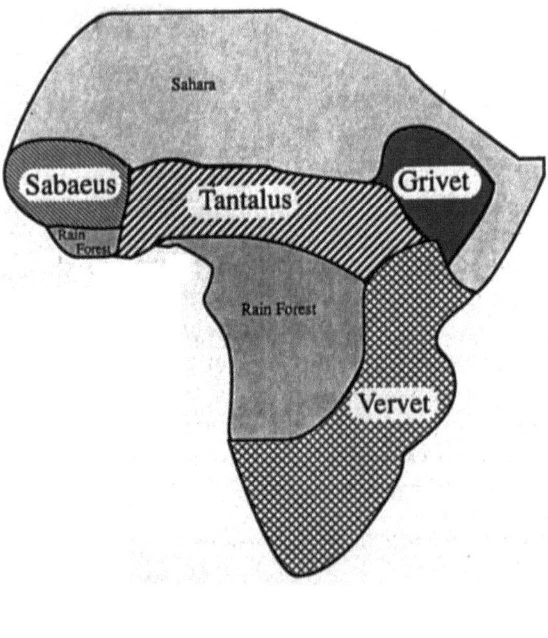

Fig. 1. Geographic distribution of the four AGM species. (Adapted from MÜLLER et al. 1993)

2 Initial Isolation of SIV_{agm}

Soon after the discovery of HIV-1 it was realized that a variety of feral Old World monkey species possess antibodies (marginally) cross-reactive with HIV-1. A sizable minority (30%–50%) of AGMs in the wild were shown to possess antibodies to SIV (KANKI et al. 1985; HENDRY et al. 1986; OHTA et al. 1988). SIV_{agm} was subsequently isolated from vervets and characterized (OHTA et al. 1988; DANIEL et al. 1988; KRAUS et al. 1989; ALLAN et al. 1990a). Isolates from the other AGM species, grivets, sabaeus, and tantalus, have been reported more recently (ALLAN et al. 1990a, 1991; MÜLLER et al 1993; JOHNSON et al. 1990a; FOMSGAARD et al. 1991). A few complete and many partial nucleotide sequences of SIV_{agm} from all four species have been reported, and biologically active clones are available (see below). The genomic structure of SIV_{agm} is typical of lentiviruses and follows the general gene order 5′ LTR-gag-pol-vif-vpx-tat-rev-env-nef-3′ LTR (FUKASAWA et al. 1988; BAIER et al. 1990).

3 SIV_{agm} Infected AGMs Remain Healthy

One of the most puzzling enigmas is why and how feral, naturally SIV-infected African monkeys remain healthy. We have perfectly healthy AGMs (vervets) in our colony that are triple retrovirus-infected, namely by simian T-lymphotropic virus, simian foamy virus, and SIV. There are now at least five different Old World monkey species that are naturally SIV-infected in the wild, and from which virus could be isolated—and none of them appears to develop immunodeficiencies! These species, besides AGM, are sooty mangabey monkeys (*Cercocebus atys*; infected by SIV_{smm}), mandrills (*Papio sphinx*; SIV_{mnd}), Syke's monkeys (*Cercopithecus mitis*; SIV_{syk}), and chimpanzees (*Pan troglodytes*; SIV_{cpz}). It is surprising how little research has been carried out with the healthy infected monkey models. Understanding the reasons for their persistent well-being despite lentivirus infection may teach us lessons about the pathogenic mechanisms of HIV-induced disease (KURTH and NORLEY 1993).

As in most pathogenic virus infections, determinants of HIV/SIV pathogenesis reside with both the virus and the host. For example, HIV is apparently not pathogenic in chimpanzees, yet causes disease in humans. Persistent SIV_{smm} infection is tolerated by the natural host, but, injected into macaques, it leads to the development of simian AIDS. In fact, due to their very close genomic homology, it has been suggested that SIV_{smm} may have recently been transmitted to rhesus monkeys (detected as SIV_{mac}) and humans (detected as HIV-2; HIRSCH et al. 1989; GAO et al. 1992). This interspecies transmission allowed the expression of the pathogenic potential of the progenitor SIV_{smm}. But what are the underlying reasons for disease susceptibility of heterologous host species?

In an attempt to answer this question we embarked on the study of the lack of disease development in SIV_{agm} infected AGM, specifically vervets. There are a number of possibilities that spring to mind when contemplating the resistance of the natural host to SIV-induced disease. These include:

- vigorous (effective) immune response of the host
- lack of immunosuppressive viral domain
- lack of cytopathogenicity
- host cell tropism different from HIV
- reduced in vivo variability compared to HIV
- reduced in vivo virus load compared to HIV
- intracellular virus suppression
- $CD8^+$ T-cell derived lymphokine suppressing SIV replication.

4 The Antiviral Immune Response

4.1 Neutralizing Antibodies

It was initially conceivable that a qualitatively different, namely vigorous and effective, immune reponse would signify the evolutionary adaptation of the healthy host species to their lentivirus infections. However, in the corresponding animal model studied in greatest detail, the SIV-infected vervet, the functional immune reponse is apparently not very pronounced (NORLEY et al. 1990a).

In our colony of long-term naturally or experimentally infected vervets, we found at best neutralizing antibody titers of 1:16–1:32 in a minority of the about 50 animals tested. Even in several homologous situations, when a given virus isolate was incubated with serial serum samples from the host animal from which the isolate originated, we observed no neutralizing antibody activity. One other laboratory has reported low AGM neutralizing antibody titers to SIV_{agm} (ROBERT–GUROFF et al. 1992). It is thus unlikely that the anti-SIV humoral immunity in AGM contributes significantly to protection from disease. Neutralizing antibodies are also absent in SIV_{smm} infected healthy sooty mangabeys (FULTZ et al. 1990). Other natural hosts have not yet been analyzed, and such analysis is urgently required to better evaluate the role of virus-neutralizing antibodies in disease prevention.

In this context, an unusual phenomenon deserves attention. Allan and his collaborators as well as our group have recently described the enhancement of SIV_{agm} infection by recombinant soluble CD4 molecules (ALLAN et al. 1990b; WERNER et al. 1990). This in vitro observation is clearly in contrast to what has been described for HIV-1 laboratory strains and may be explained by the formation of highly stable SIV_{agm}-CD4 conformational intermediates important for viral fusion that is not observed during the neutralization of HIV-1 by sCD4 (ALLAN et al. 1992). Very interestingly, the enhancing effect for SIV_{agm} infection was abrogated with antibodies from naturally infected AGM, reaching inhibitory titers of 1:2000–1:10000. These neutralizing antibody titers are an order of magnitude higher than those usually seen in humans or chimpanzees to free HIV-1 particles. In agreement with our above-mentioned studies was the observed lack of SIV_{agm} neutralization in the absence of rsCD4 (ALLAN et al. 1990b). To summarize, it is clear that AGM do not develop high titer antibodies to free SIV_{agm} particles but may be able to neutralize SIV after cell receptor binding, a defense mechanism already seen in other virus infections. This step in virus replication deserves detailed in vitro analysis for HIV and other SIV strains.

4.2 Low or Absent Anti-gag Response

The strong humoral immune response of AGM naturally infected with SIV_{agm}, as demonstrable in enzyme-linked immunosorbent assays and Western blots, is directed almost exclusively at both the external gp140 and the gp45 transmembrane virus glycoproteins. The "group-specific" p28 core antigen (gag) is rarely immunogenic. At best, low titer antibodies to p28 gag can be detected in less than 10% of the AGM. Likewise, experimentally infected AGM hardly ever develop an anti-gag response, and, if so, the response is weak and transient. In contrast, experimentally infected primates other than AGMs, such as *Macacca nemestrina*, respond well with antibodies specific for both env and gag antigens of SIV_{agm}.

AGMs are probably not genetically unable to mount an anti-gag response. Animals repeatedly immunized with purified p28 gag protein emulsified in incomplete Freund's adjuvant produce a corresponding antibody response. One must keep in mind, however, that such biochemically purified immunogen preparations always become partially denatured so that neoepitopes may well be responsible for antibody induction. The biological significance, if any, of the lack of an anti-gag response in natural infections is unclear at present.

4.3 Complement-Activating Cytotoxic Antibodies

AGMs do not develop complement-activating cytotoxic antibodies, which are also undetectable in HIV-infected individuals. This class of antibodies is readily detectable in oncoretrovirus infections of birds, rodents, or cats and in horses infected by equine infectious anemia virus (for review, see NORLEY and KURTH 1994). The antibodies recognize viral cell surface neoantigens and are effective in destroying infected cells. They require a certain cell surface neoantigen density, as at least two adjuvant antibody molecules are required to initiate the lytic complement cascade. It may well be that in HIV and SIV infections the necessary density of viral epitopes is not reached. It remains to be elucidated whether lentiviral vaccines may induce cytotoxic antibodies that may contribute to protection.

4.4 Antibody-Dependent Cellular Cytotoxicity

SIV_{agm} infected cells exposed to antisera from infected AGMs were easily lysed in the presence of effector peripheral blood lymphocytes (PBLs; NORLEY et al. 1990a). The patterns and high levels of lysis were comparable to those seen in HIV-infected individuals (NORLEY et al. 1990b). Lysis was due to true antibody-dependent cellular cytotoxicity. Natural killer cell activity of PBMC alone (i.e., in the absence of antibody) was detectable in AGMs, albeit equally directed against uninfected and infected cells.

4.5 Cytotoxic T-Lymphocyte Activity

Very little is known or published about the CTL activity of AGM. It has so far not been possible to establish autologous permanently growing AGM cell lines that could be infected or loaded with peptides to serve as targets for $CD8^+$ (or $CD4^+$) effector cells. Preliminary results from short-term PBL target cells pulsed with panels of overlapping peptides suggest that AGM possess a CTL response comparable in magnitude to the situation seen in HIV-infected humans (Siegel et al., unpublished). Yet again, the cell-mediated cytotoxic immune response of naturally infected healthy hosts urgently requires detailed investigation, as this mechanism might contribute to the persistent good health of the animals.

In summary, there is no convincing evidence that it is the immune response that keeps the AGMs healthy after SIV_{agm} infection.

5 Lack of an Immunosuppressive Domain

The immunosuppressive peptide of HIV-1 (ISU peptide) corresponds to a 17 amino acid (583–599) domain of the gp41 transmembrane glycoprotein (RUEGG et al. 1989; DENNER et al. 1994). This domain is highly conserved in the transmembrane envelope proteins of all retroviruses including type C and D viruses and lentiviruses (DENNER et al. 1994). Purified transmembrane proteins and/or corresponding ISU peptides inhibit mitogen-induced lymphoproliferation. In addition, HIV-1 ISU peptide conjugated to a protein carrier inhibited the cytopathic effect and replication of HIV-1 in human MT-4 lymphoma cells (DENNER et al. 1994).

Despite the homology of retroviral ISU domains, the corresponding sequence in the SIV_{agm} gp45 transmembrane envelope protein has diverged somewhat from the amino acid sequence of other retroviral ISU peptides (DENNER et al. 1993). In addition, proliferation of phytohemagglutinin-stimulated AGM and human lymphocytes was inhibited to a lesser degree when incubated with purified, inactivated SIV_{agm} than with HIV-1 preparations (NORLEY et al. 1990a). Preliminary data using synthetic peptides indeed indicate that the corresponding peptide of SIV_{agm} is less immunosuppressive than the HIV-1 peptide (KURTH et al. 1992). However, inhibition of lymphocyte activation appears to depend on both the species of the peripheral blood mononuclear cells (PBMC) and the virus, as the inhibitory effect of HIV-1 on human lymphocytes is greater than that observed for HIV-1 on AGM lymphocytes. Given the high concentration of HIV-1 in lymph nodes during the asymptomatic phase of HIV disease, one could easily conceive immune dysfunctional mechanisms in lymph nodes that may be absent in the lymph nodes of SIV-infected AGMs.

6 Cytopathogenicity and Tropism

Lack of pathogenicity may also be due to a lack of host cell destruction. SIV_{agm} can be propagated in a few permanently growing human $CD4^+$ T-lymphoma cells, notably in the Molt 4/clone 8 line (KRAUS et al. 1989; OHTA et al. 1988). Molt 4/clone 8 cells form syncytia after infection, underlining the potential of SIV_{agm} to fuse and thereby destroy infected cells. Freshly isolated PBMC of human or AGM origin are infectible by SIV_{agm}, including the monocyte/macrophage subpopulation (NORLEY et al. 1990a). Monocytes/macrophages have also been shown to be infected in vivo (J. Ennen and M. Dittmar, Paul-Ehrlich-Institute, Langen). Within the life span of PBMC, syncytia formation is not observed, and single-cell death is difficult to associate with virus infection or with other mechanisms such as senescence. It remains to be studied whether SIV_{agm} can induce apoptosis. SIV_{agm} isolated either from freshly or from long-term infected healthy monkeys exhibit no difference in their in vitro replicative capacity. Thus, no switch from non-syncytia-inducing to syncytia-inducing phenotypes is observed. Taken together, it does not appear that SIV_{agm} is nonpathogenic because of a lack of cytopathogenicity or because of a tropism different from HIV.

7 Diversity and Variability of SIV_{agm}

Absence of the pronounced genetic variability observed for HIV would, again, suggest an explanation for the benign nature of the SIV_{agm} infection of its natural host. Genetically stable SIV_{agm} would theoretically allow the host to mount a more efficient immune response. On the other hand, as already mentioned above, immunity to SIV_{agm} is poorly developed in AGM (caveat: the CTL response has not yet been adequately investigated). When discussing genome plasticity of SIV_{agm}, one must distinguish carefully between the heterogeneity of different SIV_{agm} prototype strains in the four different major AGM species (Table 1) and the heterogeneity that developed within one species or even within a single animal.

8 Diversity of SIV_{agm} in the Four *C. Aethiops* Species

Initially, SIVs from AGMs were described as one virus group with a novel and extremely high degree of genetic divergence (JOHNSON et al. 1990a; BAIER et al. 1990). It now appears, however, that distinct SIV_{agm} exist naturally in all four major species of AGM (Table 1) that live similar, though not identical, life-styles in geographically distinct areas of sub-Saharan Africa (Fig. 1), and that the four distinct lentiviruses coevolved with their natural hosts (ALLAN et al. 1991).

Amino acid sequence analysis of constant and variable regions of the outer envelope glycoproteins revealed that heterogeneity between SIVs from different AGM species was clearly higher than intraspecies diversity (MÜLLER et al. 1993). Amino acid identity was highest (77.5%) between isolates from vervet and sabaeus monkeys and lowest (57.1%) between tantalus and sabaeus monkeys (Fig. 2). This is somewhat surprising, as tantalus and sabaeus monkeys live in areas in West Africa that border on the Volta River, wheras the habitats of vervets and sabaeus are separated by at least 2000 miles. This observation, but more importantly the overall diversity derived from the sequencing data, supports the assumption of an SIV infection of AGMs before speciation (ALLAN et al. 1991; FOMSGAARD et al. 1991; MÜLLER et al. 1993).

The development of species-specific SIV_{agm} groups also suggests that interspecies transmission is an extremely rare event, complicating previous interpretations of recent cross-species transmission of SIV_{smm} into macaques and humans (HIRSCH et al. 1989; GAO et al. 1992).

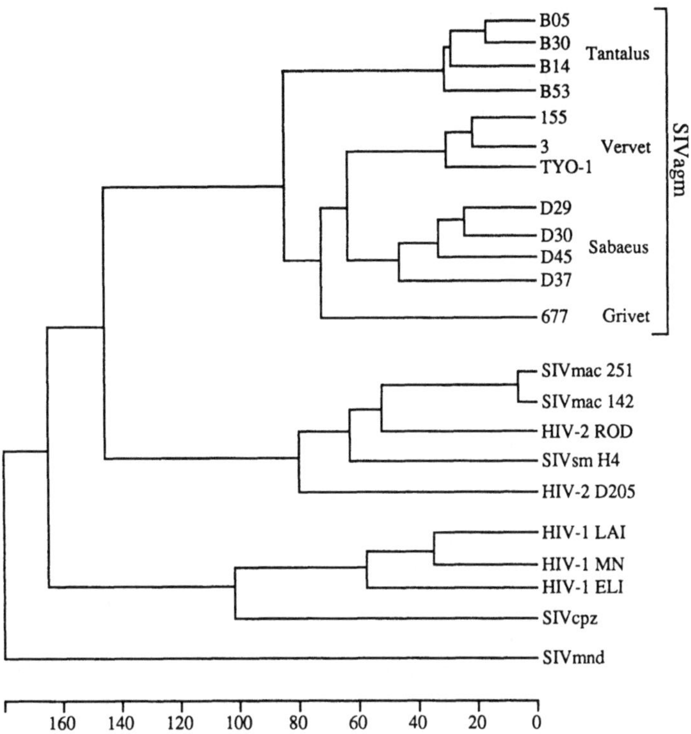

Fig. 2. Distance tree depicting the relationship in *env* of SIV_{agm} to other known SIVs and HIVs. Horizontal distances are proportional to the average number of nonsilent substitutions. (Adapted from MÜLLER et al. 1993)

9 Intraspecies Variability

$SIV_{agm/ver}$ isolates have been studied most often and extensively and therefore serve here to illustrate the pronounced intraspecies variability that can develop in single AGM species. Complete nucleotide sequence analysis revealed an unusually high degree of variability in pol (FUKASAWA et al. 1988; LI et al. 1989; BAIER et al. 1990, 1991), a high degree of divergence in rev and tat, and hypervariable regions in env (BAIER et al. 1990).

Isolates from naturally long-term infected animals also display extensive genetic variability and defectiveness, especially in the V1- and V2-like domains of their external envelope glycoproteins comparable to what is seen in HIV-infected individuals (BAIER et al. 1991). In contrast to the situation postulated for HIV, predominant major virus variants are not observed in infected vervets. In monkeys infected by a biologically active molecular virus clone, extensive variability evolved within 2 years, underlining that a viral quasispecies may rapidly develop from infection with a homogeneous, cloned viral genome (BAIER et al. 1991; CICHUTEK et al. 1992).

Lack of genome plasticity can certainly not be made responsible for the apathogenicity of SIV_{agm} strains in their natural hosts.

10 In Vivo Virus Load and Intracellular Virus Suppression

Lentivirus load in the peripheral blood of HIV-infected individuals often serves as a marker for disease progression. Persistently low virus load and/or absence of actively replicating virus might help to explain the failure of SIV_{agm} to induce an AIDS-like disease. However, results from polymerase chain reaction analysis of SIV_{agm} DNA and RNA in PBMC, in vitro titration of infected PBMC, and the detection of low-level viremia in 4 of the 17 monkeys tested all demonstrated that SIV_{agm} is continuously expressed in naturally infected monkeys, albeit at low levels that never reach the values observed in persons with disease progression (HARTUNG et al. 1992). The question remains whether AGMs have over time evolved cellular mechanisms that limit intracellular lentivirus replication, and that are not available in heterologous hosts after more or less recent transspecies transmissions. Such suppressor mechanisms require studies at the clonal cell level for both HIV and SIV infections. Unfortunately, permanently growing AGM $CD4^+$ cell lines have not yet been established.

11 Genomic Determinants of Disease

It is a well-known phenomenon in many virus infections that pathogenicity is not only a property of the virus but also the result of host susceptibility. Attempts to define determinants of pathogenicity at the viral genome level have been largely unsuccessful—notable exceptions have been elucidated in polio and influenza virus infections. The investigations of an immunodeficiency-inducing potential of SIV_{agm} for other species met with largely negative results. SIV_{agm} transiently infect *M. mulatta*, *M. nemestrina*, and *M. fascicularis* without inducing real disease (HERCHENRÖDER et al. 1989; GRAVELL et al. 1989; HARTUNG et al. 1990; HONJO et al. 1990; JOHNSON et al. 1990b). In only two studies were declines in $CD4^+$ lymphocytes (in two of eight infected rhesus monkeys; JOHNSON et al. 1990b) or death (one of four infected pig-tailed macaques; GRAVELL et al. 1989) reported. However, further passage of SIV_{agm} in pig-tailed macaques did result in the generation of a highly pathogenic, AIDS-inducing variant (P. Johnson, personal communication). The existence of pathogenic and apathogenic SIV (and HIV) isolates can be exploited, allowing the exchange of genes to establish the genomic basis of disease induction (SHIBATA et al. 1990).

12 The Value of the SIV_{agm} Model

SIV infection of the four species of *C. aethiops* represents at present the probably oldest known example of a natural lentivirus infection. More virus isolates from these species need to be analyzed to quantify more precisely viral interspecies and intraspecies divergence as well as genome plasticity in the individual animal. The selective pressure driving variant formation by favoring nonsynonymous mutations (CICHUTEK et al. 1992) is unknown at present but is unlikely to be of an immunological nature, as the immune response to SIV_{agm} does not appear to be well developed.

The animal models of SIV-infected healthy natural hosts deserve study to delineate markers of disease prevention. If it is not a particularly effective immune response that protects—what else is it? We have discussed a few other possibilities above without being able to reach firm conclusions. At the moment, we would emphasize two major areas that certainly deserve intensive investigations: (a) the CTL response in healthy natural hosts and in long-term survivors, and (b) the identification of what we have designated an immunodeficiency virus suppressing lymphokine (ISL). ISL had been described in humans as a $CD8^+$ CTL-derived soluble factor that inhibits HIV replication in vitro (WALKER et al. 1991; MACKEWICZ and LEVY 1992). We recently found ISF also produced by AGM $CD8^+$ cells and feel it worthwhile to try to isolate and characterize this putative inhibitor of HIV/SIV replication.

References

Allan JS, Kanda P, Kennedy RC, Cobb EK, Anthony M, Eichberg JW (1990a) Isolation and characterization of simian immunodeficiency virus from two subspecies of African green monkeys. AIDS Res Hum Retroviruses 6: 275–285

Allan JS, Strauss J, Buck DW (1990b) Enhancement of SIV infection with soluble receptor molecules. Science 247: 1084–1088

Allan JS, Short M, Taylor ME, Su S, Hirsch VM, Johnson PR, Shaw GM, Hahn BH (1991) Species-specific diversity among simian immunodeficiency viruses from African green monkeys. J Virol 65: 2816–2828

Allan JS, Whitehead EM, Strout K, Short M, Kanda P, Hart TK, Bugelski PJ (1992) Strong association of simian immunodeficiency virus (SIV_{agm}) envelope glycoprotein heterodimers: possible role in receptor-mediated activation. AIDS Res Hum Retroviruses 8: 2011–2020

Baier M, Garber C, Mueller C, Cichutek K, Kurth R (1990) Complete nucleotide sequence of a simian immunodeficiency virus from African green monkeys: a novel type of intragroup divergence. Virology 176: 216–221

Baier M, Dittmar MT, Cichutek K, Kurth R (1991) Development in vivo of genetic variability of simian immunodeficiency virus. Proc Natl Acad Sci USA 88: 8126–8130

Cichutek K, Merget H, Norley S, Linde R, Kreuz W, Gahr M, Kurth R (1992) Development of a quasispecies of human immunodeficiency virus type 1 in vivo. Proc Natl Acad Sci USA 89: 7365–7369

Daniel MD, Li Y, Naidu YM, Durda PJ, Schmidt DK, Troup CD, Silva DP, MacKey JJ, Kestler HW, Sehgal PK, King NW, Ohta Y, Hayami M, Desrosiers RC (1988) Simian immunodeficiency virus from African green monkeys. J Virol 62: 4123–4128

Denner J, Norley S, Cichutek K, Ennen J, Kurth R (1993) Factors determining the pathogenicity of immunodeficiency viruses. AIFO 5: 301–306

Denner J, Norley S, Kurth R (1994) The immunosuppressive peptide of HIV-1: functional domains and immune response in AIDS patients. AIDS (in press)

Fomsgaard A, Hirsch VM, Allan JS, Johnson PR (1991) A highly divergent proviral DNA clone of SIV from a distinct species of African green monkey. Virology 182: 397–402

Fukasawa M, Miura T, Hasegawa A, Morikawa S, Tsujimoto H, Miki K, Kitamura T, Hayami M (1988) Sequence of simian immunodeficiency virus from African green monkey, a new member of the HIV/SIV group. Nature 333: 457–461

Fultz PN, Stricker RB, McClure HM, Anderson DC, Switzer WM, Horaist C (1990) Humoral response to SIV/SMM infection in macaque and mangabey monkeys. J Acquir Immune Defic Syndr 3: 319–329

Gao F, Yue L, White AT, Pappas PG, Barchue J, Hanson AP, Greene BM, Sharp PM, Shaw GM, Hahn BH (1992) Human infection by genetically diverse SIV_{sm}-related HIV-2 in West Africa. Nature 358: 495–499

Gravell M, London WT, Hamilton RS, Stone G, Monzon M (1989) Infection of macaque monkeys with simian immunodeficiency virus from African green monkeys: virulence and activation of latent infection. J Med Primatol 18: 247–254

Grzimek B (1988) Enzyklopädie Säugetiere. Kinder Munich

Hartung S, Norley S, Kraus G, Werner A, Vogel M, Bergmann L, Baier M, Kurth R (1990) Infection of non-human primates with SIV_{agm} and HIV-2. In: Schellekens H, Horzinek MC (eds) Animal models in AIDS. Elsevier Amsterdam

Hartung S, Boller K, Cichutek K, Norley SG, Kurth R (1992) Quantitation of a lentivirus in its natural host: simian immunodeficiency virus in African green monkeys. J Virol 66: 2143–2149

Hendry RM, Wells MA, Phelan MA, Schneider AL, Epstein JS, Quinnan GV (1986) Antibodies to simian immunodeficiency virus in African green monkeys in Africa in 1957–62. Lancet 2: 455

Herchenröder O, Stahl-Hennig C, Lueke W, Schneider J, Schulze G, Hartmann H, Schmidt H, Tenner-Racz K, Racz P, Hayami M, Kelliher JC, Hunsmann G (1989) Experimental infection of rhesus monkeys with SIV isolated from African green monkeys. Intervirology 30 Suppl 1: 66–72

Hill WCO (1972) Evolutionary biology of the primates. Academic, London

Hirsch VM, Olmsted RA, Murphey-Corb M, Purcell RH, Johnson PR (1989) An African primate lentivirus (SIVsm) closely related to HIV-2. Nature 339: 389–392

Honjo S, Narita T, Kobayashi R, Hiyaoka A, Fujimoto K, Takasaka M, Sakakibara I, Mukai R, Ishikawa K, Ohta Y, Hayami M (1990) Experimental infection of African green monkeys and cynomolgus monkeys with a SIV_{agm} strain isolated from a healthy African green monkey. J Med Primatol 19: 9–20

Johnson PR, Fomsgaard A, Allan J, Gravell M, London WT, Olmsted RA, Hirsch VM (1990a) Simian immunodeficiency viruses from African green monkeys display unusual genetic diversity. J Virol 64: 1086–1092

Johnson PR, Goldstein S, London WT, Fomsgaard A, Hirsch VM (1990b) Molecular clones of SIV_{sm} and SIV_{agm}: experimental infection of macaques and African green monkeys. J Med Primatol 19: 279–286

Kanki PJ, Kurth R, Becker W, Dreesman G, McLane MF, Essex M (1985) Antibodies to simian T-lymphotropic retrovirus type III in African green monkeys and recognition of STLV-III viral proteins by AIDS and related sera. Lancet 1: 1330–1332

Kraus G, Werner A, Baier M, Binniger D, Ferdinand FJ, Norley S, Kurth R (1989) Isolation of human immunodeficiency virus-related simian immunodeficiency viruses from African green monkeys. Proc Natl Acad Sci USA 86: 2892–2896

Kurth R, Norley S (1993) Animal immunodeficiency viruses. In: Levy JA, Weiss R, Neu HC (eds) Focus on HIV: 1992. Churchill Livingstone, Edinburgh
Kurth R, Ennen J, Denner J, Hartung S, Binninger D, Vogel M, Norley S (1992) Immunopathogenesis of HIV/SIV infection: consequences for the development of an AIDS vaccine. In: Kurth R, Schwerdtfeger WK (eds) Current topics in biomedical research. Springer, Berlin Heidelberg New York
Li Y, Naidu YM, Daniel MD, Desrosiers RC (1989) Extensive genetic variability of simian immunodeficiency virus from African green monkeys. J Virol 63: 1800–1802
Mackewicz C, Levy JA (1992) $CD8^+$ cell anti-HIV activity: nonlytic suppression of virus replication. AIDS Res Hum Retroviruses 8: 1039–1050
Müller MC, Saksena NK, Nerrienet E, Chappey C, Herve VM, Durand JP, Legal-Campodonico P, Lang MC, Digoutte JP, Georges AJ, Georges-Courbot MC, Sonigo P, Barré-Sinoussi F (1993) Simian immunodeficiency viruses from central and Western Africa: evidence for a new species-specific lentivirus in tantalus monkeys. J Virol 67: 1227–1235
Napier JR, Napier PH (1985) The natural history of the primates. MIT Press, Cambridge, MA
Norley S, Kurth R (1994) The immune response to retroviruses. In: Levy J (ed) The retroviridae. Plenum, New York
Norley SG, Kraus G, Ennen J, Bonilla J, König H, Kurth R (1990a) Immunological studies of the basis for the apathogenicity of simian immunodeficiency virus from African green monkeys. Proc Natl Acad Sci USA 87: 9067–9071
Norley SG, Mikschy U, Werner A, Staszewski S, Helm EB, Kurth R (1990b) Demonstration of cross-reactive antibodies able to elicit lysis of both HIV-1- and HIV-2-infected cells J Immunol 145: 1700–1705
Ohta Y, Masuda T, Tsujimoto H, Ishikawa K, Kodama T, Morikawa S, Nakai M, Honjo S, Hayami M (1988) Isolation of simian immunodeficiency virus from African green monkeys and seroepidemiologic survey of the virus in various non-human primates. Int J Cancer 41: 115–122
Robert-Guroff M, Aldrich K, Muldoon R, Stern TL, Bansal GP, Mathews TJ, Markham PD, Gallo RC, Franchini G (1992) Cross-neutralization of human immunodeficiency virus type 1 and 2 and simian immunodeficiency virus isolates. J Virol 66: 3602–3608
Ruegg CL, Monell CR, Strand M (1989) Inhibition of lymphoproliferation by a synthetic peptide with sequence identity to gp41 of human immunodeficiency virus type 1. J Virol 63: 3257–3260
Shibata R, Sakai H, Kiyomasu T, Ishimoto A, Hayami M, Adachi A (1990) Generation and characterization of infectious chimeric clones between human immunodeficiency virus type 1 and simian immunodeficiency virus from an African green monkey. J Virol 64: 5861–5868
Walker CM, Erickson AL, Hsueh FC, Levy JA (1991) Inhibition of human immunodeficiency virus replication in acutely infected $CD4^+$ cells by $CD8^+$ cells involves a noncytotoxic mechanism. J Virol 65: 5921–5927
Werner A, Winskowsky G, Kurth R (1990) Soluble CD4 enhances simian immunodeficiency virus SIV_{agm} infection. J Virol 64: 6252–6256

Pathology of Simian Immunodeficiency Virus Induced Disease

A.A. LACKNER

1 Introduction

Simian immunodeficiency virus (SIV) is the general name given a large family of African nonhuman primate lentiviruses which are the closest known relatives to the human immunodeficiency virus (HIV; CHAKRABARTI et al. 1987; FRANCHINI et al. 1987; HIRSCH et al. 1989). In their natural hosts, which include sooty mangabeys (*Cercocebus atys*; FULTZ et al. 1986; LOWENSTINE et al. 1986), African green monkeys (*Cercopithecus aethiops*; DANIEL et al. 1988; KANKI et al. 1985), mandrills (*Papio sphinx*; TSUJIMOTO et al. 1988), white-crowned mangabeys (*Cercocebus torquatus lunulatus*; TOMONAGA et al. 1993), Sykes' monkeys (*Cercopithecus mitis albogularis*; HIRSCH et al.

California Regional Primate Research Center and Department of Veterinary Pathology University of California, Davis, CA 95616, USA

Current Topics in Microbiology and Immunology, Vol. 188

1993), tantalus monkeys (*Cercopithecus tantalus*; M.C. MÜLLER et al. 1993, and possibly other monkeys of African origin, the virus apparently exists as an exogenous, nonpathogenic infection (DESROSIERS 1990). Remarkably, when SIV is introduced from one of the African nonhuman primates into an Asian macaque species, as appears to have occurred in several primate centers in the United States, an AIDS-like disease results (for review see GARDNER et al. 1988).

Members of the lentivirus subfamily of retroviruses including SIV and HIV share many unique biological and genetic properties, including macrophage tropism, the presence of additional regulatory genes and the propensity to induce chronic debilitating disease associated with persistent infection in the face of a host immune response. Associated with macrophage tropism is a common organ tropism, particularly for the brain and lung first observed with the prototype visna virus in sheep (HAASE 1986).

One feature that differentiates SIV and HIV from other lentiviruses is a strong tropism for cells bearing the CD4 molecule (DANIEL et al. 1985; KLATZMANN et al. 1984). Therefore, helper/inducer lymphocytes, monocytes, macrophages, Langerhans cells' in the epidermis, and follicular dendritic cells in lymphoid organs are at risk of infection. Cells without detectable CD4 have also been reported to be infected by either SIV or HIV but this is controversial (CHENG-MAYER et al. 1987; CLAPHAM et al. 1989; KUNSCH et al. 1989).

The clinical disease associated with SIV infection in rhesus monkeys is very similar to the clinical syndrome in human AIDS patients except that the disease course is greatly shortened, with approximately 70% of the animals progressing to AIDS in 1 year (KING et al. 1990). This disease course can vary, however, depending on the host immune response, source of the virus, and virus strain. One isolate of SIV known as SIV_{smm_PBj14} (SIV from sooty mangabey monkey PBj at 14 months postinoculation) results in fatal disease in most pig-tailed macaques (*Macaca nemestrina*) within 1–2 weeks (FULTZ et al. 1989). Generally however, SIV-infected animals develop lymphadenopathy and splenomegaly early in the disease followed by generalized lymphoid depletion including thymic involution in the terminal phases of the disease. Associated with these lymphoid changes is a decrease in the number of $CD4^+$ lymphocytes in circulation and a concomitant decrease in the CD4:CD8 ratio. This decline, however, tends to be more abrupt than is commonly seen in HIV-infected humans. Blastogenic responses of peripheral blood lymphocytes to mitogens are also depressed (LETVIN et al. 1985). Many of the animals develop chronic diarrhea with progressive weight loss from which no known pathogens or opportunistic organisms can be found. A transient maculopapular rash has also been observed in up to 80% of inoculated animals (RINGLER et al. 1986, 1987).

SIVs have been isolated from several different macaque species. To differentiate the isolates a three letter subscript generally derived from either the common or scientific name of the host indicates the species of monkey from which they were originally isolated, for example, SIV_{mac} from rhesus macaques (*M. mulatta*). In addition, there are multiple subtypes of many of these isolates including various molecular clones and recombinants. Among the most notable are $SIV_{mac-251}$, the molecular clone $SIV_{mac-239}$, and $SIV_{smm\ PBj14}$, which is discussed elsewhere in this volume.

The following discussion of SIV-associated pathology focuses on that associated with infection of macaques by SIV_{mac} and SIV_{smm} (SIV derived from sooty mangabey monkeys) since most of the existing work has involved these isolates. Other investigators have described similar or identical findings using other strains of SIV in macaques of several species (BENVENISTE et al. 1988; FULTZ et al. 1989; KAHN et al. 1991; KULLER et al. 1990; LOWENSTINE et al. 1992). Significant differences between isolates are pointed out when appropriate.

The pathological sequelae of SIV infection and simian AIDS can be conceptually divided in several different ways. The most relevant is to discuss the pathology in the context of the pathogenesis of SIV by dividing it into the following four categories (initially proposed by SIMON et al. 1992): (a) pathology associated with SIV infection of resident cells, (b) pathology of uncertain etiology, (c) opportunistic infections, and, (d) lymphomas. The division between the first two categories is somewhat arbitrary and will undoubtedly change with additional information.

2 Pathology Associated with SIV Infection of Resident Cells

A number of lesions seen in the lungs, brain, and lymphoid tissue are associated with direct infection of local cells by SIV. The histopathological lesions in these organ systems are almost certainly a direct effect of SIV infection. Opportunistic pathogens are not directly or uniformly associated with these lesions.

2.1 Hemolymphatic System

Lymphoid tissues throughout the animal go through a progression of histopathological changes. The rapidity of this progression varies depending on the host immune response to the virus. Animals which mount a vigorous immune response have a prolonged disease course whereas animals that do not mount a vigorous immune response develop AIDS and severe lymphoid depletion in a matter of months. The following discussion focuses on the peripheral lymph nodes and spleen, where the changes are most thoroughly documented (CHALIFOUX et al. 1987; RINGLER et al. 1989; WYAND et al. 1989), but the thymus and diffuse lymphoid tissues of the mucosal immune system also show progressive alterations that can be correlated with disease progression (BASKERVILLE et al. 1990, 1992a,b; BASKIN et al. 1991a, b). Lymphoid changes in the intestinal tract are briefly discussed later in this chapter, and the mucosal immune system of the reproductive tract is covered in a separate chapter in this volume.

2.1.1 Peripheral Lymph Nodes and Spleen

As SIV-induced simian AIDS progresses, the lymph nodes and spleen undergo a series of histopathological alterations. These alterations have been divided into five categories which represent progressive stages of SIV infection (CHALIFOUX et al. 1987; RINGLER et al. 1989): (a) normal lymph nodes, (b) follicular hyperplasia, (c) follicular depletion with normal or expanded paracortices, (d) lymphoid depletion, and (e) giant cell lymphadenitis/splenitis.

Normal Lymph Nodes. These are seen very early in the course of infection (time course varies with viral inoculum and host immune response), and in most cases SIV antigen and nucleic acid containing cells are rare and thought to be resident macrophages and lymphocytes (RINGLER et al. 1989; WYAND et al. 1989). The extent of macrophage infection likely depends on the cellular tropism of the inoculum (DESROSIERS et al. 1991).

Follicular Hyperplasia and Dysplasia. This histological pattern is associated with generalized lymphadenopathy seen early in the course of infection. Lymphoid organs contain large dysplastic germinal centers of irregular shape that often merge. The paracortex is reduced in size relative to the cortex and has prominent high endothelial venules. Viral antigen is consistently associated with follicular dendritic cells (FDC) in the hyperplastic follicles, and the FDC network is generally intact although more evident in the outer half of the germinal center based on immunohistochemical studies (JOLING et al. 1992; RINGLER et al. 1989). Ultrastructural studies, however, have shown swelling and blunting of FDC dendrites as well as electron dense deposits that are positive for SIV core antigen by immunogold electron microscopy and rare lentiviral particles within the labyrinth of FDC processes (JOLING et al. 1992; RINGLER et al. 1989). The swelling and blunting of FDC dendrites suggests that in some animals disruption of the FDC network begins at this stage of infection. In contrast to the distribution of viral antigen, viral nucleic acid at this stage is most commonly seen in individual lymphocytes and macrophages in the paracortex (WYAND et al. 1989). Occasional animals with very high viral loads have diffuse labeling of the germinal centers for SIV nucleic acid. This lack of consistent detection of SIV nucleic acids in germinal centers may be due to procedural limitations since in recent studies of lymphoid tissue in HIV-infected humans using RNA probes viral nucleic acid could not be consistently detected without prior proteolytic digestion (FOX et al. 1991; PANTALEO et al. 1993).

Follicular Depletion with Normal or Expanded Paracortices. In this stage germinal centers are decreased in number and size and often contain hyalinized material, deposits of which can be quite extensive. The paracortex is normal or expanded and contains increased numbers of $CD8^+$ cells (CHALIFOUX et al. 1987; JOLING et al. 1992). Viral antigen and nucleic acid are present in lymphocytes and macrophages throughout the parenchyma (RINGLER et al. 1989; WYAND et al. 1989). Viral antigen can still be found on FDCs although the FDC network shows signs of fragmentation.

The pathogenesis of destruction of the FDC network is not clear. One hypothesis has been that it is the result of cytotoxic T-cells directed against FDCs infected by SIV or coated with SIV proteins. In one study examining SIV_{smm} infection in cynomolgus macaques (*M. fascicularis*), however, no evidence for this hypothesis we found, and it was proposed that virus and viral proteins play a direct role in destruction of the FDC network (JOLING et al. 1992).

Lymphoid Depletion. The majority of animals which die from SIV-induced simian AIDS have generalized lymphoid depletion at the time of necropsy. Germinal centers are virtually absent with only hyalinized remnants of follicles. The paracortex is also reduced in volume. Sinus histiocytosis is often present which can make up the bulk of the lymph node. The FDC network is fragmented and largely absent. At this stage, the number of cells containing viral antigen or nucleic acid is decreased and limited to individual cells in the parenchyma (RINGLER et al. 1989; WYAND et al. 1989). This progressive deterioration of lymphoid architecture is virutallly identical to that described in HIV-infected humans. In addition, in lymph nodes both SIV and HIV are localized primarily to cells of the macrophage and FDC lineage (EMBRETSON et al. 1993; PANTALEO et al. 1993; RINGLER et al. 1989; WYAND et al. 1989). Follicular dendritic cells appear to sequester viral proteins during follicular hyperplasia, while tissue macrophages supersede FDCs in this role in latter stages of infection.

Giant Cell Lymphadenitis/Splenitis. Approximately 25% of animals infected with SIV progress to this stage, which is characterized by the presence of multinucleated giant cells. Most of the animals which progress to this stage have disseminated giant cell disease with multinucleated giant cells in the brain (SIV encephalitis), eye, lung, intestinal tract, and other organs (BASKIN et al. 1988). Animals with giant cell lymphadenitis/splenitis have gross lymphadenopathy/splenomegaly and very high viral loads based on plasma viremia and antigenemia. Virually all of the giant cells and up to 10% of resident macrophages contain viral antigen and viral nucleic acid. It is interesting to note that while all of the lymphoid changes described are apparently directly associated with SIV infection only the giant cell lymphadenitis/splenitis is lentiviral specific. A type D retrovirus that is indigenous in macaques causes a virtually identical progression of lymphoid changes with the exception of the giant cell lymphadenitis/splenitis (ORMEROD et al. 1988; OSBORN et al. 1984).

2.1.2 Thymus

The majority of observations of thymus from SIV-infected macaques have been in terminal animals; these animals have severe thymic atrophy. Grossly, the thymus is rarely visible terminally, but thymic remnants consisting of cystic Hassall's corpuscles, macrophages, and lymphocytes can be found on histology. The progression to lymphoid depletion of the thymus generally begins within the first 2 months post infection and parallels the deterioration of the peripheral lymphoid organs described above (BASKIN et al. 1991a). Prior to lymphoid depletion some animals develop medullary expansion due to lymphocyte infiltration with formation of lymphoid follicles and germinal centers. Similar lymphofollicular infiltrates are present in multiple parenchymal organs of these animals. In extreme cases this has been referred to as a lymphproliferative syndrome (CHALIFOUX et al. 1986). Peripheral lymphoid tissue in these animals generally shows follicular hyperplasia and dysplasia. Evidence of viral infection of the thymus can be found as early as 2 weeks postinoculation and persists until death (BASKIN et al. 1991a). The distribution of virus within the thymus varies depending on viral inoculum and time postinoculation (BASKIN et al. 1991a; LACKNER et al. 1991c). In general fewer infected cells are detected in the cortex than medulla by in situ hybridization (BASKIN et al. 1991a).

2.1.3 Bone Marrow

Bone marrow hyperplasia often occurs in the absence of peripheral leukocytosis or secondary bacterial infections (BASKIN et al. 1988; KITAGAWA et al. 1991). In one study 11 of 11 monkeys with severe lymphoid depletion had hyperplastic bone marrow and yet were anemic and frequently showed lymphopenia, thrombocytopenia, or both (KITAGAWA et al. 1991). In the bone marrow SIV has been shown to be harbored in macrophages (KITAGAWA et al. 1991; WATANABE et al. 1990). In addition, bone marrow hematopoietic progenitor cell colony formation has been shown to be decreased in SIV-infected macaques (WATANABE et al. 1990). These data suggest that SIV infection of bone marrow macrophages adversely affects hematopoietic progenitor cell growth in vivo contributing to the hematological abnormalities that are commonly seen in SIV-infected macaques.

2.2 Respiratory System

Many rhesus monkeys inoculated with SIV develop pulmonary lesions. Grossly, affected lungs often have multiple 0.1- to 2-cm-diameter irregular firm tan areas scattered throughout the lung lobes that tend to coalesce. In extreme cases the lungs are consolidated, do not collapse, and are firm. Histologically the most consistent lesions are interstitial pneumonia and

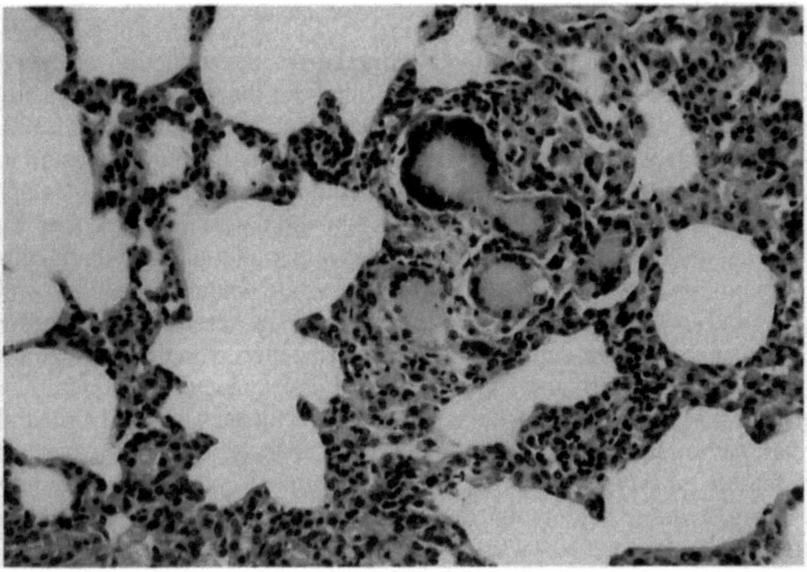

Fig. 1. Lung with multinucleated giant cells in the interstitium. Hematoxylin and eosin

multinucleated giant cells (Fig. 1). Alveolar sepatae in affected areas are thickened and contain lymphocytes, macrophages, and occasional multinucleated giant cells (BASKERVILLE et al. 1992b; BASKIN et al. 1991b). This lesion begins to develop early in infection, increasing in severity with time (BASKIN et al. 1991b). The earliest and most consistent finding in the pulmonary parenchyma is that of perivascular infiltrates of histiocytes intermixed with lymphocytes around small vessels (BASKIN et al. 1991b) similar to the early lesions seen in the central nervous system (see below). In moribund animals alveoli can be filled with fibrin, macrophages, and multinucleated giant cells, and alveolar lining cells are generally hyperplastic and hypertrophic with cuboidal metaplasia (BASKERVILLE et al. 1992b; BASKIN et al. 1991b). The majority of the multinucleated giant cells and macrophages in the pulmonary infiltrates are positive for SIV antigen and nucleic acid. Most of the animals that have a giant cell pneumonia also have giant cells in one or more other organs, in particular the brain.

Peribronchial lymphoid tissue undergoes progressive changes as described above for peripheral lymph nodes and spleen. Early in the disease when the peribronchial lymphoid tissue is hyperplastic this lesion resembles pulmonary lymphoid hyperplasia of pediatric AIDS (MURRAY and MILLS 1990a, b). Massive infiltration of alveolar septae by lymphocytes comparable to lymphocytic interstitial pneumonia in pediatric AIDS patients has not been seen (BASKIN et al. 1988; SIMON et al. 1992) except in one report (BASKERVILLE et al. 1992b).

2.3 Nervous System

Neuropathological lesions directly attributable to SIV infection are seen terminally in 30%–60% of moribund SIV-infected macaques without any evidence of opportunistic pathogens (BASKIN et al. 1992; KING et al. 1990; LACKNER et al. 1991b; SIMON et al. 1992). The primary histological lesion in the central nervous system (CNS) of SIV-infected macaques is known as SIV encephalitis (LACKNER et al. 1991a). This encephalitis is characterized by perivascular infiltrates of foamy macrophages and multinucleated giant cells in the meninges and throughout the gray and white matter of the brain and spinal cord, but the lesions are most common in the cerebral white matter (Fig. 2). As noted above, similar giant cells may be present in the lung, lymph nodes, intestine, spleen, and other organs of SIV-infected macaques. The majority of these giant cells contains SIV antigen, nucleic acid, and lentiviral particles. Although formation of syncytial giant cells throughout the body is not a feature of HIV infection of humans, identical giant cell containing lesions are a major feature of the neuropathology of HIV encephalitis (EPSTEIN 1989; PRICE et al. 1988).

Typically the CNS lesions in SIV-infected macaques consist of closely packed, multifocal perivascular infiltrates of macrophages and multinucleated giant cells with occasional lymphocytes and rare polymorphonuclear cells. In addition, small neuroglial nodules are present in some animals. Lesions are often surrounded by a zone of spongiosis. In rare instances the discrete

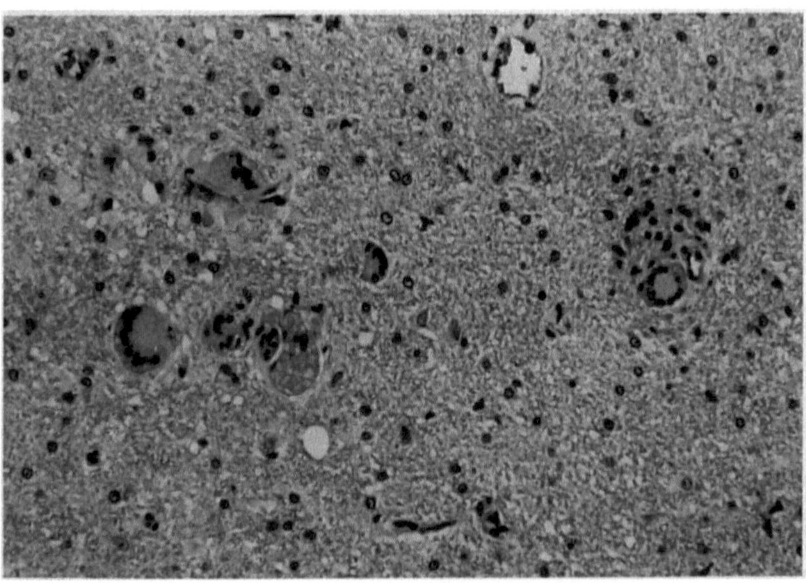

Fig. 2. Brain with scattered multinucleated giant cells. Hematoxylin and eosin

infiltrates coalesce into large, irregular masses (syncytiomas) that are visible to the naked eye. While most lesions are associated with varying degrees of parenchymal destruction, scattered multinucleated giant cells can also be found alone within white or gray matter, not associated with vessels, and without visible changes in the surrounding parenchayma. Giant cells, particularly in the larger lesions, are occasionally necrotic with hypereosinophilic cytoplasm and karryorhectic nuclei.

Similar lesions are also seen in the eye (VOGEL et al., manuscript submitted). The ocular lesions range from mild perivascular accumulations of inflammatory cells surrounding a few periaxial vessels in the optic nerve to transverse optic neuritis with multinucleated giant cells. The severely affected animals were clinically blind. In animals with severe ocular disease multinucleated giant cells were also present in the iris, ciliary body, and choroid. In general the severity of the neuropthalmic lesions mirrored the severity of inflammation in the rest of the CNS. In situ hybridization studies have demonstrated SIV but not cytomegalovirus (CMV) nucleic acids in the eyes of the affected animals. The localization of SIV to macrophages and multinucleated giant cells is identical to that seen in the brain.

Immunophenotypic studies have confirmed the morphological observation that the discrete infiltrates in the CNS consist predominantly of monocytes and macrophages with lesser numbers of T cells nearly all of which are $CD8^+$ (CHAKRABARTI et al. 1991; LACKNER et al. 1991b). In addition, these studies have demonstrated that the discrete infiltrates seen histologically are part of a much more diffuse infiltrate of monocyte/macrophages in the white matter (LACKNER et al. 1991b). This influx of monocytes and macrophages appears to be associated with expression of vascular cell adhesion molecule 1 on cerebral vasculature (SASSEVILLE et al. 1992). Vascular cell adhesion molecule 1 is relatively specific for monocytes and probably plays a major role in the trafficking of monocytes into the CNS and ultimately in the induction of SIV encephalitis.

White matter changes are generally minimal and limited to spongiosis and demyelination in the area immediately surrounding giant cell infiltrates. The reactive astrocytosis and myelin pallor commonly seen in humans infected with HIV does not occur in macaques (SHARER et al. 1991). In addition, neither the vacuolar myelopathy nor peripheral neuropathy described in AIDs patients has been seen in SIV-infected macaques although giant cell myelitis does occur, as noted above. This may be related to the shorter disease course in SIV-infected macaques and the relative infrequency of opportunistic infections in the CNS of SIV-infected macaques as compared to HIV-infected humans (LACKNER et al. 1991a).

SIV can be isolated from cerebrospinal fluid within 1–2 weeks of infection and CNS lesions have also been observed in this time frame (CHAKRABARTI et al. 1991; LACKNER et al. 1991c; SHARER et al. 1991). The earliest and most consistent lesions consist of perivascular infiltrates of lymphocytes and macrophages in the white matter and meninges (CHAKRABARTI et al. 1991;

LACKNER et al. 1991c; SHARER et al. 1991). Viral nucleic acid can be detected by in situ hybridization in these perivascular infiltrates as early as 7 days post inoculation (CHAKRABARTI et al. 1991).

In animals with SIV encephalitis viral antigen and nucleic acid are consistently detected within monocyte/macrophages, but SIV has not been detected in neurons, astrocytes, or oligodendrogliocytes. Viral antigen and nucleic acids have, however, been associated with vessels in the CNS. Individual cells or clusters of cells associated with capillaries and small vessels in the parenchyma (primarily white matter) and meninges have been shown to contain viral antigens or nucleic acids. Morphologically it is not possible at the light microscopic level to distinguish whether the viral antigen is within endothelial cells, mononuclear cells in the vessel wall or within perivascular microglia (LACKNER et al. 1991a–c).

3 Pathology of Uncertain Etiology

Several additional lesions are commonly seen in SIV-infected macaques; however, the role that SIV plays in their pathogenesis is uncertain. Some of these lesions may be the result of immune dysregulation caused by SIV infection and the direct or indirect effects of cytokines as has been proposed for the enteropathy, arteriopathy, and hepatic lesions (CHALIFOUX et al. 1992; GERBER et al. 1991; HEISE et al. 1993a). Others such as the skin lesions are probably a direct result of SIV infection of resident cells but important pieces of the puzzle remain to be found.

3.1 Intestinal Tract

Chronic diarrhea with weight loss is one of the most common clinical signs in SIV-infected animals. In a significant percentage of these animals no known enteric pathogens or opportunistic organisms can be identified to account for the diarrhea. Histologically the small intestine is characterized by villus blunting and variable crypt hyperplasia (HEISE et al. 1993a, b). Jejunal villus height to crypt depth ratios are often decreased (HEISE et al. 1993a). The lamina propria is hypercellular due to an infiltrate of lymphocytes, plasma cells, and macrophages. Occasional animals have multinucleated giant cells in the lamina propria and mesenteric lymph nodes as part of a generalized giant cell disease. Gut-associated lymphoid tissue may be in any of the histopathologic categories described for the peripheral lymph nodes and spleen depending on the stage of disease.

SIV nucleic acids have been detected in macrophages and T-lymphocytes (but not in enterocytes) throughout the intestine as early as 2 weeks postinoculation (DANDEKAR et al. 1993; HEISE et al. 1993a,). The majority of virus infected cells are in gut-associated lymphoid tissue, especially Peyer's patches in the jejunum and ileum, and lymphoid follicles of the large intestine. Lesser numbers of SIV-positive cells are present in the lamina propria and mesenteric lymph nodes (HEISE et al. 1993a, b).

In addition to morphological abnormalities, SIV-infected macaques have decreased mucosal enzyme activities for sucrase and lactase (HEISE et al. 1993a) and have D-xylose malabsorption (DANDEKAR et al. 1993). These abnormalities can be detected as early as 2 weeks postinoculation coincident with detection of SIV in the intestine (DANDEKAR et al. 1993). The depressed mucosal enzyme activity and abnormal morphometry of the jejunum are suggestive of a maturational defect in proliferating enterocytes in SIV-infected macaques (HEISE et al. 1993a). It has been postulated that this is a result of cytokines released from SIV-infected macrophages in the lamina propria of the intestine (HEISE et al. 1993a).

Enteropathies have also been associated with other immunosuppressive retroviral infections including those by feline leukemia virus in cats (REINACHER 1987), type D retrovirus in macaques (LACKNER et al. 1990), and HIV in humans (NELSON et al. 1988). In some FeLV-infected cats and macaques infected with $SIV_{smm-PBj14}$, intestinal dysfunction can be the primary manifestation of the disease (FULTZ et al. 1989; REINACHER 1987). FeLV and type D retrovirus have been shown to infect crypt epithelial cells, and thus the pathogenesis of intestinal dysfunction in these cases is probably more direct than that for SIV or HIV.

3.2 Integument

A maculopapular rash occurs in many macaques infected with SIV in the first few weeks post infection, persists for several weeks, and then regresses spontaneously (BASKIN et al. 1988; BENVENISTE et al. 1988; ISRAEL et al. 1993; RINGLER et al. 1986). Histologically, affected skin is characterized by a superficial perivascular infiltrate of mononuclear cells. Degeneration of Langerhans' cells in direct apposition to cells of a cytotoxic phenotype suggests that the Langerhans' cells are the target of a cytotoxic response (RINGLER et al. 1987). Infection of Langerhans' cells by SIV has not been proven, but in situ hybridization studies have localized viral nucleic acid to unidentified cells in the epithelium of the reproductive tract and oral mucosa (MILLER et al. 1992; SIMON et al. 1992). In humans a similar rash has been identified, as has infection of Langerhans' cells by HIV (H. MÜLLER et al. 1993; RAPPERSBERGER et al. 1988; TSCHACHLER et al. 1987). This makes it likely that the skin rash is an acute, primary manifestation of SIV infection of Langerhans' cells.

3.3 Cardiovascular System

An arteriopathy has been described in approximately 20% of SIV-infected macaques that die from simian AIDS (CHALIFOUX et al. 1992). Medium to large pulmonary arteries are most commonly affected, but similar lesions have been seen in the kidney, intestine, and other organs. The lesion is characterized by intimal thickening by smooth muscle cells, extracellular matrix, collagen, and rare macrophages (CHALIFOUX et al. 1992). The lesions range from focal eccentric intimal thickening to complete intimal obliteration of the lumen. Inflammation is not a feature of the arteriopathy. Many of the affected animals have pulmonary thrombosis with varying degrees of organization, recanalization, and pulmonary infarction (Fig. 3); (BASKERVILLE et al. 1992b; CHALIFOUX et al. 1992). These lesions are similar to those described in pediatric AIDS patients (JOSHI et al. 1987).

Two other lesions that may be related are an aseptic vegetative endocarditis (CHALIFOUX et al. 1992) and mineralization of vessels in the basal ganglia (Fig. 4); LACKNER et al. 1991a, b). Mineralization of vessels in the basal ganglia has also been described in pediatric AIDS patients (SHARER et al. 1986). As in children, this vascular mineralization is generally not associated with a cellular infiltrate. However, in time course studies, perivascular infiltrates are commonly seen in the CNS early in the infection (CHAKRABARTI et al. 1991; SHARER et al. 1991). This suggests that the mineralization is a result of transient inflammation associated with viremia during the first few weeks postinoculation.

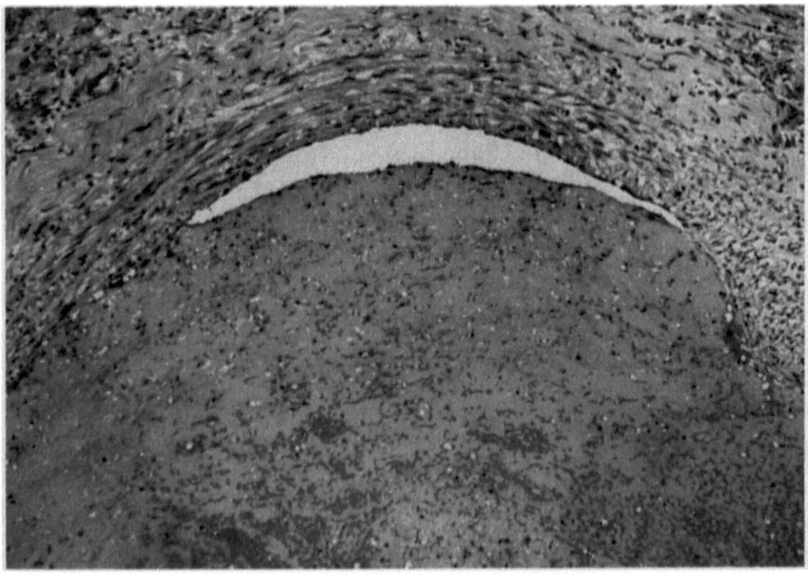

Fig. 3. A large pulmonary artery with a thrombus and recanalization. Hematoxylin and eosin

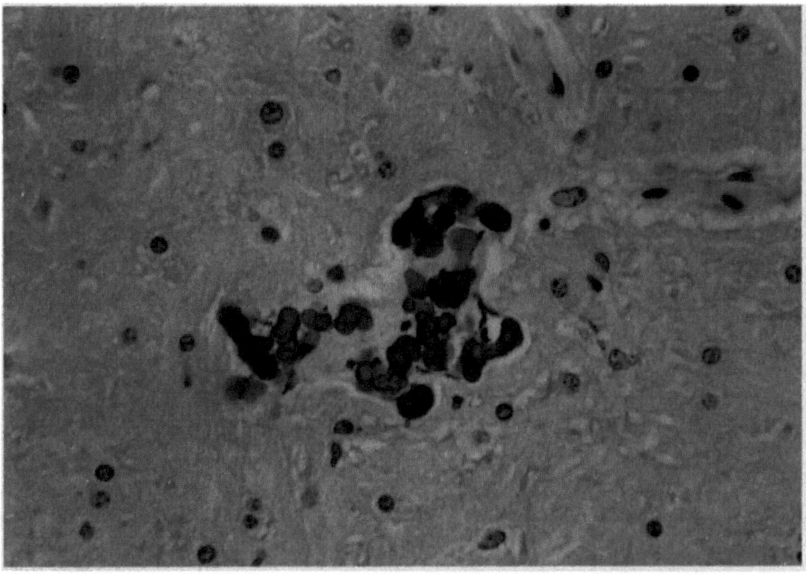

Fig. 4. Perivascular mineralization in the basal ganglia. Hematoxylin and eosin

3.4 Liver

Clinical signs associated with hepatic disease are rarely seen, but histologically many animals infected with SIV have a mild periportal hepatitis (BASKIN et al. 1988; GERBER et al. 1991). The hepatitis is characterized by proliferation of bile ductules and cellular infiltrates of lymphocytes with lesser numbers of macrophages and plasma cells. SIV nucleic acid and antigens can be detected in scattered cells within the cellular infiltrates but not in hepatocytes or bile ducts (GERBER et al. 1991). These lesions have been seen as early as 2 weeks postinoculation with SIV (GERBER et al. 1991). Similar lesions have been described in AIDS patients.

3.5 Kidney

A membranoproliferative glomerulonephritis has been described in up to 50% of SIV-infected macaques (ALPERS et al. 1988; BASKERVILLE et al. 1990; BASKIN et al. 1988). In affected animals most glomeruli are hypercellular and lobulated with hypertrophy and hyperplasia of visceral epithelial cells. Glomerular basement membranes are thickened, and mesangial matrix is increased resulting in constriction of some glomerular capillaries. Peri-

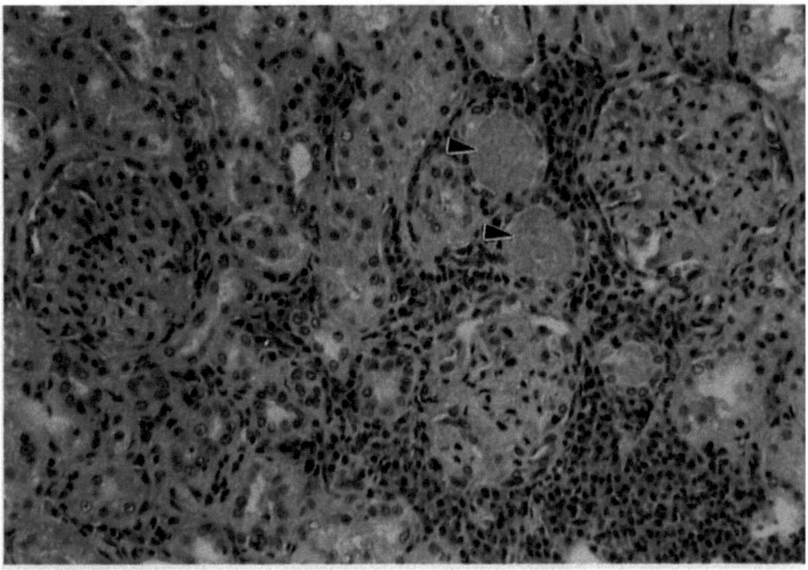

Fig. 5. Renal cortex with hypercellular, swollen glomeruli, interstitial infiltrates of mononuclear cells, and tubular casts (*arrowheads*). Hematoxylin and eosin

glomerular fibrosis, tubular atrophy, tubular casts, a lymphoplasmacytic interstitial nephritis and interstitial scaring have also been seen in association with the glomerular lesions (Fig. 5). Immune complex deposits have not been identified in affected glomeruli (LETVIN and KING 1990).

3.6 Musculoskeletal

A mild arthritis has been described in one study in animals infected with SIV_{smm} (ROBERTS and MARTIN 1991). The arthritis ranged from mild synovial cell hyperplasia to synovial hyperplasia with perivascular infiltrates of mononuclear cells in the subsynovial layer. A few animals had multinucleated giant cells in the synovium as part of a systemic giant cell disease. Viral antigens were only identified in the multinucleated giant cells. While the giant cells in the synovium are clearly related to SIV infection, it is not certain what role SIV plays in the milder lesions. This controversy also exists for the arthralgia associated with AIDS, where it is not clear whether HIV is a primary rheumatic agent or a secondary factor in producing a reactive arthritis (BUSKILA and GLADMAN 1990).

4 Opportunistic Infections

As in humans infected with HIV, monkeys infected with SIV develop a wide range of opportunistic infections. The incidence of specific opportunistic infections (OI) varies among primate facilities (BASKIN 1987; SIMON et al. 1992) and can vary over time (VOGEL et al. 1993) presumably depending on the resident flora. Among the most common opportunistic infections are *Pneumocystis carinii* (PC), CMV, and *Cryptosporidium* sp.

4.1 Viral Infections

The most common viral opportunists in decreasing order of occurrence are CMV, adenovirus, and simian virus 40 (SV-40). Interestingly, reactivated simian herpes infections have not been described despite the fact that greater than 95% of rhesus macaques are seropositive for this virus (WEIGLER et al. 1990).

Cytomegalovirus. CMV infections tend to be disseminated with involvement of multiple organs, including lung, liver, lymph nodes, spleen, heart, testicle, kidney, salivary glands, stomach, large and small intestine, and both the central and peripheral nervous systems (BASKIN 1987; SIMON et al. 1992;

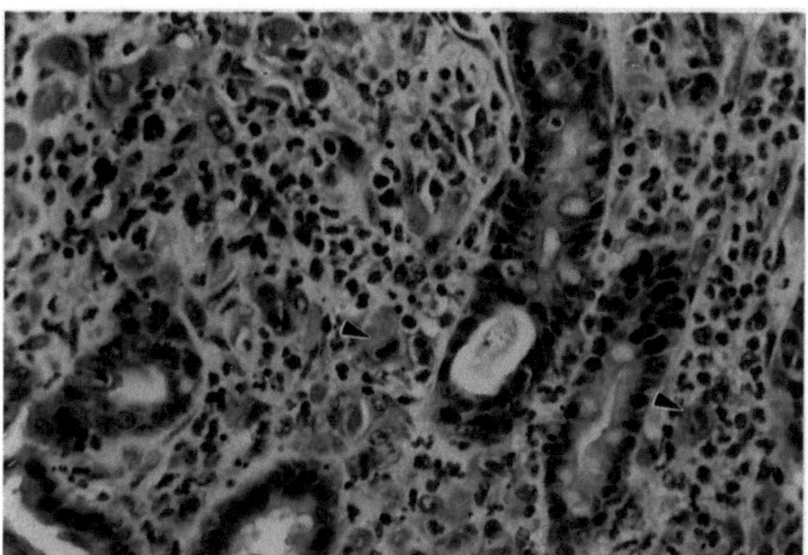

Fig. 6. Small intestine with cytomegalovirus enteritis. Numerous cytomegalic cells some of which have intranuclear inclusions are present (*arrowhead*). Hematoxylin and eosin

VOGEL et al. 1991). The lesions typically consist of foci of necrosis and neutrophilic inflammation with cytomegalic cells containing intranuclear and/or intracytoplasmic inclusions (Fig. 6). The tropism of CMV is broad, and thus numerous types of cells can be infected, including, macrophages, fibroblasts, endothelial cells, and hepatocytes. In the central and peripheral nervous systems CMV is the most common opportunistic infection, manifest most often as a polyradiculoneuritis and meningitis with surprisingly little involvement of the parenchyma of the brain and spinal cord. The occurrence of CMV in the nervous system in SIV-infected animals is intriguing because in macaques infected with type D retrovirus serotype-1, another immunosuppressive simian retrovirus, CMV infection of the nervous system is extremely rare despite a high incidence of activated systemic CMV (MARX et al. 1984; OSBORN et al. 1984).

Adenovirus. Adenoviral infections have been seen most commonly in the pancreas but disseminated infections have also been described (BASKIN et al. 1988; SIMON et al. 1992). In the pancreas the lesions consist of patchy foci of necrosis in association with large basophilic intranuclear inclusion bodies in pancreatic acinar cells (Fig. 7). Interlobular septae are thickened by edema, fibrosis, and a mononuclear cell infiltrate. Adenoviral inclusions have also been described in epithelial cells in gastric glands, Brunner's glands of the duodenum, intestine, renal tubules, and lung (BASKIN et al. 1988; SIMON

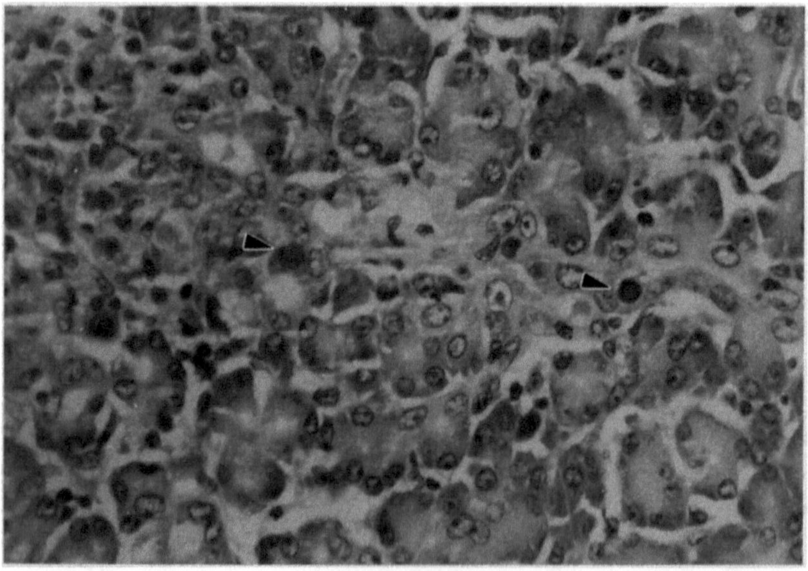

Fig. 7. Pancreas with adenoviral pancreatitis. Two acinar cells with large intranuclear inclusions are present (*arrowheads*). Hematoxylin and eosin

et al. 1992). Inflammation and necrosis are not associated with the inclusions except in the pancreas.

Papovavirus. Infection by the papovavirus SV-40 most commonly results in lesions in the kidney and brain, with rare involvement of the lungs (GRIBBLE et al. 1975; HORVATH et al. 1992; SHEFFIELD et al. 1980). Infection of the kidney appears to be a primary event while CNS infection occurs only after reactivation of SV-40 in severely immunosuppressed individuals (HORVATH et al. 1992). Infection in the CNS leads to progressive multifocal leukoencephalopathy. This lesion is characterized by multiple foci of demyelination, necrosis, and dense infiltrates of macrophages with gemistocytic astrocytes around the periphery of the lesions (Fig. 8). Scattered oligodendrocytes within the lesions have intranuclear inclusions and can be shown to contain SV-40 by in situ hybridization, immunofluorescence and/or electron microscopy (GRIBBLE et al. 1975; HOLMBERG et al. 1977; HORVATH et al. 1992). The lesions occur primarily in the subcortical white matter of the cerebrum, commonly near the boundary with gray matter but have also been described in the spinal cord, optic nerve, and retina (GRIBBLE et al. 1975). In the kidney infection by SV-40 is associated with chronic interstitial nephritis. Intranuclear inclusions containing typical viral particles and SV-40 nucleic acid are present within tubular epithelial cells (HORVATH et al. 1992).

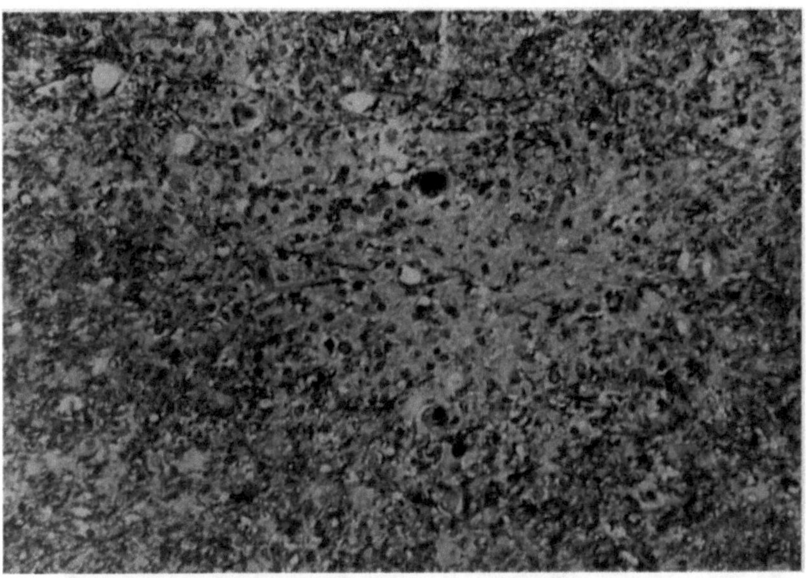

Fig. 8. Progressive multifocal leukoencephalopathy. Cerebral white matter with a focal area of demyelination. Luxol fast blue/ cresyl echt violet

4.2 Bacterial and Fungal Infections

Immunosuppressed animals develop infections by a variety of bacterial organisms, many of which can cause disease in normal animals. The most important bacterial opportunist is *Mycobacterium avium/intracellulare* (MAI). Infections by MAI are generally uncommon in experimentally infected macaques, but in one spontaneous epizootic in a group of *M. arctoides* (stump-tailed macaques) at the California Regional Primate Research Center (CRPRC) in the late 1970s, 18 of 37 (49%) macaques with evidence of infection by SIV (SIV_{stm}, SIV from stump-tailed macaques) died with MAI infection (KAHN et al. 1991; LOWENSTINE et al. 1992). This is probably related to the fact that these animals were housed outdoors, unlike most experimentally infected animals.

Infections by MAI in SIV-infected macaques tend to be disseminated; however, the intestines are most frequently involved (HOLMBERG et al. 1982). Grossly, the intestinal wall is thickened and lymphatics are corded and mesenteric lymph nodes are greatly enlarged (Fig. 9). The mucosal surface is often granular and corrugated. Histologically the lesion consists of large numbers of monomorphic macrophages with abundant foamy cytoplasm in the lamina propria of the intestine. The infiltrates in the intestine and regional nodes efface normal architecture (Fig. 10). Similar, small lesions are often present in the liver, spleen, and occasionally lung, bone marrow, kidney, and other organs (HOLMBERG et al. 1985). Macrophages in the lesions contain large numbers of acid fast bacilli. This is in contrast to infection with *M.*

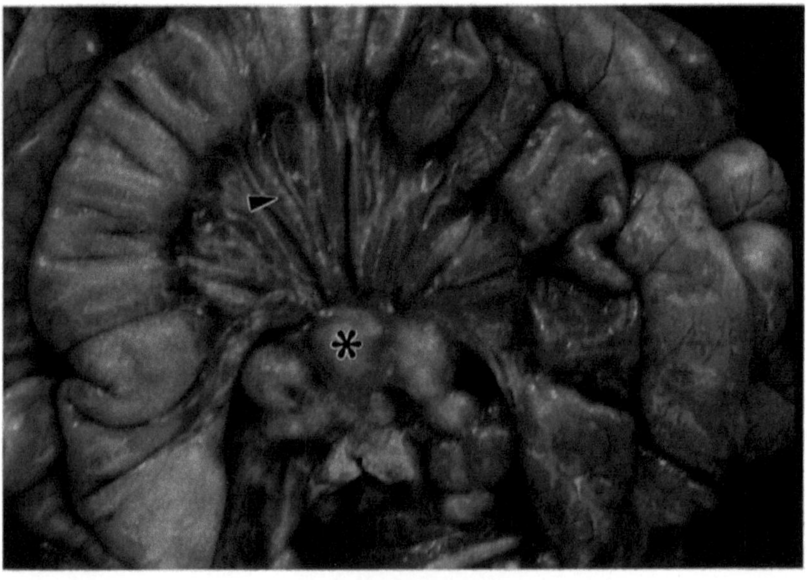

Fig. 9. Intestine of an animal infected with *Mycobacterium avium.* Note the enlarged mesenteric lymph nodes (*asterisk*) and cording of the lymphatics (*arrowhead*)

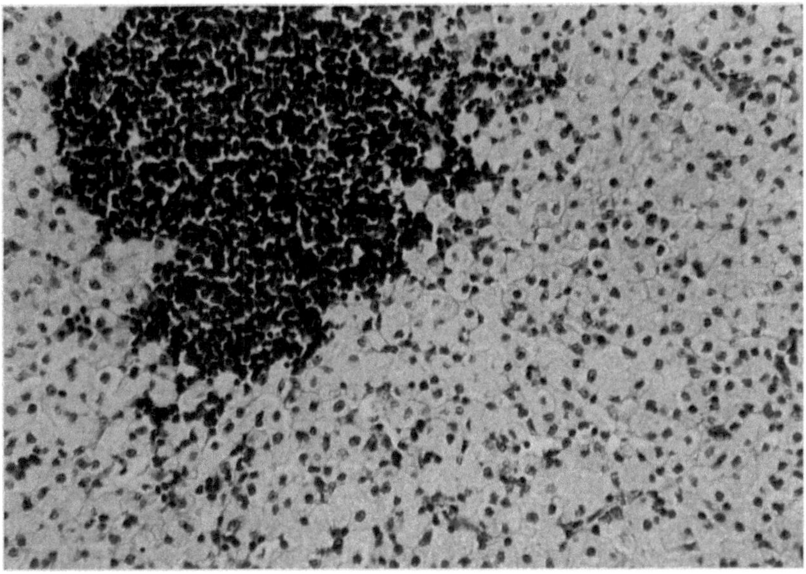

Fig. 10. Mesenteric lymph node effaced by sheets of pale foamy macrophages as a result of infection by *Mycobacterium avium*. Hematoxylin and eosin

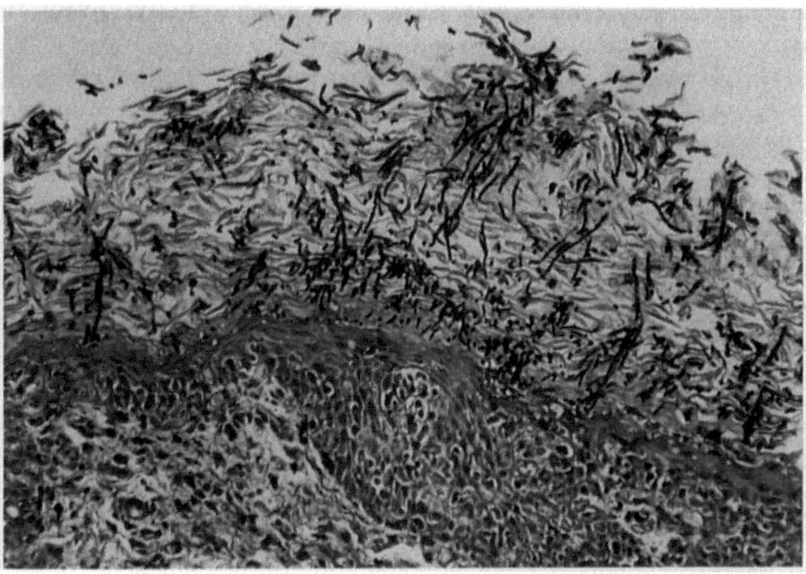

Fig. 11. Candidiasis of the esophagus. Note the sheets of squamous epithelial cells woven together by pseudohyphae of *Candida* sp. Periodic acid–Schiff

tuberculosis in which the lesions are necrotizing and contain few acid fast organisms. It is interesting to note that MAI infections have not been reported in macaques immunosuppressed due to infection with type D retrovirus (OSBORN et al. 1984). This is presumably due to the more profound suppression of cell-mediated immunity caused by SIV as compared to type D retrovirus.

The most important fungal infection is that by *Candida* sp. Infection is most commonly seen in the oral cavity and esophagus. Lesions appear as pale gray plaques on the mucosa. Histologically, the lesions consist of sheets of squamous epithelial cells woven together by pseudohyphae of *Candida* sp (Fig. 11). Inflammation in the underlying mucosa is minimal. Rare cases of dermatophilosis and mucormycosis have also been recognized.

4.3 Protozoal Infections

PC pneumonia is a major life threatening opportunistic infection in AIDS patients (MURRAY and MILLS 1990a,b) and is one of the Most common opportunistic infections in SIV-infected macaques (BASKERVILLE et al. 1991a; SIMON et al. 1992; VOGEL et al. 1993). In a study of 107 SIV-infected macaques at the CRPRC over a 4-year period PC was detected by immunohistochemistry in the lungs of 52% (44 of 85) of terminally ill SIV-infected animals but in only 9% (2 of 22) of clinically normal SIV-infected culls and in 0% (0 of 10) of normal SIV-uninfected control animals (VOGEL et al. 1993). Typical findings included: (a) variable numbers of lightly basophilic trophozoites enmeshed in an eosinophilic foamy matrix within alveoli, (b) type II pneumocyte hyperplasia, and (c) variable degrees of interstitial thickening due to fibrosis, edema, and minimal inflammatory cell infiltrates. Lesions of severe PC pneumonia in rhesus were very similar to those seen in human AIDS patients (MURRAY and MILLS 1990a,b). Infection by PC was centered on terminal airways. As the infection became more severe, alveolar filling tended to extend outward from terminal airways, producing more diffuse pulmonary consolidation (Fig. 12). Extrapulmonary dissemination of PC was not seen in this study but PC has been detected in the pulmonary interstitium, lymphatics, and veins by other investigators, suggesting that dissemination may occur (BASKERVILLE et al. 1991a; SIMON et al. 1992). Over the 4-year period of the study at the CRPRC clinically important PC pneumonia increased from 0% in the first 2 years to over 50% during the final year. In addition, it was found that the infection rate varied significantly between rooms and was never detected in SIV-infected macaques held in high-efficiency particulate air filtered isolators. The epidemiological data in conjunction with the orientation of the lesions around terminal airways suggests that horizontal spread of PC infection is more important in the epidemiology of PC pneumonia in immunosuppressed macaques than reactivation of latent infections (VOGEL et al. 1993).

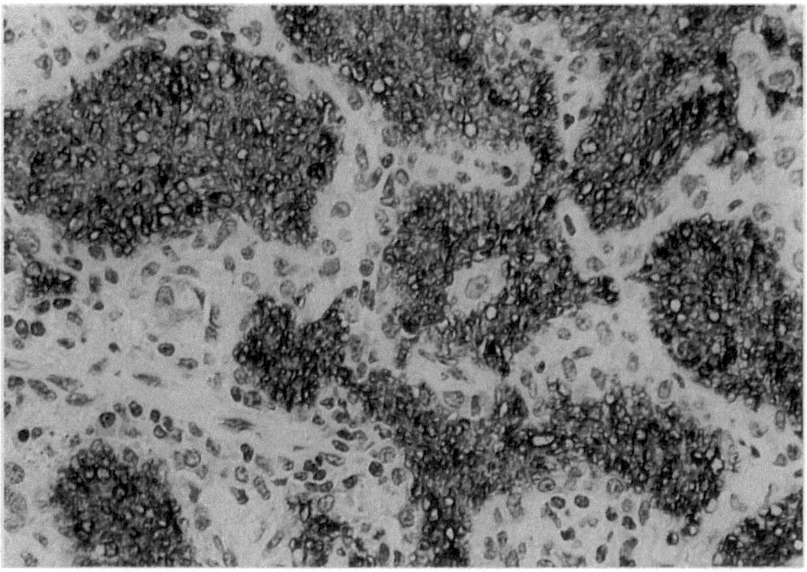

Fig. 12. Lung with pneumonia caused by *Pneumocystis carinii*. Immunohistochemistry with antibody 3F6 (DAKO, Santa Barbara, CA)

Cryptosporidiosis. Rhesus monkeys with cryptosporidiosis generally have a profuse watery diarrhea associated with dehydration, particularly in infants (COCKRELL et al. 1974; MILLER et al. 1990a,b; WILSON et al. 1984). The intestinal tract of affected animals is variably distended with fluid and gas. In animals that have involvement of the gall bladder and bile ducts mural thickening may be present and the bile tends to be scant, pale and mucoid. In macaques, coccidian parasites of the genus *Cryptosporidium* are most commonly seen in the small intestine, gall bladder, biliary and pancreatic ducts (BASKERVILLE et al. 1991b; BLANCHARD et al. 1987; COCKRELL et al. 1974; KOVATCH and WHITE 1972; WILSON et al. 1984). Involvement of the large intestine is uncommon, and infection of the gastric mucosa has not been reported. Cryptosporidiosis involving the respiratory tract has also been described in two SIV-infected macaques (BLANCHARD et al. 1987).

In the small intestines the lesion consists of mild to moderate blunting and fusion of villi, flattening of surface enterocytes, necrosis of individual enterocytes, and increased mitotic figures. The various stages of *Cryptosporidium* can be seen attached to the apical surface of epithelial cells (Fig. 13). The organisms are 1–4 µm in diameter, round to oval, and basophilic. The number of organisms present is quite variable. Most of the organisms are present on the mucosal surface, but occasional infected crypts are also present. Inflammation, consisting of moderate numbers of lymphocytes,

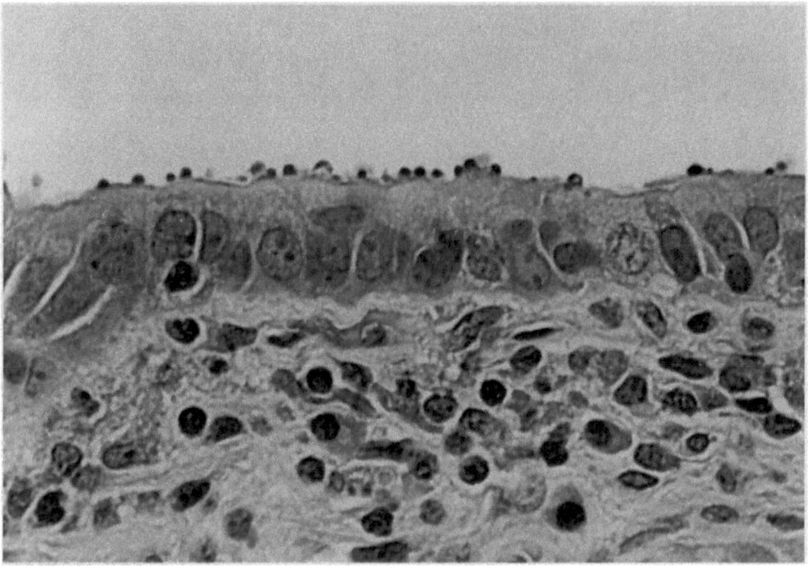

Fig. 13. Gall bladder with numerous cryptosporidia attached to the apical surface of epithelial cells. Hematoxylin and eosin

plasma cells, and macrophages, is generally present. In the intrahepatic and extrahepatic bile ducts, gall bladder, and largest pancreatic ducts infection of the mucosa is characterized by hyperplasia rather than atrophy (BASKERVILLE et al. 1991b; BLANCHARD et al. 1987; KOVATCH and WHITE 1972). The hyperplasia can be particularly severe at the hepatopancreatic ampulla where the pancreatic and bile ducts empty into the duodenum. In extreme cases the hyperplasia can resemble adenomas (BASKERVILLE et al. 1991b). The pancreatic and biliary ducts can become partially occluded by folds of mucosa protruding into the ductal lumen. Sloughed epithelial cells, cellular debris, and occasional inflammatory cells present in the lumina of affected ducts also contributes to ductal occlusion. In one report this was associated with chronic pancreatitis and cirrhosis of the liver (BASKERVILLE et al. 1991b). The stroma of the gall bladder, biliary and pancreatic ducts often contains a moderate infiltrate of mixed inflammatory cells.

Trichomonads and Other Protozoa. Trichomonads have been seen in approximately 15% of SIV-infected macaques in the crypts of the cecum and colon and less commonly in gastric pits (BASKIN et al. 1988; SIMON et al. 1992). When present the affected crypt or pit is often mildly dilated with attenuation of the epithelium. Crypt or pit abscesses are often present. Trichomonads are occasionally present in the lamina propria around crypt abscesses and in rare cases invasive trichomoniasis has been observed (Fig. 14). In most instances, however, the significance of trichomonads as an opportunistic pathogen in

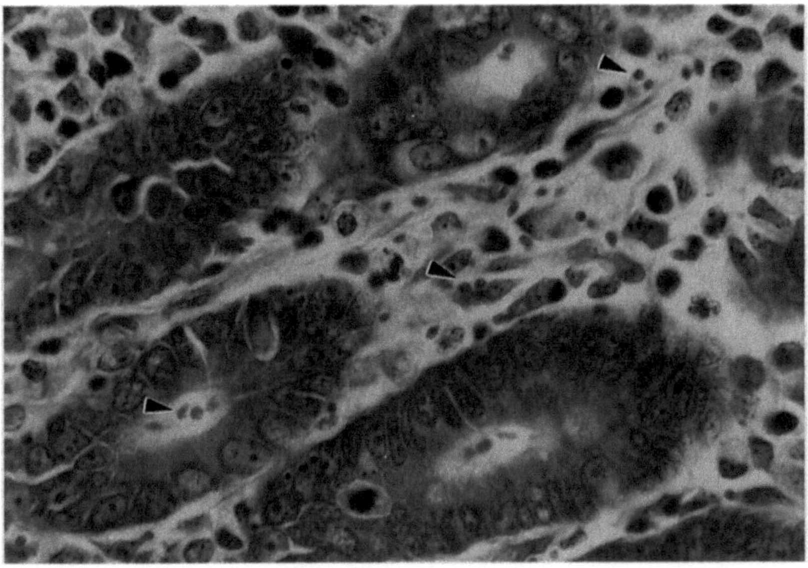

Fig. 14. Intestine with invasive trichomoniasis. Note that numerous organisms are present in crypts and the lamina propria of the intestine (*arrowheads*). Giemsa

macaques is questionable, as organisms are commonly found in fecal smears of normal animals. This is also true of *Balantidium coli* but in rare SIV-infected macaques this organism can be seen to invade the lamina propria of the intestinal tract. Other protozoal infections include toxoplasmosis of the CNS (LOWENSTINE et al. 1992), giardiasis, and amebiasis, which have only rarely been reported (BASKIN et al. 1988; SIMON et al. 1992).

5 Lymphomas

Lymphomas have occurred sporadically in SIV-infected macaques (BASKERVILLE et al. 1992b; BASKIN et al. 1986, 1988; BENVENISTE et al. 1986; DANIEL et al. 1985), with two notable exceptions: a recent outbreak of lymphoma in 9 of 24 *M. fascicularis* experimentally infected with SIV at the Karolinska institute (FEICHTINGER et al. 1990) and an epizootic of lymphoma involving 45 macaques (42 *M. mulatta* and 3 *M. arctoides*) beginning in 1969 at the CRPRC (MANNING and GRIESEMER 1974; STOWELL et al. 1971; TERRELL et al. 1980). While the outbreak at the CRPRC cannot be associated unequivocally with SIV infection, there is substantial circumstantial evidence that it was (TERRELL et al. 1980). Furthermore, virological evidence suggests that at least some of the animals were infected with SIV (LOWENSTINE et al. 1992).

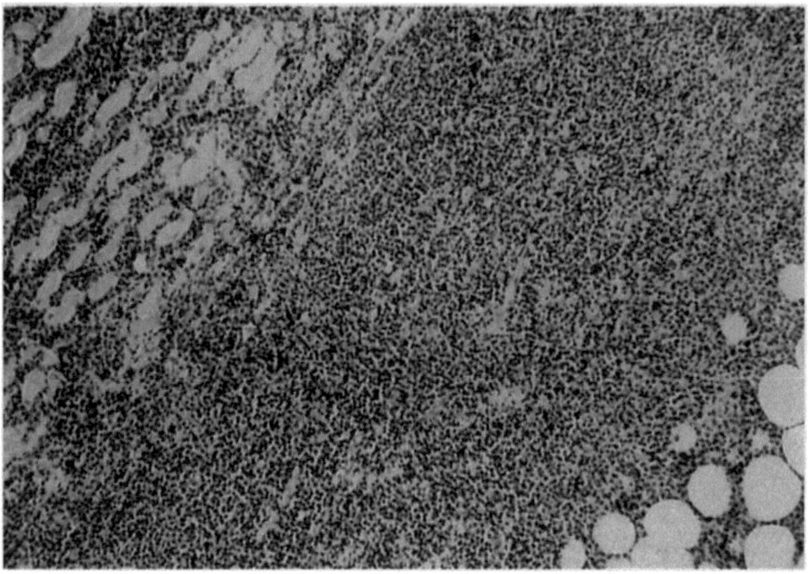

Fig. 15. A retrobulbar lymphoma with infiltration of retrobulbar adipose tissue and extraocular muscles. Hematoxylin and eosin.

The role of SIV in the development of the lymphomas is unclear. A B-lymphotropic virus related to Epstein-Barr virus (EBV) has been isolated from a lymphoma in one SIV-infected macaque (RANGAN et al. 1986), and preliminary studies indicate integration of EBV-like sequences in genomic DNA of other lymphomas (FEICHTINGER et al. 1990). However, in two other cases, no EBV-related DNA could be detected (DESROSIERS et al. 1987).

The lymphomas described to data are all of B-cell origin and tend to be disseminated with mostly extranodal and visceral involvement (BASKIN et al. 1986; FEICHTINGER et al. 1990; RAMSAY et al. 1991; TERREL et al. 1980). Histologically, most of the tumors are lymphoblastic, with a Burkitt-like appearance (Fig. 15). Infiltration of the tumors by macrophages and $CD8^+$ lymphocytes is a common feature. Occasionally, the infiltrating macrophages are positive for SIV antigen or nucleic acid, but tumor cells have not been shown to be positive (FEICHTINGER et al. 1990; Lackner, unpublished).

6 Discussion

The SIV/macaque model of AIDS is the preeminent model for examining the pathogenesis of HIV infection. Studies of viral determinants of disease, vaccines, mucosal transmission of SIV, maternal-fetal transmission of SIV, and tissue distribution of SIV over time are examples of the key role that this model has played and will continue to play in elucidating the pathogenesis of AIDS.

The strength of the model is derived from the close evolutionary, biological, and genomic relationship between HIV and SIV. The disease resulting from infection of humans with HIV or macaques with SIV are strikingly similar from clinical, immunological, virological, and pathological perspectives. In addition, the ability to uniformly induce AIDS in SIV-infected macaques in a time frame appropriate for laboratory investigation further strengthens this model and contributes to its importance in the development of interventive stratagies and vaccines.

Acknowledgments. This work was made possible by Public Health Service grants RR00169 from the National Center for Research Resources and NS30769 from the National Institute of Neurologic Disorders and Stroke, National Institutes of Health. Special thanks to Robert Munn for photograpic assistance.

References

Alpers CE, Tsai C-C, Thouless M, Kueller L, Morton WR (1988) Focal and segmental glomerulosclerosis in primates infected with simian immunodeficiency virus. Lab Invest 58: 3A

Baskerville A, Ramsay A, Cranage P, Cook N, Cook RW, Dennis MJ, Greenaway PJ, Kitchin PA, Stott EJ (1990) Histopathologic changes in simian immunodeficiency virus infection. J Pathol 162: 67–75

Baskerville A, Dowsett AB, Cook RW, Dennis MJ, Cranage MP, Greenaway PJ (1991a) Pneumocystis carinii pneumonia in simian immunodeficiency virus infection: immunohistological and scanning and transmission electron microscopical studies. J Pathol 164: 175–184

Baskerville A, Ramsay A, Millwar-Sadler G, Cook R, Cranage M, Greenaway P (1991b) Chronic pancreatitis and biliary fibrosis associated with cryptosporidiosis in simian AIDS. J Comp Pathol 105: 415–421

Baskerville A, Cook RW, Dennis MJ, Cranage MP, Greenaway PJ (1992a) Pathological changes in the reproductive tract of male rhesus monkeys associated with age and simian AIDS. J Comp Pathol 107: 49–57

Baskerville A, Ramsay AD, Addis BJ, Dennis MJ, Cook RW, Cranage MP, Greenaway PJ (1992B) Interstitial pneumonia in simian immunodeficiency virus infections. J Pathol 167: 241–247

Baskin GB (1987) Disseminated cytomegalovirus infection in immunodeficient rhesus monkeys. Am J Pathol 129: 345–352

Baskin GB, Martin LN, Rangan SRS, Gormus BJ, Murphey-Corb M, Wolf RH, Soike KF (1986) Transmissible lymphoma and simian acquired immunodeficiency syndrome in rhesus monkeys. J Natl Cancer Inst 77: 127–139

Baskin GB, Murphey-Corb M, Watson EA, Martin LN (1988) Necropsy findings in rhesus monkeys experimentally infected with cultured simian immunodeficiency virus (SIV)/delta. Vet Pathol 25: 456–467

Baskin GB, Murphey-Corb M, Martin LN, Davison-Fairburn B, Hu F-S, Kuebler D (1991a) Thymus in simian immunodeficiency virus-infected rhesus monkeys. Lab Invest 65: 400–407

Baskin GB, Murphey-Corb M, Martin LN, Soike KF, Hu F-S, Kuebler D (1991b) Lentivirus-induced pulmonary lesions in rhesus monkeys (*Macaca mulatta*) infected with simian immunodeficiency virus. Vet Pathol 28: 506–513

Baskin GB, Murphey-Corb M, Roberts ED, Didier PJ, Martin LN (1992) Correlates of SIV encephalitis in rhesus monkeys. J Med Primatol 21: 59–63

Benveniste RE, Arthur LO, Tsai C-C, Sowder R, Copeland TD, Henderson LE, Oroszlan S (1986) Isolation of a lentivirus from a macaque with lymphoma: comparison with HTLV-III/LAV and other lentiviruses. J Virol 60: 483–490
Benveniste RE, Morton WR, Clark EA, Tsai C-C, Ochs HD, Ward JM, Kuller L, Knott WB, Hill RW, Gale MJ, Thouless ME (1988) Inoculation of baboons and macaques with simian immunodeficiency virus/Mne, a primate lentivirus closely related to human immunodeficiency virus type 2. J Virol 62: 2091–2101
Blanchard JL, Baskin GB, Murphey-Corb M, Martin LN (1987) Disseminated cryptosporidiosis in simian immunodeficiency virus/delta-infected rhesus monkeys. Vet Pathol 24: 454–456
Buskila D, Gladman D (1990) Musculoskeletal manifestations of infection with human immunodeficiency virus. Rev Infect Dis 12: 223–235
Chakrabarti L, Guyader M, Alizon M, Daniel MD, Desrosiers RC, Tiollais P, Sonigo P (1987) Sequence of simian immunodeficiency virus from macaque and its relationship to other human and simian retroviruses. Nature 328: 543–547
Chakrabarti L, Hurtrel M, Maaire M, Vazeux R, Dormont D, Montagnier L, Hurtrel B (1991) Early viral replication in the brain of SIV-infected rhesus monkeys. Am J Pathol 139: 1273–1280
Chalifoux LV, King NW, Daniel MD, Kannagi M, Desrosiers RC, Sehgal PK, Waldron LM, Hunt RD, Letvin NL (1986) Lymphoproliferative syndrome in an immunodeficient rhesus monkey naturally infected with an HTLV-III-like virus (STLV-III). Lab Invest 55: 43–50
Chalifoux LV, Ringler DJ, King NW, Sehgal PK, Desrosiers RC, Daniel MD, Letvin NL (1987) Lymphadenopathy in macaques experimentally infected with the simian immunodeficiency virus (SIV). Am J Pathol 128: 104–110
Chalifoux LV, Simon MA, Pauley DR, MacKey JJ, Wyand MS, Ringler DJ (1992) Arteriopathy in macaques infected with simian immunodeficiency virus. Lab Invest 67: 338–349
Cheng-Mayer C, Rutka JT, Rosenblum ML, McHugh T, Stites DP, Levy JA (1987) Human immunodeficiency virus can productively infect cultured human glial cells. Proc Natl Acad Sci USA 84: 3526–3530
Clapham PR, Weber JN, Whitby D, McIntosh K, Dalgleish AG, Maddon PJ, Deen KC, Sweet RW, Weiss RA (1989) Soluble CD4 blocks the infectivity of diverse strains of HIV and SIV for T cells and monocytes but not for brain and muscle cells. Nature 337: 368–370
Cockrell BY, Valerio MG, Garner FM (1974) Cryptosporidiosis in the intestines of rhesus monkeys. Lab Anim Sci 24: 881–887
Dandekar S, Heise C, Vogel P, Lackner A (1993) SIV-associated intestinal abnormalities and cellular targets of SIV in the intestinal tract of rhesus macaques (Abstr. Q412). J Cell Biochem Suppl 17E: 61
Daniel MD, Letvin NL, King NW, Kannagi M, Sehgal PK, Hunt RD, Kanki PJ, Essex M, Desrosiers RC (1985) Isolation of T-cell tropic HTLV-III-like retrovirus from macaques. Science 228: 1201–1204
Daniel MD, Li Y, Naidu YM, Durda PJ, Schmidt DK, Finger CD, Silva DP, MacKey JJ, Kestler HW, Sehgal PK, King NW, Hayami M, Desrosiers RC (1988) Simian immunodeficiency viruses from African green monkeys. J Virol 62: 4123–4128
Desrosiers RC (1990) The simian immunodeficiency viruses. Annu Rev Immunol 8: 557–578
Desrosiers RC, Letvin NL, King NW, Hunt RD, Blake BJ, Arthur LO, Daniel MD (1987) Three retroviruses infecting macaques at the New England Regional Primate Research Center. In: Gallo RC, Haseltine W, Klein G, zurHausen H (eds) Viruses and human cancer. Liss, New York, pp 451–466
Desrosiers RC, Hansen-Moosa A, Mori K, Bouvier DP, King NW, Daniel MD, Ringler DJ (1991) Macrophage-tropic variants of SIV are associated with specific AIDS-related lesions but are not essential for the development of AIDS. Am J Pathol 139: 29–35
Embretson J, Zupancic M, Ribas JL, Burke A, Racz P, Tenner-Racz K, Haase AT (1993) Massive covert infection of helper T lymphocytes and macrophages by HIV during the incubation period of AIDS. Nature 362: 359–362
Epstein LG (1989) Lentiviral encephalitides in the immature host: a comparison of human immunodeficiency virus type 1 (HIV-1) and simian immunodeficiency virus (SIV) brain infection. Brain Dev 11: 353–359
Feichtinger H, Putkonen P, Parravicini C, Li SL, Kaaya EE, Böttiger D, Biberfeld G, Biberfeld P (1990) Malignant lymphomas in cynomolgus monkeys infected with simian immunodeficiency virus. Am J Pathol 137: 1311–1315

Fox CH, Tenner-Racz K, Racz P, Firpo A, Pizzo PA, Fauci AS (1991) Lymphoid germinal centers are reservoirs of human immunodeficiency virus type 1 RNA. J Infect Dis 164: 1051–1057
Franchini G, Gurgo C, Guo HG, Gallo RC, Collalti E, Fargnoli KA, Hall LF, Wong-Staal F, Reitz MSJ (1987) Sequence of simian immunodeficiency virus and its relationship to the human immunodeficiency viruses. Nature 328: 539–543
Fultz PN, McClure H, Anderson DC, Swenson RB, Anand R, Srinivasan A (1986) Isolation of a T-lymphotropic retrovirus from naturally infected sooty mangabey monkeys (*Cercocebus atys*). Proc Natl Acad Sci USA 83: 5286–5290
Fultz PN, McClure HM, Anderson DC, Switzer WM (1989) Identification and biologic characterization of an acutely lethal variant of simian immunodeficiency virus from sooty mangabeys (SIV/SMM). Aids Res Hum Retroviruses 5: 397–409
Gardner MB, Luciw P, Lerche N, Marx P (1988) Non-human primate retrovirus isolates and AIDS. In: Perk K (ed) Immunodeficiency disorders and AIDS. Academic Press, New York, pp 171–226
Gerber M, Chen M, Hu F, Baskin G, Petrovich L (1991) Liver disease in rhesus monkeys infected with simian immunodeficiency virus. Am J Pathol 139: 1081–1088
Gribble DH, Haden CC, Schwartz LW, Henrickson RV (1975) Spontaneous progressive multifocal leukoencephalopathy (PML) in macaques. Nature 254: 602–604
Haase AT (1986) Pathogenesis of lentivirus infections. Nature 322: 130–136
Heise C, Vogel P, Miller CJ, Halsted CH, Dandekar S (1993a) Simian immunodeficiency virus infection of the gastrointestinal tract of rhesus macaques: functional, pathological and morphological changes. Am J Pathol 142: 1–13
Heise C, Vogel P, Miller CJ, Lackner AA, Dandekar S (1993b) Distribution of SIV infection in the gastrointestinal tract of rhesus macaques at early and terminal stages of AIDS. J Med Primatol (in press)
Hirsch VM, Olmsted RA, Murphey-Corb M, Purcell RH, Johnson PR (1989) An African primate lentivirus (SIV_{sm}) closely related to HIV-2. Nature 339: 389–392
Hirsch VM, Dapolito GA, Goldstein S, McClure H, Emau P, Fultz PN, Isahakia M, Lenroot R, Myers G, Johnson PR (1993) A distinct African lentivirus from Sykes' monkeys. J Virol 67: 1517–1528
Holmberg CA, Gribble DH, Takemoto KK, Howley PM, Espana C, Osburn BI (1977) Isolation of simian virus 40 from rhesus monkeys (*Macaca mulatta*) with spontaneous progressive multifocal leukoencephalopathy. J Infect Dis 136: 593–596
Holmberg CA, Henrickson RV, Malaga C, Schneider R, Gribble D (1982) Nontuberculous mycobacterial disease in rhesus monkeys. Vet Pathol 19 Suppl 7: 9–16
Holmberg CA, Henrickson R, Lenninger R, Anderson J, Hayashi L, Ellingsworth L (1985) Immunologic abnormality in a group of Macaca arctoides with high mortality due to a typical mycobacterial and other disease processes. Am J Vet Res 46: 1192–1196
Horvath CJ, Simon MA, Bergsagel DJ, Pauley DR, King NW, Garcea RL, Ringler DJ (1992) Simian virus 40-induced disease in rhesus monkeys with simian acquired immunodeficiency syndrome. Am J Pathol 140: 1431–1440
Israel ZR, Dean GA, Maul DH, O'Neil SP, Dreitz MJ, Mullins JI, Fultz PN, Hoover EA (1993) Early pathogenesis of disease caused by SIVsmmPBj14 molecular clone 1.9 in macaques. Aids Res Hum Retroviruses 9: 277–286
Joling P, Van Wichen DF, Parmentier HK, Biberfeld P, Böttiger D, Tschopp J, Rademakers LHPM, Schuurman HJ (1992) Simian immunodeficiency virus (SIVsm) infection of cynomolgus monkeys: effects on follicular dendritic cells in lymphoid tissue. Aids Res Hum Retroviruses 8: 2021–2030
Joshi VV, Path FRC, Pawel B, Connor E, Sharer L, Oleske JM, Morrison S, Marin-Garcia J (1987) Arteriopathy in children with acquired immune deficiency syndrome. Pediatr Pathol 7: 261–275
Kahn A, Galvin T, Lowenstine L, Jennings M, Gardner M, Buckler C (1991) A highly divergent simian immunodeficiency virus (SIV_{stm}) recovered from stored stump-tailed macaque tissues. J Virol 65: 7061–7065
Kanki PJ, Alroy J, Essex M (1985) Isolation of T-lymphotropic retrovirus related to HTLV-III/LAV from wild-caught african green monkeys. Science 230: 951–954
King NW, Chalifoux LV, Ringler DJ, Wyand MS, Sehgal PK, Daniel MD, Letvin NL, Desrosiers RC, Blake BJ, Hunt RD (1990) Comparative biology of natural and experimental SIV_{mac} infection of macaque monkeys: a review. J Med Primatol 19: 109–118

Kitagawa M, Lackner AA, Martfeld DJ, Gardner MB, Dandekar S (1991) Simian immunodeficiency virus infection of bone marrow: correlation of viral infection with disease progression and hematologic abnormalities. Am J Pathol 138: 921–930
Klatzmann D, Barre-Sinoussi F, Nugeyre MT, Danguet C, Vilmer C, Griscelli C, Brun-Vézinet F, Rouzioux C, Gluckman JC, Cherman JC, Montagnier L (1984) Selective tropism of lymphadenopathy associated virus (LAV) for helper-induced T-lymphocytets. Science 225: 59–63
Kovatch RM, White JD (1972) Cryptosporidiosis in two juvenile rhesus monkeys. Vet Pathol 9: 426–440
Kuller L, Morton WR, Benveniste RE, Tsai CC, Clark EA, Gale MJ, Hu SL, Thouless ME, Katze MG (1990) Inoculation of *Macaca fascicularis* with simian immunodeficiency virus, SIVmne: immunologic, serologic, and pathologic changes. J Med Primatol 19: 367–380
Kunsch C, Hartle HT, Wigdahl B (1989) Infection of human fetal dorsal root ganglion glial cells with human immunodeficiency virus type 1 involves an entry mechanism independent of the CD4 T4A epitope. J Virol 63: 5054–5061
Lackner AA, Moore PF, Marx PA, Munn RJ, Gardner MB, Lowenstine LJ (1990) Immunohistochemical localization of type D retrovirus serotype 1 in the digestive tract of rhesus monkeys with simian AIDS. J Med Primatol 19: 339–349
Lackner AA, Dandekar S, Gardner MB (1991a) Neurobiology of simian and feline immunodeficiency virus infections. Brain Pathol 1: 201–212
Lackner AA, Smith MO, Munn RJ, Martfeld DJ, Gardner MB, Marx PA, Dandekar S (1991b) Localization of simian immunodeficiency virus in the central nervous system of rhesus monkeys. Am J Pathol 139: 609–621
Lackner AA, Vogel P, Hoogenboom E, Dandekar S, Kluge JD, Marthas M (1991c) Pathogenic (SIV_{mac}-239) and nonpathogenic (SIV_{mac}-1A11) molecular clones of SIV have distinct tissue distributions that vary with length of infection. 9th Annual Symposium on Nonhuman Primate Models for AIDS, Nov 6–9, Seattle
Letvin NL, King NW (1990) Immunologic and pathologic manifestations of infection of rhesus monkeys with simian immunodeficiency virus of macaques. J Acquir Immune Defic Syndr 3: 1023–1040
Letvin NL, Daniel MD, Sehgal PK, Desrosiers RC, Hunt RD, Waldron LM, MacKey JJ, Schmidt DK, Chalifoux LV, King NW (1985) Induction of AIDS-like disease in macaque monkeys with T-cell tropic retrovirus STLV-III. Science 230: 71–73
Lowenstine LJ, Pedersen NC, Higgins J, Pallis KC, Uyeda A, Marx PA, Lerche NW, Munn RJ, Gardner MB (1986) Seroepidemiologic survey of captive old-world primates for antibodies to human and simian retroviruses and isolation of a lentivirus from sooty mangabeys (Cercocebus atys). Int J Cancer 38: 563–574
Lowenstine LJ, Lerche NW, Yee JL, Uyeda A, Jennings MB, Munn RJ, McClure HM, Anderson DC, Fultz PN, Gardner MB (1992) Evidence for a lentiviral etiology in an epizootic of immune deficiency and lymphoma in stump-tailed macaques (*Macaca arctoides*). J Med Primatol 21: 1–14
Manning JS, Griesemer RA (1974) Spontaneous lymphoma of the nonhuman primate. Lab Anim Sci 24: 204–210
Marx PA, Maul DH, Osborn KG, Lerche NW, Moody P, Lowenstine LJ, Henrickson RV, Arthur LO, Gravell M, London WT, Sever JL, Levy JA, Munn RJ, Gardner MB (1984) Simian AIDS: isolation of a type D retrovirus and disease transmission. Science 223: 1083–1086
Miller CJ, Vogel P, Alexander NJ, Sutjipto S, Hendrickx AG, Marx PA (1992) Localization of SIV in the genital tract of chronically infected female rhesus macaques. Am J Pathol 141: 655–660
Miller RA, Brondson MA, Kuller L, Morton WR (1990a) Clinical and parsitologic aspects of cryptosporidiosis in nonhuman primates. Lab Anim Sci 40: 42–46
Miller RA, Brondson MA, Morton WR (1990b) Experimental cryptosporidiosis in a primate model. J Infect Dis 161: 312–315
Müller H, Weier S, Kojouharoff G, Grez M, Berger S, Kappus R, Shah PM, Stutte HJ, Schmidts HL (1993) Distribution and infection of Langerhans cells in the skin of HIV-infected healthy subjects and AIDS patients. Res Virol 144: 59–67
Müller MC, Saksena NK, Nerrienet E, Chappey C, Hervé VMA, Durand JP, Legal-Compodonico P, Lang MC, Digoutte JP, Georges AJ, Georges-Courbot MC, Sonigo P, Barré-Sinoussi F (1993) Simian immunodeficiency viruses from central and western Africa. J Virol 67: 1227–1235

Murray JF, Mills J (1990a) Pulmonary infectious complications of human immunodeficiency virus infection. Part I. Am Rev Respir Dis 1: 1356–1372
Murray JF, Mills J (1990b) Pulmonary infectious complications of human immunodificiency virus infection. Part II. Am Rev Respir Dis 141: 1582–1598
Nelson JA, Wiley CA, Reynolds-Kohler C, Reese CE, Margetten W, Levy JA (1988) Human immunodeficiency virus detected in bowel epithelium from patients with gastrointestinal symptoms. Lancet: 259–262
Ormerod LD, Osborn KG, Lowenstine LJ, Meyer PR, Parker JW, Smith RE, Taylor CR (1988) Lymphoid immunohistochemistry of macaque primates. Clin Exp Immunol 74: 435–442
Osborn KG,Prahalada S, Lowenstine LJ, Gardner MB, Maul DH, Henrickson RV (1984) The pathology of an epizootic of acquired immunodeficiency in rhesus macaques. Am J Pathol 114: 94–103
Pantaleo G, Graziosi C, Demarest JF, Butini L, Montroni M, Fox CH, Orenstein JM, Kotler DP, Fauci AS (1993) HIV infection is active and progressive in lymphoid tissue during the clinically latent stage of disease. Nature 362: 355–358
Price RW, Brew B, Sidtis J, Rosenblum M, Scheck AC, Cleary P (1988) The brain in AIDS: central nervous system HIV-1 infection and AIDS dementia complex. Science 239: 586–592
Ramsay AD, Giddings J, Baskerville A, Cranage MP (1991) Phenotypic analysis of malignant lymphoma in simian immunodeficiency virus infection using anti-human antibodies. J Pathol 164: 321–328
Rangan SRS, Martin LN, Bozelka BE, Wang N, Gormus BJ (1986) Epstein-Barr virus-related herpesvirus from a rhesus monkey (*Macaca mulatta*) with malignant lymphoma. Int J Cancer 38: 425–432
Rappersberger K, Gartner S, Schenk P, Stingl G, Groh V, Tschachler E, Mann DL, Wolff K, Konrad K, Popvic M (1988) Langerhans' cells are an actual site of HIV-1 replication Intervirology 29: 185–194
Reinacher M (1987) Feline leukemia virus-associated enteritis-a condition with features of feline panleukopenia. Vet Pathol 24: 1–4
Ringler DJ, Murphy GF, King NW (1986) An erythematous maculopapular eruption in macaques infected with an HTLV-III-like virus (STLV-III). J Invest Dermatol 87: 674–677
Ringler DJ, Hancock WW, King NW, Letvin NL, Daniel MD, Desrosiers RC, Murphy GF (1987) Immunophenotypic characterization of the cutaneous exanthem of SIV-infected rhesus monkeys: apposition of degenerative Langerhans cells and cytotoxic lymphocytes during the development of acquired immunodeficiency syndrome. Am J Pathol 126: 199–207
Ringler DJ, Wyand MS, Walsh DG, MacKey JJ, Chalifoux LV, Popovic M, Minassian AA, Sehgal PK, Daniel MD, Desrosiers RC, King NW (1989) Cellular localization of simian immunodeficiency virus in lymphoid tissues. I. Immunohistochemistry and electron microscopy. Am J Pathol 134: 373–383
Roberts E, Martin L (1991) Arthritis in rhesus monkeys experimentally infected with simian immunodeficiency virus (SIV/delta). Lab Invest 65: 637–643
Sasseville VG, Newman WA, Lackner AA, Smith MO, Lausen NC, Beall D, Ringler DJ (1992) Elevated vascular cell adhesion molecule-1 in AIDS encephalitis induced by simian immunodeficiency virus. Am J Pathol 141: 1021–30
Sharer LR, Epstein LG, Cho E-S, Joshi VV, Meyenhofer MF, Rankin LF, Petito CK (1986) Pathologic features of AIDS encephalopathy in children: evidence for LAV/HTLV-III infection of the brain. Hum Pathol 17: 271–284
Sharer LR, Michaels J, Murphey-Corb H, Hu F-S, Kuebler DJ, Martin LN, Baskin GB (1991) Serial pathogenesis study of SIV brain infection. J Med Primatol 20: 211–217
Sheffield WD, Strandberg JD, Braun L, Shah K, Kalter SS (1980) Simian virus 40-associated fatal interstitial pneumonia and renal tubular necrosis in a rhesus monkey. J Infect Dis 142: 618–622
Simon MA, Chalifoux LV, Ringler DJ (1992) Pathologic features of SIV-induced disease and the association of macrophage infection with disease evolution. Aids Res Hum Retroviruses 8: 327–337
Stowell RC, Smith EK, Espana C, Nelson VG (1971) Outbreak of malignant lymphoma in rhesus monkeys. Lab Invest 25: 476–479
Terrell TG, Gribble DH, Osburn BI (1980) Malignant lymphoma in macaques: a clinicopathologic study of 45 cases. J Natl Cancer Inst 64: 561–568
Tomonaga K, Katahira J, Fukasawa M, Hassan MA, Kawamura M, Akari H, Miura T, Goto T, Nakai M, Suleman M, Isahakia M, Hayami M (1993) Isolation and characterization of simian

immunodeficiency virus from African white-crowned mangabey monkeys (Cercocebus torquatus lunulatus). Arch Virol 129: 77–92
Tschachler E, Groh V, Popvic M, Mann DL, Konrad K, Safai B, Eron L, diMarzo-Veronese F, Wolff K, Stingl G (1987) Epidermal Langerhans cells: a target for HTLV-III/LAV infection. J Invest Dermatol 88: 233–237
Tsujimoto H, Cooper RW, Kodama T, Fukasawa M, Miura T, Ohta Y, Ishikawa K-I, Frost E, Roelents GE, Nakai M, Roff J, Hayami M (1988) Isolation and characterization of simian immunodeficiency virus from mandrills in Africa and its relationship to other human and simian immunodeficiency viruses. J Virol 62: 4044–4050
Vogel P, Lackner AA, Alcendor DJ, Dandekar S (1991) Distribution of rhesus cytomegalovirus (CMV) and simian immunodeficiency virus (SIV) in lungs of rhesus macaques (Absh). Lab Invest 64: 90A
Vogel P, Miller CJ, Lowenstine L, Lackner A (1993) Evidence of horizontal transmission of *Pneumocystis carinii* pneumonia in SIV-infected rhesus macaques. J Infect Dis (in press)
Watanabe M, Ringler DJ, Nakamura M, DeLong PA, Letvin NL (1990) Simian immunodeficiency virus inhibits bone marrow hematopoietic progenitor cell growth. J Virol 64: 656–663
Weigler BJ, Roberts JA, Hird DW, Lerche NW, Hilliard JK (1990) A cross sectional survey for B virus antibody in a colony of group housed rhesus macaques. Lab Anim Sci 40: 257–261
Wilson DW, Day PA, Brummer MEG (1984) Diarrhea associated with Cryptosporidium spp in juvenile macaques. Vet Pathol 21: 447–450
Wyand MS, Ringler DJ, Naidu YM, Mattmuller M, Chalifoux LV, Sehgal PK, Daniel MD, Desrosiers RC, King NW (1989) Cellular localization of simian immunodeficiency virus in lymphoid tissues. II. In situ hybridization. Am J Pathol 134: 385–393

$SIV_{smmPBj14}$: An Atypical Lentivirus

P.N. FULTZ

1 Introduction

Among the various retrovirus-animal systems currently in use or in development as models for infection and disease induced by the human immunodeficiency virus (HIV), the most relevant model is infection of macaque monkeys with simian immunodeficiency viruses (SIVs). In various macaque species the majority of SIV isolates exhibit a pathogenicity which varies from development of AIDS-like disease and death within 6–12 months, such as $SIV_{DeltaB670}$ (ZHANG et al. 1988) and SIV_{mac239} (KESTLER et al. 1990), to long-term asymptomatic infection with slowly progressive disease and survival for as long as 4–5 years, such as SIV_{smm9} infection of rhesus monkeys (MCCLURE et al. 1989). These model systems have multiple parallels to HIV infection of humans and have therefore been useful not only for gaining insight into the natural history and pathogenesis of the primate lentiviruses but also for testing antiviral therapies and vaccines. In addition, infection of macaques with pathogenic molecularly cloned SIVs has been particularly valuable in identifying possible roles of accessory genes in disease development (KESTLER et al. 1991; LANG et al. 1993) and regions of the genome associated with distinct tropisms and disease states (DESROSIERS et al. 1991; MORI et al. 1992).

Department of Microbiology, University of Alabama at Birmingham, 845 19th Street So. Birmingham, AL 35294, USA

Current Topics in Microbiology and Immunology, Vol. 188

Although lentiviruses are classically associated with slowly progressive diseases, which usually affect the central nervous system, lungs, or immune system, the acute phase of infection with HIV-1 is often accompanied by a mononucleosis-like disease syndrome that can be severe, but symptoms usually subside as the immune system effectively limits virus replication (LEVY 1993). The most extreme case of acute disease associated with a primate lentivirus is that induced by $SIV_{smmPBj14}$ (SIV-PBj14), which induces an acute disease syndrome that culminates in death within 6–10 days after intravenous inoculation (FULTZ et al. 1989a). This particular isolate evolved from a prototypic sooty mangabey (*Cercocebus atys*) SIV, SIV_{smm9}, during 14 months of infection of a pig-tailed macaque (*Macaca nemestrina*) with the identifier PBj. Even though the virulence of SIV-PBj14 is atypical for lentiviruses, it and its biologically and molecularly cloned derivatives can provide insight into lentivirus-host interactions that are relevant to the pathogenesis of AIDS. Furthermore, the rapid course of the acute disease syndrome will facilitate some aspects of AIDS-related research, such as those associated with evaluating potential vaccines and therapies for prevention of disease.

2 $SIV_{smmPBj14}$ Disease Syndrome

SIV-PBj14 induces the same disease syndrome in pig-tailed, cynomolgus (*M. fascicularis*), and rhesus (*M. mulatta*) macaques as well as in SIV-seronegative sooty mangabey monkeys, but pig-tailed macaques appear to be more uniformly susceptible to the acute lethal effects of the virus (FULTZ et al. 1989a; ROSENBERG et al. 1991; LEWIS et al. 1992). Within 4–6 days after intravenous inoculation of approximately 10^4 50% tissue culture infectious doses ($TCID_{50}$) macaques develop anorexia and depression which is often associated with a profuse bloody diarrhea. Despite intravenous or subcutaneous fluid therapy, most animals become moribund or die 1–2 days after onset of diarrhea. Within this time both marked lymphadenopathy and splenomegaly develop, and a maculopapular rash is often noted.

Hematological evaluation of peripheral blood samples obtained during acute infection shows a decrease in all lymphocyte subsets coincident with neutrophilia and monocytosis (DEWHURST et al. 1990; LEWIS et al. 1992; ISRAEL et al. 1993). In animals that survive the acute syndrome, all lymphocyte subsets gradually increase to levels higher than those at the time of virus inoculation. During the acute phase, increases in serum and tissue levels of cytokines, several of which are associated with acute inflammatory reactions, have been detected. The most prevalent cytokines appear to be interleukin (IL)6 and tumor necrosis factor-α (TNF-α; BIRX et al. 1993; P.N. Fultz and R. Schwiebert, unpublished).

In pig-tailed macaques the pathogenic sequelae of SIV-PBj14 and one of its biologic clones (bcl3) are essentially the same, irrespective of whether animals are inoculated by blood transfusion, with tissue culture-propagated cell-free virus, tissue homogenates from an animal that died during the acute phase (FULTZ et al. 1989a; LEWIS et al. 1992; NOVEMBRE et al. 1991), or virus-infected lymphocytes from an animal that died on day 7 from SIV-PBj14-bcl3 infection (P.N. Fultz and R. Schwiebert, unpublished data). In vivo titration of serial tenfold dilutions of cell-free SIV-PBj14 in two pig-tailed macaques each resulted in a delay of approximately 48 h of both onset of symptoms and death. This pattern extended to a 1000-fold dilution, or to approximately 10 $TCID_{50}$, a dose that resulted in death of animals on days 15 and 16 postinfection (FULTZ et al. 1989b).

The responses of rhesus macaques are more variable. Approximately 30%–50% of animals of this species succumb to the acute disease syndrome; the remainder of infected rhesus macaques exhibit no obvious evidence of disease and often survive for several years with no apparent immunodeficiency. Although the exact reason for the wide variation in susceptibility of rhesus macaques to SIV-PBj14 is not known, a correlation between elevated IL-6 levels and acute disease induction has been noted (BIRX et al. 1993).

In contrast to marked depletion of lymphocytes in the peripheral blood, lymphoid tissues show a dramatic increase in the number of lymphocytes, resulting in gross enlargement, particularly in the gut-associated lymphoid tissue and throughout the lamina propria of the gastrointestinal (GI) tract (FULTZ et al. 1989a; LEWIS et al. 1992; P. Zack personal communication). Peyer's patches become greatly enlarged, due to large numbers of lymphoblasts expressing both CD4 and CD8 cell-surface antigens and activation markers, such as CD25 (IL-2 receptor), HLA-DR, and CD71 (transferrin receptor; P.N. Fultz and R. Schwiebert, unpublished), as detected by immunohistochemistry and immunofluorescent flow cytometry. Immunohistochemistry and electron microscopic examination also revealed SIV antigens and large numbers of virions throughout these tissues. Another characteristic pathological finding in peripheral lymph nodes was a marked reduction in B-cell areas with expansion of T-cell compartments, in which more $CD8^+$ than $CD4^+$ lymphocytes and fewer SIV antigen positive cells were detected (LEWIS et al. 1992).

In all animals that died acutely there was generalized edema of the GI mucosa with villus blunting and fusion of the superficial mucosal epithelium. However, no viral antigens were detected in epithelial cells. Histopathological lesions were not ulcerative, but were consistent with a diffuse lymphohistiocytic gastroenteritis. Only occasionally have other pathogenic organisms been detected in tissues of animals that died acutely, despite extensive examination of tissues by several groups of investigators.

While dehydration, electrolyte imbalance, and metabolic acidosis may contribute to death in some animals, more recently animals have been observed to deteriorate rapidly (within a few hours) after onset of anorexia and depression, with little or no diarrhea or fluid loss but with symptoms that suggest a shocklike syndrome. The presence of high levels of IL-6 and TNF-α in serum of animals that failed to develop diarrhea yet died within 7–8 days after virus inoculation supports this diagnosis. In addition, elevation in levels of other acute-phase inflammatory reactants, such as IgM, IgA, C3 and C4 complement components, haptoglobin, and IL-1, have been noted (M. Lewis and P. Zack, personal communication). Thus, the virulence of SIV-PBj14 appears to result from the production of extremely high levels of virus, which is accompanied by induction of inflammatory cytokines and other acute-phase reactants.

The acute disease syndrome, including anorexia, diarrhea, rash, depression, and lymphadenopathy, has also been observed uniformly in pig-tailed macaques inoculated with a molecular clone of SIV-PBj14 (mcl1.9; ISRAEL et al. 1993). However, in contrast to infection with the original SIV-PBj14 isolate and SIV-PBj14-bcl3, these symptoms generally resolved, and all but one macaque infected with SIV-PBj14-mcl1.9 has survived long-term. During the acute symptomatic phase animals infected with this molecular clone are viremic and antigenemic, and a high percentage of circulating $CD4^+$ lymphocytes and monocyte/macrophages appear to be infected when tested for the presence of intracellular p27 Gag antigen by FACScan analysis. As in acute HIV-1 infection, viremia and numbers of infected peripheral blood cells decrease coincident with detection of anti-SIV antibodies. The similarities between acute infection with HIV-1 and SIV-PBj14-mcl1.9 and, to a certain extent, SIV-PBj14 are striking, indicating that this virus may provide a useful model for defining early events in infection by primate lentiviruses. Similarities between acute infection with HIV-1 and SIV-PBj14-mcl1.9 include:

- anorexia
- erythematous rash
- diarrhea and gastrointestinal problems
- immune activation
- lymphopenia
- transient loss of $CD4^+$ lymphocytes
- lymphadenopathy and splenomegaly
- high-level viremia and antigenemia that resolve as antibodies become detectable.

3 Biological Properties of $SIV_{smmPBj14}$

In addition to its extreme pathogenicity, SIV-PBj14 acquired multiple in vitro biological properties that distinguish it not only from its parent virus, SIV_{smm9}, but also from other primate lentiviruses. Unique biological properties of SIV-PBj14 are

- replication in chimpanzee PBMC
- efficient replication in quiescent macaque PBMC
- activation of macaque lymphocytes
- induction of macaque lymphocyte proliferation
- cytopathicity for sooty mangabey $CD4^+$ lymphocytes (relative to natural SIV_{smm} isolates)
- induction of acute disease and death.

Consistent with isolates of HIV-1 obtained from persons in late stages of disease (FENYO et al. 1988; TERSMETTE et al. 1989; CONNOR et al. 1993), SIV-PBj14 has an expanded host range compared to that of SIV_{smm9}. SIV-PBj14, but not SIV_{smm9}, replicates in several human T-cell lines and in chimpanzee peripheral blood mononuclear cells (PBMC). In fact, SIV-PBj14 is the only SIV isolate identified to date that replicates in chimpanzee PBMC, albeit at a much lower rate compared to its replication in human and macaque PBMC (FULTZ et al. 1989a). Another property of the acutely lethal virus that may have been an important factor in its evolution is that it is not neutralized by serum obtained from pig-tailed macaque PBj at any time during infection. In contrast, the parental SIV_{smm9} isolate is neutralized by all serum samples obtained from macaque PBj at 6 months after infection or thereafter (FULTZ et al. 1989a), suggesting that immune selection may have been important in SIV-PBj14 evolution. Neutralization escape mutants have been identified in macaques infected with other SIV isolates (BURNS et al. 1993) and in HIV-1-infected chimpanzees (NARA et al. 1990).

Among the other biological differences between SIV-PBj14 and SIV_{smm9}, those properties most likely to be directly related to acute virulence are the ability to replicate in resting or quiescent lymphocytes and the ability to activate and induce proliferation of both $CD4^+$ and $CD8^+$ lymphocytes (FULTZ 1991; P.N. Fultz and R. Schwiebert, unpublished). These two properties may be linked, and both may contribute to pathogenesis, but replication of SIV-PBj14 is not required for activation and proliferation of lymphocytes. Heat-inactivated purified SIV-PBj14 virions induce macaque lymphocytes to incorporate [^{3}H]thymidine after only 2–4 days of in vitro incubation (FULTZ 1991). These data also demonstrate that proliferation is not mediated by cytokines induced as a result of virus replication. It is not known whether cytopathicity for mangabey $CD4^+$ T cells is required for virulence in pig-tailed macaques, but it probably contributes to the virus' pathogenicity for mangabeys. SIV isolates from naturally infected sooty mangabeys are not cytopathic in vitro for mangabey $CD4^+$ lymphocytes and usually do not cause an AIDS-like disease in this host species (FULTZ et al. 1990b).

Since SIV-PBj14 is the most acutely virulent lentivirus identified to date, it seemed worthwhile to document the evolution of the virus from the minimally pathogenic SIV_{smm9}. Sequential isolates, designated SIV-PBj3, -PBj6, -PBj8, etc., were obtained at various times (PBjX, where X is months after infection) between inoculation of SIV_{smm9} and recovery of SIV-PBj14 from macaque PBj PBMC; these isolates were obtained by cocultivation of PBj PBMC with normal human PBMC and were cryopreserved as cell-free culture supernatants in liquid nitrogen vapor. Passage 1 virus stocks were made from each of the PBjX viruses, and these were tested for each of SIV-PBj14 novel in vitro properties (see above; B. Tao and P.N. Fultz unpublished). The results are consistent with previous studies (FULTZ et al. 1989a) which indicated that a major change in the quasispecies phenotype occurred between 6 and 10 months after infection of PBj. Biological clones of SIV-PBj6, -PBj8, and -PBj10 are being generated and characterized biologically and molecularly. SIV-PBj8 is of particular interest because this isolate possesses in vitro properties indicative of both SIV_{smm9} and SIV-PBj14; this quasispecies does not replicate in resting macaque PBMC, induces minimal proliferation of resting macaque PBMC, yet efficiently kills mangabey $CD4^+$ lymphocytes.

4 Correlation Between In Vitro and In Vivo Properties of $SIV_{smmPBj14}$

The ability of heat-inactivated SIV-PBj14 to activate and induce macaque $CD4^+$ and $CD8^+$ lymphocytes to proliferate has been demonstrated convincingly in [^{3}H]thymine-incorporation assays. Activation of both lymphocyte subsets was also shown to occur during active virus replication in cultures of primary macaque PBMC. Using double-label immunofluorescence and FACScan analysis, both $CD4^+$ and $CD8^+$ cells expressed increasing levels of CD25 and CD71 antigens as SIV-PBj14 replicated and accumulated in culture media. More importantly, both lymphocyte populations acquired these same cell-surface activation markers during in vivo infection. Lymphocytes from peripheral blood and lymph node biopsies before infection and at different times after SIV-PBj14 inoculation were evaluated for evidence of activation using double-label immunofluorescence (P.N. Fultz and R. Schwiebert, unpublished). These results indicated that activation of both lymphocyte subsets occurs during infection of macaques (Fig. 1), consistent with the presence of lymphoblasts observed in tissue sections by histology and electron microscopy. Furthermore, the validity of using in vitro experiments to define potential pathogenic mechanisms was demonstrated.

Based on the hypothesis that the ability to replicate in and to activate and induce proliferation of primary macaque lymphocytes was required for induction of disease by SIV-PBj14, various biologically and molecularly cloned viruses derived from SIV-PBj14, for which the pathogenesis for pig-tailed macaques was also known, were characterized for these properties. All of the viruses tested, with the exception of SIV_{smm9} and SIV-PBj14-bcl1,

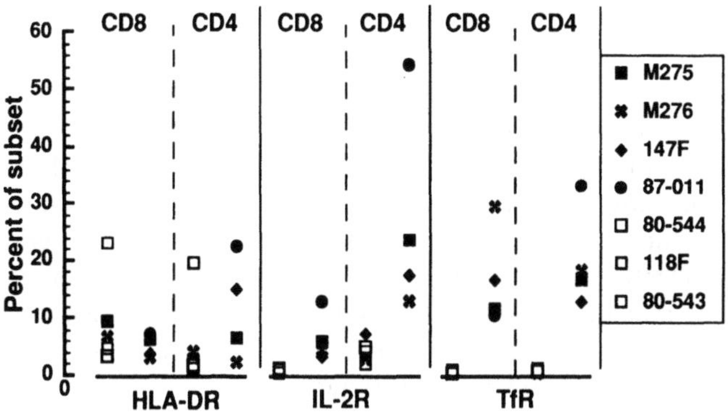

Fig. 1. Coexpression of activation markers on CD4$^+$ and CD8$^+$ peripheral blood lymphocytes before inoculation of SIV-PBj14 (first value in each group) and at necropsy 7–8 days later (second value in each group). Data are for four individual infected pig-tailed macaques. *Open symbols*, values for normal, uninfected control pig-tailed macaques

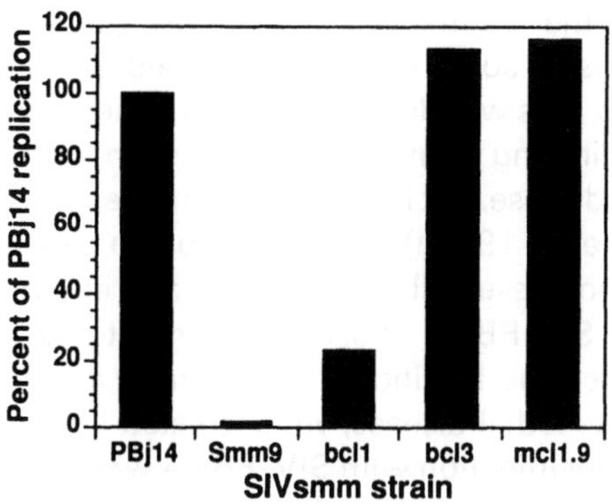

Fig. 2. Comparative replication of SIV_{smm9}, SIV-PBj14, and biologically and molecularly cloned viruses derived from SIV-PBj14 in resting peripheral blood lymphocytes from an uninfected pig-tailed macaque. Values were calculated from reverse transcriptase activity in culture supernatants on day 14 after infection and were relative to that for SIV-PBj14 (531 260 cpm). Culture medium was RPMI-1640 containing 10% fetal bovine serum, glutamine, and antibiotics

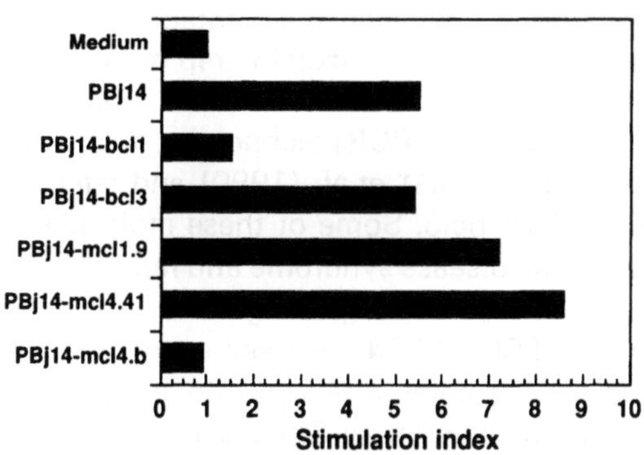

Fig. 3. Induction of proliferation of resting pig-tailed macaque PBMC by viruses derived from SIV-PBj14. Stimulation index is based on incorporation of [^{3}H]thymidine (average of quadruplicate samples) after incubation of 10^5 PBMC with 10 ng of the different viruses in 0.2 ml RPMI-1640 medium containing 10% human AB serum. Viruses are as described in the text, except for the acutely pathogenic molecular clone SIV-PBj14-mcl4.41 (Dewhurst et al. 1990) and SIV-PBj14-mcl4.b, which is a recombinant virus containing the 5′ half of SIV-PBj14-bcl3 and the 3′ half of SIV-PBj14-bcl1 (Dewhurst et al. 1992)

replicated in and induced proliferation of resting pig-tailed macaque PBMC (Figs. 2, 3). In addition, these same viruses, but not SIV_{smm9} or SIV-PBj14-bcl1, were cytopathic for mangabey $CD4^+$ lymphocytes (not shown). Thus, these in vitro properties correlated exactly with in vivo pathogenesis, that is, with the ability to induce the acute disease syndrome. Using chimeric molecular clones, NOVEMBRE et al. (1993) also found that the ability to induce proliferation of macaque PBMC correlates with acute disease and death. Although these correlations have not been proven definitively, there is strong circumstantial evidence to support the hypothesis.

5 Molecular Characterization of $SIV_{smmPBj14}$

Like other primate lentivirus isolates, the original SIV-PBj14 was a quasispecies, composed of individual viruses, some of which retained all the properties of the parental SIV_{smm9} virus. This was demonstrated by generating biologically cloned viruses by serial limiting dilutions; the first such virus, SIV-PBj14-bcl1, did not induce acute disease, but led to persistent infections and slow progression to AIDS (FULTZ et al. 1989a). Thus, attempts to generate infectious molecular clones that possessed all of the in vitro and in vivo properties observed with the original SIV-PBj14 quasispecies might never have succeeded. Therefore, to enhance the likelihood of obtaining a biological clone that reproduces the acute lethal disease, virus isolated from a pig-tailed macaque that died 7 days after infection with SIV-PBj14 was used as the virus source for limiting dilution analyses. This approach was based on the assumption that virulent members of the quasispecies would be amplified preferentially in vivo because of a replicative advantage. The first biologically cloned virus so generated, SIV-PBj14-bcl3, reproduced all aspects of the acute lethal disease and also exhibited the in vitro properties of SIV-PBj14. Using polymerase chain reaction (PCR) technology, infectious molecular clones were generated by DEWHURST et al. (1990) and later by NOVEMBRE et al. (1993) from SIV-PBj14-bcl3. Some of these molecularly cloned viruses also reproduced the acute disease syndrome and rapid death, thus confirming that this lentivirus is the sole etiological agent.

Because of the extreme virulence of SIV-PBj14 the complete nucleotide sequences of multiple clones of the parental SIV_{smm9}, the nonacutely pathogenic SIV-PBj14-bcl1, and the acutely lethal SIV-PBj14-bcl3 were obtained (DEWHURST et al. 1990, 1992; COURGNARD et al. 1992). Comparisons of consensus sequences for the various viruses revealed numerous point mutations and two duplications; one duplication encompassed 22 bp in the U3 region of the long terminal repeat, resulting in a second nuclear factor κB (NF-κB) site; the second duplication was in the V1 region of the *env* gene and was 15 bp. This latter duplication was not found in any of the clones from the parental isolate SIV_{smm9} but was found in all clones generated from both SIV-PBj14 and SIV-PBj14-bcl1. Interestingly, several biological clones of the SIV-PBj10 sequential isolate had duplications or insertions of 27 bp in the V1 region of *env*. Thus, since SIV-PBj14-bcl1 is not acutely pathogenic, this mutation is unlikely to represent the sole viral determinant responsible for the acute disease syndrome; however, it may contribute to some unique aspects of the SIV-PBj14 phenotype.

Because SIV-PBj14 was the only known virus belonging to the SIV_{smm}/SIV_{mac}/HIV-2 group of the primate lentiviruses with two NF-κB binding sites in the enhancer region, there was speculation that this mutation played a major role in the acutely lethal disease syndrome. However, two independent lines of evidence indicate that this 22-bp duplication is not sufficient. First, NOVEMBRE et al. (1991, 1993) generated recombinant virus clones using an acutely lethal molecularly cloned virus and an apathogenic (relative to SIV-PBj14) molecular clone, SIV_{sm} H4 (HIRSCH et al. 1989). In vivo analysis of the pathogenesis of six recombinant viruses showed that viruses containing two NF-κB sites did not induce acute disease or rapid death. These investigators also found that only 25% of clones obtained directly from intestinal tissue of a pig-tailed macaque that died acutely from infection with SIV-PBj14 had two NF-κB sites (NOVEMBRE et al. 1991), confirming the quasispecies nature of SIV-PBj14.

Second, PCR amplification and nucleotide sequence data have been obtained for viruses cloned biologically from sequential isolates recovered from macaque PBj before death at 14 months. There appears to be no correlation between the presence of two NF-κB sites and in vitro biological properties identical to those of SIV-PBj14 (Table 1). Although neither the sequential isolates nor their biologically cloned derivatives have been tested for pathogenicity in vivo, it is unlikely that viruses which do not replicate in resting PBMC or induce proliferation of pig-tailed macaque PBMC will elicit the acute disease syndrome and death in susceptible animals.

The conclusion from the molecular analyses performed to date with SIV-PBj14 and the sequential PBj virus isolates is that determination of the acutely lethal phenotype is multifactorial and is likely to involve more than one region of the viral genome. One region that has not been evaluated is the region of *env* encoding the transmembrane portion of the glycoprotein. Both nonacutely pathogenic viruses, SIV_{smm9} and SIV-PBj14-bcl1, contain premature stop codons in TM whereas all other SIV-PBj14-derived viruses have full-length TM. In addition, COURGNAUD et al. (1992) identified a maximum of 36 nonsynonymous amino acid differences between SIV-PBj14 and SIV_{smm9}; 64% of these differences were in Pol (12) or Env (11). An additional five amino acid changes were in Nef. It is possible, therefore, that regions of the genome other than those examined to date are important for induction of acute disease and death.

While it may be difficult to localize viral determinants of acute pathogenicity using chimeric recombinant viruses, the sequential SIV-PBj isolates

Table 1. Genetic analysis of PBj sequential SIV_{smm} isolates and biological clones

Virus strain	Phenotype	Genotype			
		LTR (22-bp dupl)		Env (15- or 27-bp dupl)	
		smm9	PBj14	smm9	PBj14
PBj6	smm9	5/5			
PBj6-bclA-1	smm9	1/1		3/3	
PBj6-bclA-2	smm9	1/1		2/2	
PBj10	PBj14	5/6	1/6		
PBj10-bclA-1	smm9		10/10		7/7
PBj10-bclB-2	PBj14	2/2			9/9
PBj10-bclC-1	PBj14	1/1			4/4

Values given are number of PCR-amplified clones with nucleotide sequences resembling that of SIV_{smm9} or SIV-PBj14 in a specific region/number of clones sequenced. Within one virus strain, no clones were identical in sequence.

and their biological clones (see above) may prove of value in identifying genomic regions that confer some of the unique individual biological properties of SIV-PBj14. Preliminary nucleotide sequence data for limited regions of some of the SIV-PBj6 and SIV-PBj10 biological clones, which exhibit the in vitro phenotypes of either SIV_{smm9} or SIV-PBj14, also reveal the complexity of the acutely lethal phenotype (Table 1). However, if SIV-PBj8 biological clones can be obtained that reproduce the SIV-PBj8 phenotype (see above), it may be possible to localize viral genetic determinants for those properties that segregate, for example, cytopathicity for mangabey $CD4^+$ lymphocytes and the ability to replicate in and induce proliferation of quiescent pig-tailed macaque PBMC.

6 Summary and Conclusions

SIV-PBj14 is atypical for a lentivirus in that infection of pig-tailed macaques usually does not result in long-term progressive disease; however, this model may potentially provide valuable information about the pathogenesis of HIV and the development of AIDS. By capitalizing on some of the unique properties of the virus and the model system discussed above, new insights may be gained in: (a) understanding pathogenic mechanisms of acute lentiviral infections, (b) dissecting lentivirus-host cell interactions, (c) evaluating the role(s) of cytokines in lentivirus-induced disease, (d) rapidly assessing therapeutic and prophylactic benefits of new drugs and vaccines, and (e) identifying regions of the viral genome that influence specific biological properties. While it is unlikely, but possible, that HIV-1 variants as virulent as SIV-PBj14 will become a threat to humans, SIV-PBj14 may possess properties important to the development of lentivirus-induced disease. An understanding of all possible virus-host interactions, from the most virulent to the most benign, may be required to make a significant, positive impact on the HIV-1 pandemic.

Acknowledgments. The author wishes to acknowledge the contributions of past and present collaborators and support from NIH (AI27136 and AI32377).

References

Birx DL, Lewis MG, Vahey M, Tencer K, Zack PM, Brown CR, Jahrling PB, Tosato G, Burke D, Redfield R (1993) Association of interleukin-6 in the pathogenesis of acutely fatal $SIV_{smm/PBj14}$ in pigtailed macaques. AIDS Res Hum Retroviruses 9: 1123–1129

Burns DPW, Collignon C, Desrosiers RC (1993) Simian immunodeficiency virus mutants resistant to serum neutralization arise during persistent infection of rhesus monkeys. J Virol 67: 4104–4113

Connor RI, Mohri H, Cao Y, Ho DD (1993) Increased viral burden and cytopathicity correlate temporally with $CD4^+$ T-lymphocyte decline and clinical progression in human immunodeficiency virus type 1-infected individuals. J Virol 67: 1772–1777

Courgnaud V, Laure F, Fultz PN, Montagnier L, Brechot C, Sonigo P (1992) Genetic differences accounting for evolution and pathogenicity of simian immunodeficiency virus from sooty mangabey monkey after cross-species transmission to a pig-tailed macaque. J Virol 66: 414–419

Desrosiers RC, Hansen-Moosa A, Mori K, Bouvier DP, King NW, Daniel MD, Ringler DJ (1991) Macrophage-tropic variants of SIV are associated with specific AIDS-related lesions but are not essential for the development of AIDS. Am J Pathol 139: 29–35

Dewhurst S, Embretson JE, Anderson DC, Mullins JI, Fultz PN (1990) Sequence analysis and acute pathogenicity of molecularly cloned $SIV_{smm\text{-}PBj14}$. Nature 345: 636–640

Dewhurst S, Embretson JE, Fultz PN, Mullins JI (1992) Molecular clones from a non-acutely pathogenic derivative of $SIV_{smmPBj14}$: characterization and comparison to acutely pathogenic clones. AIDS Res Hum Retroviruses 8: 1179–1187

Fenyo EM, Morfeldt-Manson L, Chiodi F, Lind B, von Gegerfelt A, Albert J, Olausson E, Asjo B (1988) Distinct replicative and cytopathic characteristics of human immunodeficiency virus isolates. J Virol 62: 4414–4419

Fultz PN (1991) Replication of an acutely lethal simian immunodeficiency virus activates and induces proliferation of lymphocytes. J Virol 65: 4902–4909

Fultz PN, McClure HM, Anderson DC, Swenson RB, Anand R, Srinivasan A (1986) Isolation of a T-lymphotropic retrovirus from naturally infected sooty mangabey monkeys (*Cercocebus atys*). Proc Natl Acad Sci USA 83: 5286–5290

Fultz PN, McClure HM, Anderson DC, Switzer WM (1989a) Identification and biologic characterization of an acutely lethal variant of simian immunodeficiency virus from sooty mangabeys (SIV/SMM). AIDS Res Hum Retroviruses 5: 397–409

Fultz PN, Mullins JI, Dewhurst S, McClure HM, Ansari A, Anderson DC (1989b) Virus-host interactions in SIV/SMM infection of macaque and mangabey monkeys In: Girard M, Valette L (eds) Retroviruses of human AIDS and related animal diseases. Merieux, Lyon, pp 161–165

Fultz PN, Anderson DC, McClure HM, Dewhurst S, Mullins JI (1990a) SIV_{smm} infection of macaque and mangabey monkeys: correlation between in vivo and in vitro properties of different isolates. Dev Biol Stand 72: 253–258

Fultz PN, Gordon TP, Anderson DC, McClure HM (1990b) Prevalence of natural infection with SIV_{smm} and STLV-I in a breeding colony of sooty mangabey monkeys. AIDS 4: 619–625

Hirsch VM, Olmsted RA, Murphey-Corb M, Purcell RH, Johnson PR (1989) An African primate lentivirus (SIV_{sm}) closely related to HIV-2. Nature 339: 389–392

Israel ZR, Dean GA, Maul DH, O'Neil SP, Dreitz MJ, Mullins JI, Fultz PN, Hoover EA (1993) Early pathogenesis of disease caused by SIV_{smm}PBj14 molecular clone 1.9 in macaques. AIDS Res Hum Retroviruses 9: 277–286

Kestler H, Kodama T, Ringler D, Marthas M, Pedersen N, Lackner A, Regier D, Sehgal P, Daniel M, King N, Desrosiers R (1990) Induction of AIDS in rhesus monkeys by molecularly cloned simian immunodeficiency virus. Science 248: 1109–1112

Kestler HW, Ringler DJ, Mori K, Panicali DL, Sehgal PK, Daniel MD, Desrosiers RC (1991) Importance of the *nef* gene for maintenance of high virus loads and for development of AIDS. Cell 65: 651–662

Lang SM, Weeger M, Stahl-Hennig C, Coulibaly C, Hunsmann G, Muller J, Muller-Hermelink H, Fuchs D, Wachter H, Daniel MM, Desrosiers RC, Fleckenstein B (1993) Importance of *vpr* for infection of rhesus monkeys with simian immunodeficiency virus. J Virol 67: 902–912

Levy JA (1993) Pathogenesis of human immunodeficiency virus infection. Microbiol Rev 57: 183–289

Lewis MG, Zack PM, Elkins WR, Jahrling PB (1992) Infection of rhesus and cynomolgus macaques with a rapidly fatal SIV ($SIV_{smm/PBj}$) isolate from sooty mangabeys. AIDS Res Hum Retroviruses 8: 1631–1639

McClure HM, Anderson DC, Fultz PN, Ansari AA, Lockwood E, Brodie A (1989) Spectrum of disease in macaque monkeys chronically infected with SIV/SMM. Vet Immunol Immunopathol 21: 13–24

Mori K, Ringler DJ, Kodama T, Desrosiers RC (1992) Complex determinants of macrophage tropism in *env* of simian immunodeficiency virus. J Virol 66: 2067–2075

Nara PL, Smit L, Dunlop N, Hatch W, Merges M, Waters D, Kelliher J, Gallo RC, Fischinger PJ, Goudsmit J (1990) Emergence of viruses resistant to neutralization by V3-specific antibodies in experimental human immunodeficiency virus type 1IIIB infection of chimpanzees. J Virol 64: 3779–3791

Novembre FJ, Hirsch VM, McClure HM, Johnson PR (1991) Molecular diversity of $SIV_{smm/PBj}$ and a cognate variant, $SIV_{smm/PGg}$. J Med Primatol 20: 188–192

Novembre FJ, Johnson PR, Lewis MG, Anderson DC, Klumpp S, McClure HM, Hirsch VM (1993) Multiple viral determinants contribute to pathogenicity of the acutely lethal simian immunodeficiency virus SIV_{smmPBj} variant. J Virol 67: 2466–2474

Rosenberg Y, White BD, Papermaster SF, Zack P, Jahrling PB, Eddy GA, Burke DS, Lewis MG (1991) Variation in T-lymphocyte activation and susceptibility to $SIV_{PBj\text{-}14}$-induced acute death in macaques. J Med Primatol 20: 206–210

Tersmette M, Gruters RA, deWolf F, de Goede REY, Lange JMA, Schellekens PTA, Goudsmit J, Huisman HG, Miedema F (1989) Evidence for a role of virulent human immunodeficiency virus (HIV) variants in the pathogenesis of acquired immunodeficiency syndrome: studies on sequential HIV isolates. J Virol 63: 2118–2125

Zhang J, Martin LN, Watson EA, Montelaro RC, West M, Epstein L, Murphey-Corb M (1988) Relationship of antibody response and viral antigenemia to SIV/Delta-induced immunodeficiency disease in the rhesus monkey. J Infect Dis 158: 1277–1286

Conserved Framework of Envelope Glycoproteins Among Lentiviruses

G. PANCINO[1], H. ELLERBROK[3], M. SITBON[2], and P. SONIGO[1]

1 Introduction

The envelope glycoproteins mediate important aspects of host-virus relationships and lentivirus pathogenic properties: they govern the cell tropism by interacting at the cell membrane with cellular receptors, they mediate membrane fusion to allow virus entry, and they constitute a major target of the immune response. In the case of HIV-1, as well as that of other lentiviruses, comparisons of amino acid sequences among various isolates

[1] Génétique des virus (CNRS UPR0415), Institut Cochin de Génétique Moléculaire, 22 rue Méchain, 75014 Paris, France
[2] Oncologie cellulaire et moléculaire (INSERM U363), Institut Cochin de Génétique Moléculaire, 22 rue Méchain, 75014 Paris, France
[3] AIDS Zentrum am Bundesgesundheitsamt, Nordufer 20, 13353 Berlin, Germany

Current Topics in Microbiology and Immunology, Vol. 188

have shown that envelope glycoproteins consist of a succession of variable domains which are bordered by conserved domains, most likely determining a conserved spatial framework organization (ALIZON et al. 1986; MYERS et al. 1990)

The existence of a more general framework, conserved between envelopes of different lentiviruses, remains an open question. Indeed, a high variation rate of lentiviral genomes leads to rapid divergence of amino acid sequences. However, in view of the similar pathogenic strategies of lentiviruses it seems that major lentiviral functions are conserved. Some data support the idea that common functions of the envelope glycoproteins of genetically divergent lentiviruses such as feline immunodefiency virus (FIV), equine infectious anemia virus (EIAV), and Human Immunodeficiency Virus (HIV) rely on a conserved structural framework that tolerates extensive variations in amino acid sequence (BALL et al. 1992; PANCINO et al. 1993b). The delineation of this framework will help to clarify structure-function relationships within lentiviral envelope glycoproteins.

In this chapter we deal with selected aspects of variation and conservation within envelope proteins of lentiviruses. We focus on the following features, presented in the order of appearance during translation: (a) signal peptide, (b) surface glycoprotein framework, (c) immunodominant domain of transmembrane glycoprotein, and (d) C-terminal domain of transmembrane glycoprotein.

2 The Signal Peptides of Lentiviral Envelopes

The 5′ end of the retroviral *env* gene encodes a class I signal peptide (SP) which ensures translocation in the endoplasmic reticulum (ER). Translocation occurs after the docking to the ER of a complex which includes the ribosome and the *env* mRNA, the nascent protein, and a ribonucleoprotein designated as signal recognition particle (Fig. 1). The translocation function of SP, which is universally conserved as it also ensures protein secretion in prokaryotes, is not based upon sequence homologies, but rather on a general organisation in three domains of various sizes. These domains are arranged from the amino-to the carboxy-termini into: a rather slightly charged region (n-region), a hydrophobic core (h-region) composed of at least six consecutive hydrophobic amino acids and a small stretch of more polar amino acids (NOTHWEHR and GORDON 1990; VON HEIJNE 1986) which defines the cleavage site (*c*-region) between the SP and the amino terminus of the mature protein. Class I SP, such as SP of lentiviral envelopes, are co-translationally cleaved by a cellular signal peptidase, leaving the protein NH_2-terminus into the ER lumen, which corresponds to the extracellular side of the cell membrane.

Further maturation occurs through the ER and Golgi apparatus and includes the different stages of glycosylation and cleavage leading to formation of the Env extracellular surface (SU) and transmembrane (TM) components (HUNTER and SWANSTROM 1990). In some cases the cleavage site between the retroviral SP and the mature Env has been formally identified by direct sequencing of the amino-terminus of the extracellular SU (KALYANARAMAN et al. 1990; KNOWLES et al. 1991). Identification of SP and delineation of the c-region has considerably gained from the predictive model established by VON HEIJNE (1986). Comparison of model predictions with the data obtained from direct sequencing of the SU amino-terminus has confirmed the validity of the model, although the predicted scores for retroviral Env SP are often very low and would not lead one to suspect the presence of an SP if these sequences were placed in a different gene context. A remarkable feature of many retroviral SP which is likely to contribute to this poor scoring is the presence of a large n-region with accumulation of positively (Arg, Lys, His) and negatively (Asp, Glu) charged residues. This feature is the most striking in FIV strains which generally harbor putative SP exceeding 140 amino acids. Such a large n-region is not limited to the Env SP from the lentivirinae subfamily, since certain viruses from the oncovirinae (Rous sarcoma virus, mouse mammary tumor virus), and the Spumavirinae subfamilies (human and simian foamy viruses) also harbor SP over 60 amino acids (ELLERBROK et al. 1992).

When applying the von Heijne predictive model, we use the PSIGNAL program from PCgene (Intelligenetics) to analyze the *env* sequences of a representative set of primate and nonprimate lentiviruses. Three distinct groups could be distinguished according to the characteristics of their SP (Table 1).

The Env SP from the first group, which includes exclusively and all of the nonprimate lentiviruses tested, are more than 60 amino acids long and have an exceptionally elevated number of charged residues in their n-region (16–47 residues representing 19%–27% of their entire SP). For example, in a compilation of 185 unrelated eukaryotic nonretroviral SP which are generally less than 25 amino acids long, none have more than five charges, and over 150 have a single or no charged residue in the n-region (ELLERBROK et al. 1992; WATSON 1984). In lentiviruses from the first group of Table 1, variable length of the SP amino-terminus also constitutes the first exon of Rev. The second group represented in Table 1 includes the HIV-1 strains and two SIVs: SIV_{mnd} and SIV_{cpz} isolated from mandrills (TSUJIMOTO et al. 1988) and chimpanzees (HUET et al. 1990), respectively. Although significantly longer than most eukaryotic cellular and other viral SP, the Env SP of this second group have an intermediate size, comprised of 30–45 amino acids, and also have a significant accumulation of charges (seven to nine residues representing 17%–23% of their entire SP). Viruses from this group are characterized by an overlap between the Env SP coding frame and a gene

Table 1. Characteristics of the envelope signal peptides of a representative set of lentiviruses

Lentivirus strain	Total length	n-region			n-region charges / Total length × 100
		Length	Total charges	Positive charges	
FIV_{Wo}	172	149	52	25	30
FIV_{peta}	172	149	47	22	27
OMVV	97	82	26	16	27
EIAV	86	74	16	10	19
CAEV	83	66	19	12	23
BIV	66	46	17	5	26
SIV_{mnd}	45	30	9	6	20
HIV-1_{bru}	30	16	7	6	23
SIV_{cpz}	30	10	7	5	23
SIV_{smm}	22	7	0	0	0
SIV_{agm}	20	3	1	1	5
SIV_{mac}	19	7	0	0	0
HIV-2_{rod}	19	4	0	0	0
HIV-2_{D205}	19	7	1	1	5

The n-region is the sequence comprised between the initiation Met and the last charged residue before the hydrophobic region of the SP (h-region). Total charges include positive (Arg, Lys His) and negative (Asp, Glu) residues. All sequences were retrived from the EMBL data bank and analyzed using the PCgene programs (Intelligenetics). FIV_{Wo} and FIV_{peta}, feline immunodeficiency virus Wo and Petaluma strains, respectively; OMVV, ovine Maedi Visna virus; EIAV, equine infectious anemia virus; CAEV, caprine arthritis and encephalitis virus; BIV, bovine immunodeficiency virus; HIV-1 and -2, type 1 and type 2 human immunodeficiency viruses, respectively; SIV_{mnd}, SIV_{cpz}, SIV_{smm}, SIV_{agm}, and SIV_{mac}, simian immunodefiency viruses from mandrill, chimpanzee, sooty mangabey monkey, African green monkey, and macaque, respectively.

translated in a different coding frame. These sequences code for Vpu in HIV-1 and SIV_{cpz} or for the first exon of Rev in SIV_{mnd}.

The SP of lentiviruses in the third group of Table 1 are very similar to most eukaryotic SP. They are composed of 19–22 residues with only one or no charge present in the n-region (Table 1). This group includes all HIV-2 strains and the related SIVs isolated from sooty mangabey monkeys (SIV_{smm}) and macaques (SIV_{mac}). This third group also includes SIVs isoloted from African green monkeys (SIV_{agm}), which sequences are divergent from the other members of this group (KIRCHHOFF et al. 1990). However, scoring of the SIV_{agm} SP was below threshold (0 in the PSIGNAL program), whereas scoring of the other members of this group had elevated scores, similar to those of canonic SP (between 7 and 8 in the PSIGNAL program; VON HEIJNE 1986). No alternative coding abilities have been identified in the region coding for the SP of the viruses from this third group.

2.1 Functional Studies on Lentiviral SP

Only few studies have been reported on retroviral SP in general (HUNTER and SWANSTROM 1990) and on lentiviral SP in particular. Concerning lentiviral SP, STEPHENS et al. (1992) used a vaccinia-based vector in fibroblasts to express a secreted SU component of FIV in the absence of TM, whereas in a second study, an in vitro translational system was used to produce the entire HIV Env precursor in the presence of microsomes, which reproduce the ER stages of protein maturation (ELLERBROK et al. 1992). Both studies showed that the charged residues in the n-region are not necessary for efficient translocation and maturation. However, deletion of the entire n-region, with the exception of the initiation codon, leads to the formation of an unstable and uncompletely matured product, resulting in undetectable secretion of the FIV SU, although translocation did not seem to be altered (STEPHENS et al. 1992). Concerning the HIV-1 Env SP, only slight inhibition of translocation was noted upon deletion of the n-region, suggesting that deletion of the n-region does not significantly alter the ER steps of Env maturation (ELLERBROK et al. 1992). In the latter work several mutations within the h-region have also been studied. Substitutions similar to those described to prevent SP function in both eukaryotes and prokaryotes still allow significant levels of translocation and maturation of the HIV-1_{bru} Env. These mutations include introduction in the h-region of one or two charged residues (Arg) or one α-helix breaking residue (Pro) for residues located at positions -8 and -9 from the SP cleavage site (Leu and Gly in HIV-1_{bru}). Such a functional tolerance to substitutions has also been observed upon dramatic mutations of the amino-acids located immediately upstream either from the h-region or from the cleavage site of the HIV-1 SP (Ellerbrok and Sitbon, unpublished). Nevertheless, deletion of the h-region completely abolished translocation, underlining that yet no other structure could substitute for the SP function. Most of these results on the HIV-1 Env SP observed in an in vitro translational system have recently been confirmed in cell cultures using CD4-expressing fibroblasts with an HIV-based Env expression vector (Ellerbrok and Sitbon unpublished). Taken together, these first studies indicated a remarkable functional tolerance of lentiviral SP to mutations.

2.2 Alternative Functions for Lentiviral SPs

The biological significance of the conservation of an unusually long n-region in the Env SP of certain lentiviruses remains to be elucidated. Several clues are left to further experimentation and a better understanding of the role of these structures during infection of the natural host. One clue is given by the homogeneous clustering of lentiviruses in three groups, as shown in Table 1. Grouping of lentiviruses based on SP structural features, such as the length and charge-content of the n-region, in the absence of any significant nucleotide or amino-acid homologies corresponds to certain phylogenic branching established on sequence homologies in other parts of the genome. Since distinctive features such as a long and highly charged n-region do not seem to influence dramatically the primary translocation function of the SP (ELLERBROK et al. 1992; HUNTER and SWANSTROM 1990; STEPHENS et al.; 1992; VON HEIJNE 1986), alternative roles for such sequences might be postulated in relationship to selective pressure for their conservation. Certain viral functions for these sequences have indeed been described. For instance, in viruses from the first group in Table 1, the sequence coding for the first exon of Rev is included in the amino-most amino acids of the SP. In viruses from the second group the sequence coding for the carboxy-terminus of Vpu or the first exon of Rev also overlaps within a different coding frame with the sequence coding the amino-most amino acids of the SP. Thus, 42 codons out of the 46 of the n-region of BIV Env SP also constitutes the first *rev* exon (OBERSTE et al. 1991), and the SP coding region of SIV_{cpz} and HIV-1 encodes, in a different coding frame, approximately the 30 carboxy-terminal aminoacids of *vpu*. However, in the case of the SP of FIV and SIV_{mnd}, over 70 and 27 amino acids, respectively, are located before the h-region and do not overlap with any demonstrated alternative feature.

Although the corresponding sequences might be part of alternative genes yet to be identified, other types of function may explain conservation of this peculiarly long n-region. Type 1 SP, such as Env SP, are cleaved during translation (Fig. 1), and it has been shown that they may follow a specific processing pathway before association to MHC class I molecules and presentation at the cell surface (HENDERSON et al. 1992). Interestingly, a peptide derived from the SP of the viral glycoprotein precursor GP-C of the murine lymphocytic choriomeningitis virus, an arena-virus, has been shown to be a T immunodominant epitope (BUCHMEIER and ZINKERNAGEL 1992). Therefore, it is most likely that peptides derived from these long retroviral SP also associate with MHC molecules. As such, and by analogy to the T immunodominant epitope of the GP-C SP, they might also be recognized by the cellular immune response. In favor of this prediction is the fact that FIV and HIV-1 SP have been observed to contain hypervariable sequences in their c-region indicative of the presence of epitopes escaping the immune response through "selection for change" (PANCINO et al. 1993a). Alternatively, association of SP-derived peptides with MHC molecules may contribute to immunosuppressive mechanisms allowing viral persistence (see OLDSTONE 1991, for a review), for instance, by inhibiting presentation of peptides which might be better targets for an efficient immune response or by altering directly MHC expression. In this case, positive selection for change may ensue directly from binding requirements with a wide range of MHC molecules.

In addition to help in testing these hypotheses, elucidation of the precise degradation processes of lentiviral SP and their association with the MHC will warrant a better understanding of lentivirus/host-specific interactions.

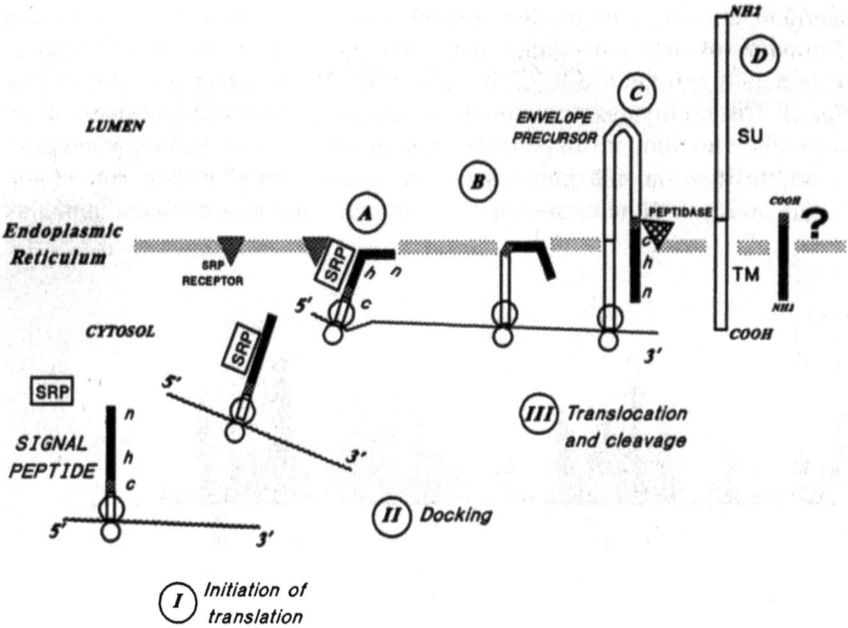

Fig. 1. Model of translocation in the endoplasmic reticulum (ER) and synthesis of the envelope precursor glycoprotein. *I*, Translation is initiated at the AUG initiation codon of the leader sequence coding for the signal peptide (SP). The amino-terminus (*n*), hydrophobic (*h*), and cleavage (*c*) regions of the SP and their respective orientation during translocation is indicated. *II* When enough of the h-region emerges from the ribosome interaction between the signal recognition particle (*SRP*), the nascent SP and the ribosome allows the docking of this complex to the ER through recognition of the SRP receptor. *III*, *A*, Charges in the n-region are thought to help orientation of the SP and subsequent translocation of the h-region into the lumen of the ER. *B* Further elongation of the nascent protein occurs into the lumen of the ER orienting the SP with its n-region in the cytosol. *C* Cleavage of the SP by an ER-associated peptidase occurs cotranslationally, releasing the NH_2-terminus of the envelope precursor into the lumen. *D* After cleavage of the SP, anchorage into the membrane is ensured by a hydrophobic stretch of the transmembrane component (*TM*) of the envelope precursor leaving the carboxy-terminus of the TM as a cytoplasmic tail and the amino-terminus of the precursor corresponding to the future surface extracellular component (*SU*) entirely present in the lumen. Exact fate of the cleaved SP remains unclear

3 A Framework of Lentiviral SU?

The variability of envelope glycoproteins is considered as the result of random mutations introduced by error-prone reverse transcriptase, followed by selective fixation of some of these mutations in the viral population. Selective pressures result from diverse elements such as functional constraints or escape from the host immune response. Consequently, variability is not randomly distributed, and its distribution reflects structural constraints and immunological pressures. Precise identification of such highly conserved and divergent sequences might then provide useful information on the structure of lentiviral envelopes in relation to their biological properties. We compared the variability profile of SIV_{agm} SU with that of HIV-1 and with that of FIV (Fig. 2). This comparison pointed out analogies in location and distribution of genetic variations in the surface glycoprotein. These analogies encompassed the three variable domains termed V3 , V4, and V5 in SIV, HIV-1, and FIV and suggested the existence of a common structure between lentiviral

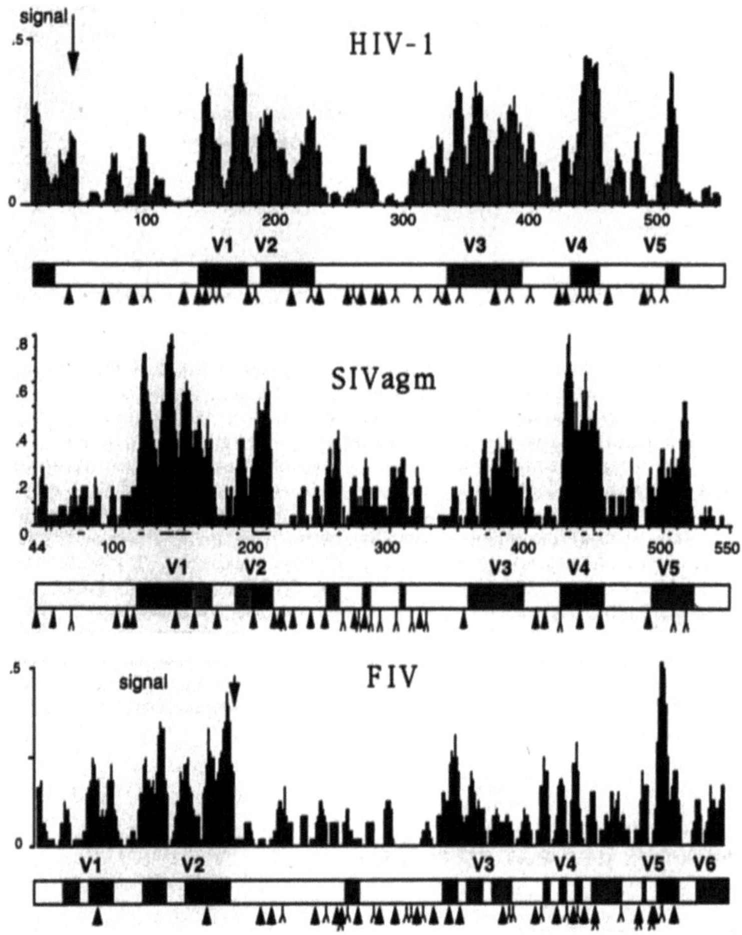

Fig. 2. Variable and conserved domains of SU in HIV-1 established from 20 sequences (CHAPPEY and HAZOUT 1992), SIV_{agm} established from 5 sequences (MÜLLER et al. 1993), and FIV established from 9 sequences (PANCINO et al. 1993b). The variability profiles are based on the frequency of amino acid substitutions relative to the consensus sequence and smoothed (value indicated on the *y axis*). Amino acids are numbered on the *x axis*. Below each variability profile, the conserved regions are *boxed in white*, the variable regions *in grey*, and the hypervariable regions in *black*. *Arrow*, cysteine residues; *forks*, putative N-glycosylation sites. (Kindly provided by C. Chappey)

envelope glycoproteins, not apparent at the level of the amino acid sequences. Indeed, faint amino acid sequence homologies still allow alignment between HIV-1 and SIV_{agm} SU, but sequence homology is lost between FIV SU and any of the primate lentivirus SU. Thus, when observing Fig. 2 from top to bottom, the similarities of the variation profiles are lost more slowly than those of amino acid sequences. This can be observed especially in the C-terminal halves.

To further assess these similarities, a schematic model of the spatial folding of FIV envelope glycoproteins was constructed (PANCINO et al. 1993b) and compared to structural models of surface envelope proteins previously proposed for HIV-1 (LEONARD et al. 1990; MODROW et al. 1987) and EIAV (BALL et al. 1992). The FIV SU model exhibited a conserved domain 1 and the three variable domains 3, 4, and 5. The variation, exposition, and folding of domains 3 and 4 of FIV may bear homologies to the domain of EIAV SU (BALL et al. 1992) and to the V3 loop of HIV-1 SU (JAVAHERIAN et al. 1989).

4 Conserved Framework of Retroviral TM

The TM of retroviruses presents common conserved structural features. These similarities, first observed when comparing sequences from type C, B, and D retroviruses (PATARCA and HASELTINE 1984), allowed construction of a general model for retrovirus TM proteins, including lentiviruses, based on the influenza virus hemagglutinin glycoprotein HA_2 (GALLAHER et al. 1989; Fig. 3). The conserved framework is organized in three functional domains. The N-terminal extracellular domain associates to the SU, the major hydrophobic domain anchors the protein into the lipid bilayer, while the function of the C-terminal domain, also designated as cytoplasmic, remains to be defined. Below we focus on two structures of the TM: the immunodominant domain which contains two neighboring cysteines at the apex of the extracellular domain and the unusually long cytoplasmic domain that follows the hydrophobic anchor domain at the C-terminus.

5 The Immunodominant Domain

5.1 A Conserved Cysteine Loop

Approximately in the middle of the external domain of the TM there is a highly conserved region containing two cysteine residues in lentiviruses and type B retroviruses and three cysteines residues in type C and D retroviruses

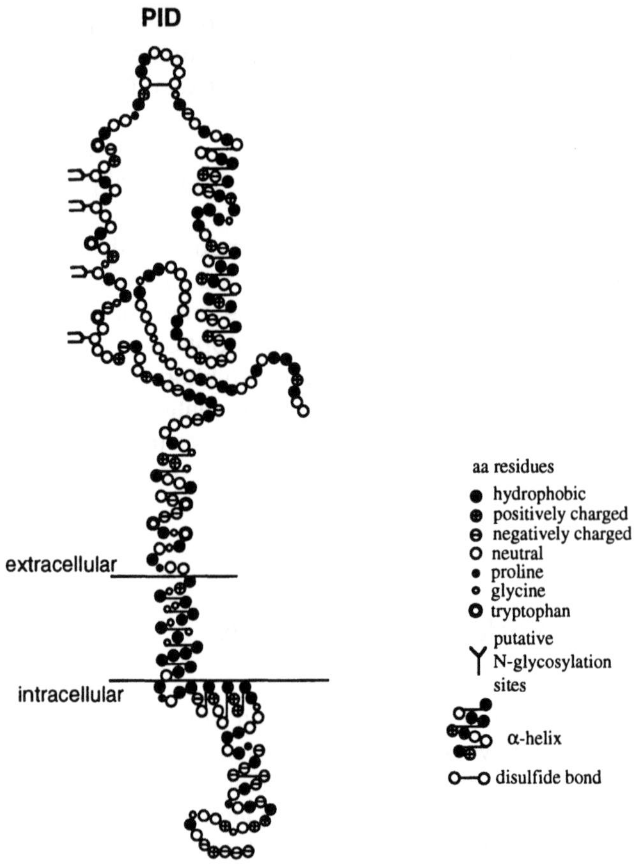

Fig. 3. Structural model of TM of FIV, based on predictive algorithms for secondary structures (PANCINO et al. 1993b). The model illustrates the framework for retroviral TM, defined by (GALLAHER et al. (1989). *PID*, Principal immunodominant domain

(Table 2). This region has been described as containing an immunosuppressive sequence in murine oncoviruses and in human T-cell leukemia virus (HTLV-1 and 2; CIANCIOLO et al. 1985) and a strongly immunogenic peptide in lentiviruses (GNANN et al. 1987a,c; WANG et al. 1986; see below).

An amino acid sequence in this region has been found to be significantly conserved between type C and D oncoviruses (CIANCIOLO et al. 1985; SONIGO et al. 1986). Conversely, the corresponding domain does not present primary sequence homology between different lentiviruses. However, despite the high variability observed in other regions of lentiviral envelopes, this domain, and particularly the sequence comprised between the two cysteines, is conserved among the different isolates of a given lentivirus. For example, the sequence is more than 99% conserved in isolates of HIV-1. A perfect

Table 2. Sequence alignment of principal immunodominant domains of TM from diverse lentiviruses (from MYERS et al. 1990; RUSHLOW et al. 1986; SALTARELLI et al. 1990; SONIGO et al. 1985; TALBOTT et al. 1989)

SIV_{mm142}	NAWG**CAFRQVC**HTTV
SIV_{smmPBj}	NSWG**CAFRQVC**HTTV
$HIV\text{-}2_{rod}$	NSWG**CAFRQVC**HTTV
SIV_{agm}	NAWG**CAWKQVC**HTTV
HIV_{bru}	GIWG**CSGKLIC**TTAV
HIV_{mn}	GFWG**CSGKLIC**TTV
HIV_{mal}	GMVG**CSGKHIC**TTFV
Visna	QELD**CWHYQHYC**VTST
CAEV	QELD**CWHYHQYC**VTST
FIV	QELG**CNQNQFFC**KVPS
EIAV	NLIG**CIERTHVFC**HTGH

The cysteine residues are conserved. Amino acids located between the cysteines are highly conserved among isolates of a given lentivirus (e.g., SIV and HIV-2) while they are divergent among different lentiviruses (e.g., HIV-1 versus HIV-2). Sequences likely folded as a cysteine loop are in boldface.

conservation of this sequence between different isolates of the same lentivirus is also found in HIV-2, in all SIV (MYERS et al. 1992) and in FIV (PANCINO et al. 1993b; Table 2). Irrespective of divergence of primary amino acid sequences the conservation of similarly spaced cysteines in all retroviruses argues in favor of a conservation of the secondary and tertiary structural features (GALLAHER et al. 1989; SCHULZ et al. 1992) and suggests a functional role of this domain.

No direct evidence has yet been reported on the existence of a disulfide bond between the two cysteines present in the TM of all retroviruses, including lentiviruses. However immunological studies, showing a preferential or a selected reactivity of human monoclonal antibodies and sera from HIV-1 infected patients with synthetic peptides containing both cysteines, suggest that such a cysteine loop is formed (GNANN et al. 1987c; OLDSTONE et al. 1991; XU et al. 1991). Analysis by proton nuclear magnetic resonance spectroscopy has shown that the immunologically reactive form of the heptapeptide CSGKLIC of HIV-1 contains a disulfide bond and is probably folded in a type I reverse turn (OLDSTONE et al. 1991). Site-directed mutagenesis has been used to analyze the role of the two cysteines in the ectodomain of HIV-1 TM (DEDERA et al. 1992; SYU et al. 1991). Both cysteines appear to be critical for correct envelope processing. Substitution of either cysteine leads to production of noninfectious virions, impairing the cleavage of envelope precursor into SU gp120 and TM gp41 subunits, yet not preventing TM oligomerization (DEDERA et al. 1992; SYU et al. 1991).

5.2 Possible Functions of the Immunodominant Domain

5.2.1 SU/TM Association

This apical region of TM, including the putative cysteine loop, is thought to be involved in the association of the SU and TM proteins. In HIV-1 the sCD4 binding to SU gp120 at the surface of HIV-1 infected cells, weakening the association SU/TM, increases the exposition of a TM epitope centered on the cysteine loop (SATTENTAU and MOORE 1991). Computer models have been proposed supporting a "knob-and-socket" mechanism, with the knob, formed by the TM loop fitting in the socket, formed by the C-terminal region of the SU (SCHULZ et al. 1992). This view leads to an important issue: the existence of a general mechanism, conserved in all retroviruses, by which the conformational changes due to the binding of the surface unit of the envelope to its cellular receptor may be transduced from the SU to TM in the envelope complex, and so activate the fusiogenic activity of the TM (SCHULZ et al. 1992) through the so-called fusion peptide which is located at the amino terminus of the TM.

5.2.2 Immunosuppressive or Immunodominant?

A peptide (the "immunosuppressive peptide") corresponding to a conserved 26 amino acid stretch of the TM ectodomain of C and D retroviruses has been shown to inhibit the proliferation of an interleukin-2 dependent murine cytotoxic cell line, the blastogenic response of murine and human lymphocytes in mixed leukocyte reactions (CIANCIOLO et al. 1985), and the immunoglobulin secretion of activated feline B lymphocytes (MITANI et al. 1987). The minimal sequence required for the immunosuppressive activity has been further mapped to the octapeptide LQNRRGLD (RUEGG et al. 1989a).

Although inhibition of human and murine lymphoproliferation by a HIV-1 peptide corresponding to the immunosuppressive segment of oncoviruses has been reported (CIANCIOLO et al. 1988; RUEGG et al. 1989b), by far prominent is the role of the major B-cell determinant played by this domain in lentiviruses. Using synthetic peptides, a region centered around the two conserved cysteines of the HIV-1 TM protein has been shown to be reactive with most, if not all, sera from HIV-1 infected patients (GNANN et al. 1987c; WANG et al. 1986). The main immunoreactive sequences are the seven amino acids CSGKLIC that form the cysteine loop (BUGGE et al. 1990; GNANN et al. 1987b; OLDSTONE et al. 1991) and the immediately N-contiguous sequence QLLGIWG (XU et al. 1991). Interestingly, monoclonal antibodies reacting with the loop structure irrespective of the primary sequence have been described (OLDSTONE et al. 1991). Corresponding regions of HIV-2, and SIV_{mac} are also immunodominant B epitopes (GNANN et al. 1987a; NORRBY et al. 1987). The most broadly reactive peptides in EIAV envelopes correspond, as for HIV-1, to two regions, preceeding or including the cysteine loop (CHONG et al. 1991). In FIV envelopes, the major antibody binding domain of TM mapped to the QNQFF sequence within the cysteine loop (PANCINO et al. 1993a). Based on this immunodominance and conservation, the respective peptides of HIV-1, HIV-2, SIV, and FIV have been used for diagnosis of the lentiviral infections in serological immunoassays (AVRAMEAS et al. 1993; BAILLOU et al. 1991; GNANN et al. 1987a; NORRBY et al. 1989).

The presence of antibodies to the putative immunosuppressive peptide of HIV-1 in sera of HIV-1 seropositive individuals has been associated with the absence of disease (KLASSE et al. 1988). This observation has led to speculation on a possible role of the antibody response against this region in preventing its immunosuppressive effects (KLASSE et al. 1988; RUEGG et al. 1989a). However, subsequent studies did not substantiate the described association (LANGE et al. 1989). The absence of anti-TM antibodies in late stages of disease appears more likely to be due to the general failure of the immune system.

Few works have reported on the serological reactivity of the corresponding region of oncoviruses. Using synthetic peptides and sera of HTLV-I infected patients, a region corresponding to the immunosuppressive peptide was recognized by only 18% of sera (PALKER et al. 1989). Similarly, the feline leukemia virus immunosuppressive domain was found to be poorly antigenic (FONTENOT et al. 1992). Studies on the respective role of the cysteine loop in relation to the immunosuppressive peptide are not available. It thus remains possible that different determinants in this region account for the B epitopic recognition and for immunosuppressive mechanisms.

5.3 Possible Implications for Pathogenesis

What role is played by the immunological response directed against the immunodominant epitope, which arises early during the lentiviral infection (AVRAMEAS et al. 1993; SCHRIER et al. 1988), and which accounts for the most prevalent part of the anti-viral circulating antibodies (NEURATH et al. 1990; REDFIELD et al. 1991; XU et al. 1991)? Why do persistent viruses such as HIV and FIV, skilled in immunological escape and continuous variation of envelope sequences, keep constant their most immunogenic peptide?

As expected, no neutralizing activity of antibodies to the immunodominant epitope of HIV-1 or proximate epitopes has been found (BUGGE et al. 1990; MATTHEWS et al. 1986). A low neutralizing activity of rabbit sera raised against the immunodominant epitope (antipeptide antibody titer 1:163 840, neutralizing titer 1:10), reported by SCHRIER et al. (1988), constitutes an exception. Conversely, ROBINSON et al. (1990, 1991) have described an enhancing activity of HIV-1 infection in vitro by four human monoclonal antibodies mapping to the immunodominant region of the TM protein. Similar results have been obtained using rabbit antisera to HIV-1 peptides

(JIANG et al. 1991). Antibody enhancement of HIV-1 and SIV infections in vitro has also been described using HIV-1 infected patients sera or SIV-infected rhesus macaques plasma (HOMSY et al. 1988; MONTEFIORI et al. 1990; ROBINSON et al. 1988; TAKEDA et al. 1988). Two main mechanisms have been implicated: one involves Fc receptors present on target cells (HOMSY et al. 1989; JOUAULT et al. 1989; TAKEDA et al. 1988), and the other involves the C3 component of complement and requires both complement receptor type 2 and CD4 receptor (ROBINSON et al. 1989, 1990).

Using sera from HIV-1 infected individuals at different clinical stages, HOMSY et al. (1990) has reported that antibody enhancing activity was more frequent than neutralizing activity, and that enhancement increased over time toward disease. Whether enhancing antibodies, assayed in vitro, play a role in virus spreading in vivo and retroviral disease progression, as demonstrated for other viral infections (PORTERFIELD 1986), is not yet clear. Vaccination experiments based on the viral envelope of caprine arthritis and encephalitis virus and feline leukemia virus, showing disease enhancement (ISSEL et al. 1992; MCGUIRE et al. 1986; PEDERSEN et al. 1986), support rather the existence of antibody-mediated enhancement of viral infectivity. Antibodies raised to HIV-1 TM, may also explain the absence of protection against infection of chimpanzees vaccinated with the complete envelope precursor (rgp160) whereas chimpanzees vaccinated with SU (rgp120) alone were protected (BERMAN et al. 1990).

Extracellular HIV-1 virions found at early stages of infection in patients lymphoid tissues, accumulated between follicular dendritic cells villi, are coated by antibodies and complement (ARMSTRONG et al. 1985; EMBRETSON et al. 1993; FOX et al. 1991; PANTALEO et al. 1993; TENNER-RACZ et al. 1988). These immune complexes keep infectious virus available to infect monocyte/macrophages and $CD4^+$ lymphocytes in transit. Whether nonneutralizing antibodies and complement have a more active role, such as mediating virus attachment and entry in susceptible cells remains to be determined.

If the humoral response to some epitopes is useful to virus infection of host cells, in spite of being inhibitory, it is reasonable to think that the immunodominant/enhancing structure and function are preferentially fixed in the virus population and adapted to the host. This could explain the contrast between the high conservation of the immunodominant loop sequence between various isolates of a given lentivirus and the total sequence divergence of the loop sequences between different lentiviruses. The different and apparently opposite role of this domain described in vitro for oncogenic viruses and for lentiviruses address questions yet unanswered on retrovirus evolution and pathogenesis. In vivo studies of viruses mutated in this domain or manipulations of the immune response against it might open new and different scenarios.

6 The C-Terminal "Cytoplasmic" Domain of SIV and Other lentiviruses

6.1 A Premature Stop Codon in SIV and HIV-2 *env*

A typical feature of the lentiviral envelope is the unusual length of the cytoplasmic domain of the TM, due to presence of 100–200 additional amino acids absent from onco- or spuma-retroviruses. Indeed, TM of oncoviruses such as Rous sarcoma virus and Mason-Pfizer monkey virus carry a cytoplasmic domain of 20–40 amino acids whereas the cytoplasmic domain contains 140–160 in HIVs and SIVs, 130 in visna, and 220 in EIAV. FIV constitutes an exception among lentiviruses since infectious molecular clones encode a cytoplasmic domain of approximately 50 amino acids. It must be noted that no definitive evidence has been brought as to whether the long C-terminal of lentiviruses is indeed entirely cytoplasmic or cross-backs the cell membrane. In the following sections we designate this domain CD, for C-terminal or cytoplasmic domain of TM.

Attention was drawn to the CD of SIV after communication of the first sequences of infectious molecular clones of SIV_{mac} (isolates 142 and 251, the latter initially described as STLV-3_{agm} and HTLV-4 (CHAKRABARTI et al. 1987; FRANCINI et al. 1987; HIRSCH et al. 1987). All these proviral clones carried an in-frame TAG nonsense codon (stop codon) in the *env* open reading frame, located immediately downstream from the anchor domain, at amino acid position 736 or 737. This position corresponds to the splice acceptor site for the second exon of *tat* and *rev*, encoded in other reading frames. This stop codon interrupts the envelope gene and shortens cytoplasmic sequences to a degree equivalent to that observed in other retroviral subfamilies (Fig. 4).

This stop codon does not abolish viral infectivity. In its presence, viral progeny of molecular clones remain infectious in vitro (NAIDU et al. 1988). Interestingly, a premature TAG stop codon was found at exactly the same position in one of the clones derived from the HIV-2_{rod} isolate (GUYADER et al. 1987). Apart from the SIV_{mac}/HIV-2 group, similar truncations have been observed in most other lentiviruses: in primates, such as in SIV from sooty mangabey, African green monkeys, and mandrills (MYERS et al. 1990), and in ungulates, such as in visna (SONIGO et al. 1985) and EIAV (BEISEL et al. 1993).

However, as data accumulated, the presence and position of the stop codon appeared more variable than that initially observed in the first isolats of SIV_{mac} and HIV-2. The TM of a SIV_{mac239} infectious molecular clone is not truncated. One isolate of SIV_{agm} TM is truncated 30 amino acids downstream of the stop codon initially described in SIV_{mac} and HIV-2 (MYERS et al. 1990). In the case of HIV-1 a spontaneous C-terminal truncation of the TM has been described only once (SHIMIZU et al. 1990).

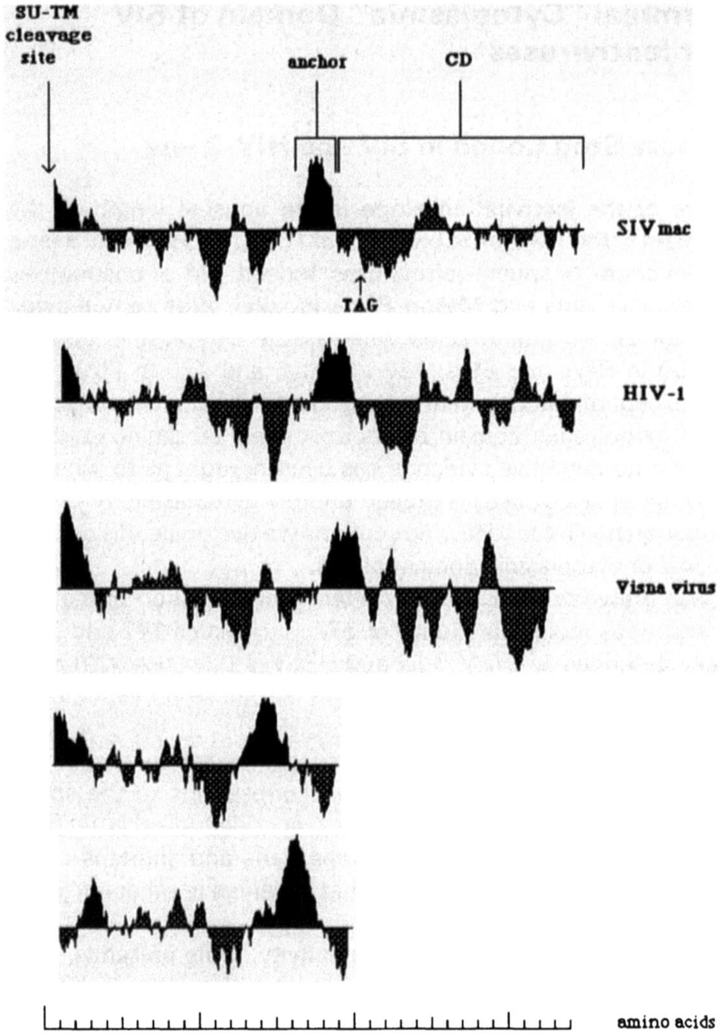

Fig. 4. Hydrophobicity profile of TM for various retroviruses: SIV_{mac} (CHAKRABARTI et al. 1987), HIV-1, and visna virus (BRAUN et al. 1987; SONIGO et al. 1985) are representative of lentiviruses; Mason–Pfizer monkey virus (SONIGO et al. 1986) and Rous sarcoma virus (SCHWARZ et al. 1983) are representatives of oncoviruses. (Adapted from CHAKRABARTI et al. 1989)

As mentioned above, all described molecular infectious clones of FIV contain a short CD (PHILLIPS et al. 1990) although the *env* open reading frame could be translated further, providing a frameshift occurred in a C-terminal orf, designated G (Fig. 5). Therefore, it could be wondered whether isolates of FIV with a long CD exist.

These observations raise a series of questions about the CD in lentiviruses. The first concerns the significance of the premature truncation of SIV *env*.

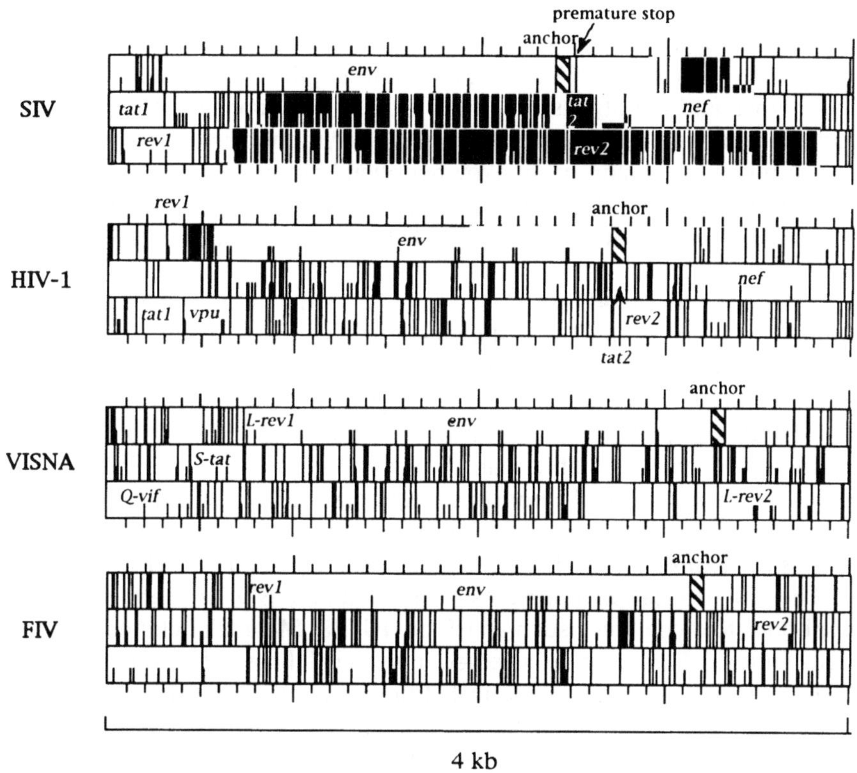

Fig. 5. Open reading frame (orf) organization of the 3′ end of lentiviral genomes. *Box*, the location of sequences encoding the hydrophobic anchor domain within the *env* orf. The sequences encoding the CD of TM are separated from the rest of the *env* orf by a premature stop codon in some isolates of SIV (SIV_{mac142} is represented here; CHAKRABARTI et al. 1987). In SIV, HIV-1, and visna, 3′ terminal sequences of *env* overlap with the second exons of *tat* and/or *rev*. In sequenced isolates of FIV the CD is truncated by the presence of multiple-stop codons.

First, we discuss the experiments that were initially performed to address this point, from which it has been concluded that the truncation is a consequence of in vitro culture. Then we deal with more general aspects, which also apply to other lentiviruses, concerning the function of the CD in the viral life cycle and its possible mechanisms of action.

6.2 The Truncated TM is Selected in a Cell-Specific Manner In Vitro

A first series of data has been obtained by mutagenesis of SIV_{mac} infectious clones from various isolates (CHAKRABARTI et al. 1989; HIRSCH et al. 1989; KODAMA et al. 1989). Introduced mutations consist in the "removal" of the premature stop codon (TAG changed to CAG) in initially truncated TMs (isolates SIV_{mac142} and SIV_{mac251}) or in the introduction of a similar stop codon (CAG changed to TAG) in initially full-length TMs (isolate SIV_{mac239}). Phenotypes and genotypes of virions produced by transfection have been studied in vitro after infection of either various lymphoblastoid cell lines from human origin, or human or monkey peripheral blood lymphocytes (PBLs). Some mutants have also been studied in vivo after infection of rhesus monkeys (CHAKRABARTI et al. 1989; HIRSCH et al. 1989; KODAMA et al. 1989).

It has been observed that the presence of a stop codon always correlates with a shorter size of the TM or the envelope glycoprotein precursor: a short TM (28–32 kDa) is detected in the presence of the premature stop codon whereas a full-length TM (41 kDa) is detected in its absence (CHAKRABARTI et al. 1989; (HIRSCH et al. 1989; KODAMA et al. 1989). This indicates that ribosomes do not detectably read through the premature termination, via frameshifting or the use of a nonsense suppressor tRNA, as is the case at the *gag-pol* junction. Thus, indeed, the stop codon abolishes or dramatically reduces the expression of the CD as a fused extension of the TM.

The TM truncation correlates with culturing SIV on certain human lymphoblastoid cell lines, such as H9 or HUT78. Molecular clones carrying a full-length TM are counter-selected in such cell lines. After infection with full-length TM viruses, spontaneously truncated mutants predominate in the culture after 2–3 weeks. Indeed, the truncated TM genotype is associated with a higher infectivity in these cell lines (CHAKRABARTI et al. 1989). It is important to note that simian lymphoblastoid cell lines are not available to perform similar cultures. In contrast, in animals infected with viruses carrying a truncated TM, mutants carrying a full-length TM appear and predominate. This observation was made by sequencing proviruses present in blood samples after polymerase chain reaction amplification. Whether such a reversion in vivo to a full-length TM occurs rapidly remains to be determined since the analysis was performed 10–13 months after inoculation (HIRSCH et al. 1989; KODAMA et al. 1989). Similar analysis has been made using human and monkey PBLs grown in vitro. It appeared that the short TM is selected in human PBLs, whereas the long TM is selected in monkey PBLs.

From these results, KODAMA et al. (1989), and HIRSCH et al. (1989) initially proposed that the stop codon constitutes an artifactual adaptation of SIV to human cells. They also proposed that the similar truncation observed in HIV-2 may reflect the recent simian origin of HIV-2.

However this initial interpretation was a matter of controversy, for two main reasons. First, a full-length TM is generally observed in HIV-2, although it was introduced in humans at least 30 years ago (MYERS et al. 1992) (and possibly much earlier). Second, the truncated TM was not selected in all human cell lines, as demonstrated with SIV initially isolated from macaques or sooty mangabey, or African green monkeys and productively cultured in CEMX174 or Molt4/8 human cell lines (WERNER et al. 1990). An alternative hypothesis was therefore proposed: the CD contains a negative regulatory element whose expression is selected "on and off" by the cellular context, via the absence or presence, respectively, of the stop codon. According to this hypothesis, truncated TMs might be selected in vivo in some organs or cell types of the simian or human host. However, to our knowledge, such naturally truncated forms have never been described when using polymerase chain reaction amplification directly in blood samples, either from SIV-infected monkeys (Fossati and Sonigo, unpublished) or from HIV-2 infected humans. Nevertheless, it might be of interest to study viral sequences present in different cell types and organs in vivo.

In conclusion, we would rather favor the idea that the full-length TM consititutes the natural predominant form in vivo, at least in PBLs. Conversely, the truncated TM would be selected by the culture conditions in vitro. This selection event is cell specific and not species specific.

6.3 The CD is not Essential for Viral Replication In Vitro but Influences Viral Dissemination In Vivo

As mentioned above, most of the initial interest attended the question of the significance of the premature stop codon in the *env* gene. The proposed interpretation that the TM truncation is mainly a cell culture artifact led to ignoring the associated question of the function of the CD. Indeed, initial experiments have indicated that the full-length TM is not strictly required for viral replication in vitro.

However, mutagenesis experiments consisting of the introduction or replacement of the original stop codon, leave downstream sequences intact. The possible existence in primate lentiviruses of proteins other than the full-length TM—which include the C-terminal domain of Env, due, for example, to alternative mRNA splicing or proteolytic cleavage of the C-terminus of the TM—remains a central issue. In visna and caprine arthritis encephalitis virus such proteins exist; the C-terminal orf constitutes an exon for the equivalent of Rev (SALTARELLI et al. 1990). In EIAV a 27-kDa protein (Ttm), corresponding to the CD has recently been described (BEISEL et al. 1993) In HIV-1 and SIV_{mac} transcripts including the C-terminus of *env* have also been documented, but the corresponding proteins, which could use alternate CUG or GUG initiation codons, have not been characterized (discussed in BEISEL et al. 1993). The hypothesis of an independent function of CD is also supported bý the fact that (except for FIV) the sequences encoding the CD, downstream from the spontaneously appearing premature stop codon, remain open for translation; no further accumulation of stop codons or other obvious genetic changes such as insertions, deletions, or frameshifts have been mentioned. Thus one can not exclude from the introduction of a single stop codon that the CD is indeed required, but as an element of an independent and yet unidentified protein. To address this point, deletion or additional nonsense mutations were introduced downstream to the original stop codon in isolate SIV_{mac142}. None of the modifications introduced in the C-terminal domain in the presence of the original stop codon has an effect on the viral phenotype in vitro. Finally, this confirmed that the C-terminal domain is not required for viral replication in vitro, either as an extension to the TM or in a possible independent protein (CHAKRABARTI et al. 1989).

Conversely, in vivo experiments showing that SIV_{mac} genomes encoding a truncated TM are overgrown by full-length TM revertants (HIRSCH et al. 1989; KODAMA et al. 1989) strongly suggest that the CD is an important element of viral dissemination in vivo and support the inclusion of the long CD of SIV (and possibly of other lentiviruses) in the category of genetic elements of virulence in vivo, which are not required for viral replication in vitro.

However, the viral phenotype in vivo that could result of a genetic inactivation of the CD in the absence of reversion has not yet been reported. Indeed, the introduction of deletions is hampered by multiple gene overlap in the 3′ part of the genome. The first half of the sequence encoding the CD overlaps with the second exon for *tat* and *rev*, and the second half overlaps with *nef* (Fig. 5). Such reading frame overlaps may also create structural constraints on the nucleotide sequences encoding the C-terminal domain, unrelated to the conservation of its function. Such constraints might limit the spontaneous apparition of additional deletion or stop codons in the CD, downstream from the truncation initially described around position 736.

6.4 Evaluation of the CD Functions In Vitro

Evaluation of the functions of the CD confronts the general problem of studying virulence genes of pathogenic micro-organisms in vitro. Paradoxically, most of the mutagenesis studies performed in SIV or HIV to address the functions of the CD have been performed in vitro, in the same cell culture systems in which the CD of SIV has been shown to be non-essential for viral replication.

This also raises the problem of spontaneous mutations that might occur in nonessential genes in vitro due to the high error rate of lentiviral polymerases and the absence of selection for their function in vitro. The premature stop codon of SIV *env* may be only the most "visible" part of such a process. Additional point mutations leading to amino acid changes may also have occurred in isolates maintained in culture. As no functional assay yet exists for the CD, it is difficult to control its activity before undertaking mutagenesis. As a recent illustration of these problems, we mention the spontaneous occurrence of stop codons in the *nef* gene in vitro, (RATNER et al. 1985) and the contradictory results initially obtained concerning its role: was it or was it not a negative regulator of viral replication? These results were obtained by initial studies performed in cell culture systems in which *nef* was non essential for viral replication. Progress in our understanding was made possible when experiments were performed in vivo using the SIV_{mac} model; these revealed a major positive effect of *nef* for viral dissemination in the infected organism (KESTLER et al. 1991).

In the case of the CD the problem may be even more obscured because of its attachment to the envelope, a function which is essential for viral infectivity. Any mutation in the CD that would modify the overall stability, transport, processing, functional efficacy, etc. of the envelope precursor or the TM would influence viral phenotype. Many laboratories have analyzed the CD of HIV-1 (DUBAY et al. 1992; GABUZDA et al. 1992; WILK et al. 1992; YU et al. 1993) or SIV_{mac} (CHAKRABARTI et al. 1989) by introduction of mutations, stop codons, or deletions. These studies have indicated that certain domains of the CD of HIV-1 or SIV_{mac} participate in envelope stability, processing, or function. Such results must be interpreted with caution. Obviously they do not mean that the only function of the CD is to modulate envelope stability, processing, or function. Another study demonstrates that the CD is strongly associated with cell membranes (HAFFAR et al. 1991). Concerning the role of the CD in HIV-1 envelope incorporation into virions, via a possible interaction with the nucleocapsid, contradictory results have been reported (DUBAY et al. 1992; GABUZDA et al. 1992; WILK et al. 1992; YU et al. 1993). In view of the production of infectious particles by clones carrying a truncated TM it is most likely that the long lentiviral CD does not play a major role in viral assembly.

The most constant result of mutagenesis studies of the lentiviral CD is paradoxical. Complete truncation of the domain does not inhibit and sometimes enhances viral infectivity whereas partial truncation is able to completely abolish it (CHAKRABARTI et al. 1989; DUBAY et al. 1992; GABUZDA et al. 1992; WILK et al. 1992; YU et al. 1993). Recalling our preceding remarks, we think that truncations of the CD which do not affect viral infectivity, clearly described for HIV-1 (WILK et al. 1992), SIV_{mac} (CHAKRABARTI et al. 1989), HIV-2 (MULLIGAN et al. 1992), and EIAV (RICE et al. 1990), are more informative than those which do. The former mutations confirm the non-essential characteristic of the CD, at least as a fused extension of the TM, whereas the latter will always be suspected of altering the overall structure and functional efficacy of the envelope precursor. Structural prediction based on sequence analysis suggests that the CD may be constituted by two amphipathic helices (VENABLE et al. 1989). Amphipathic helices contain alternatively hydrophilic and hydrophobic amino acid residues that are located on opposite sides of the helix. Such helices may require to be associated by pairs via their hydrophobic sides to allow correct positioning of the envelope relative to the hydrophobic lipid bilayer of the membrane. This requirement for pair association provides a possible explanation for the paradoxical consequences of mutations in the CD. Schematically, a completely truncated TM lacking both amphipathic helices can still associate in a correct structure whereas a partially truncated one, containing an isolated helix, cannot. These hypothesis are supported by experiments showing the association of the CD with cellular membranes (HAFFAR et al. 1991).

As already indicated, no definitive evidence has been obtained that the long CD of lentiviruses indeed has an entirely cytoplasmic localization. For example, no explanation can be provided concerning the existence of an accessible neutralizing B epitope, reported in the CD of HIV-1 (BROLIDEN et al. 1992), which suggests exposition of this CD epitope at the virion surface. Thus, defining the exact structure and position of the CD, especially relative to the cell membranes, remains a central issue.

6.5 A Possible Common Origin with *nef*

Analysis of 3′ ends of primate lentiviruses genomes by dot matrix analysis reveals a repeated motif containing the consensus tetrapeptide Arg-Arg-Ile-Arg. We found three occurrences of this motif in the HIV-2/SIV group: the first in the anchor domain of TM, the second in the CD of the TM, and the third in Nef (Chappey and Sonigo, unpublished). In addition to influencing persistence in vivo, both *nef* (KESTLER et al. 1991) and the CD have the same subcellular localization, at the inner side of the plasma membrane, the latter as the C-terminal extremity of a type 1 transmembrane protein and the former by virtue of its myristylated N-terminal residue. These elements and the fact that the CD of TM and *nef* are tandemly located in the genome may constitute a vestige of ancient gene duplications. A duplication was proposed to explain the origin of *vpr* and *vpx* (TRISTEM et al. 1990). Similarly, successive duplications may have given rise to the long CD in the ancestor of all lentiviruses and then to *nef* in the ancestor of primate lentiviruses.

6.6 The Lentiviral CD as an Independent Protein In Vivo?

The hypothetical existence of a CD protein, performing a non-essential function independently of the TM is supported by the results of the already reviewed genetic analysis. Since no standard AUG initiation codon is present in the CD, this independent protein might be generated by different mechanisms: mRNA splicing and the use of alternate CUG or GUG initiation codons or proteolytic cleavage at the C-terminus, as observed for murine oncoviruses (reviewed in HUNTER and SWANSTROM 1990).

Interestingly, a recent analysis of EIAV viral transcripts directly in liver tissues from an acutely infected horse revealed the existence of a transcript that was not observed in vitro in cell culture systems (BEISEL et al. 1993). This transcript encodes a 27-kDa protein, called Ttm, containing the CD. This protein is located in the endoplasmic reticulum or Golgi apparatus of infected cells. Unlike HIV-1 Tev, which is a fusion protein between Tat, Rev, and/or

Env, Ttm does not contain sequences involved in transactivation (BEISEL et al. 1993). Due to the known sensitivity of mRNA splicing to the cellular context, it may be of interest to analyze lentiviral mRNA directly in vivo, especially when looking for virulence functions that are not essential for viral replication in vitro. Proteins equivalent to Ttm of EIAV, generated by a splicing event occurring mainly or specifically in vivo, are good candidates for carrying the independent function encoded by CD sequences in other lentiviruses.

7 Conclusion: Variation, Conservation, and Innovation in Lentiviruses

We have reviewed above the literature concerning three conserved elements of lentiviruses. The signal peptide of retroviral envelopes presents specific features absent from classical type I eukaryotic cellular signal peptides. The immunodominant domain of the TM and the long cytoplasmic domain should be added to previously proposed models (GALLAHER et al. 1989) as additional elements of the lentiviral TM framework. It is also possible that lentiviral SUs present a common folding pattern, especially in their C-terminal halves.

Is there a common biological significance of these peculiar features of lentiviral envelopes? The answer may lie in persistent lentivirus populations having found a compromise to solve apparently incompatible requirements: on the one hand, a dynamic phenotypic and antigenic adaptability, useful particularly in facing intense selection by the immune system (discussed in SONIGO et al. 1990), which is based on a high genomic instability due to the infidelity of genome replication, and on the other, the necessary conservation of functional structures allowing viral replication.

The envelope glycoproteins are key elements of this compromise. They are the main target of the immune response, and their structural requirements appear less stringent than those for *gag* or *pol* products. Accordingly, lentiviral envelopes are built according to a conserved framework responsible for correct envelope synthesis and function, and robust enough to tolerate a high number of mutations necessary for immunological escape and persistence. This conclusion, well accepted for a given lentivirus, could thus be considered as a general rule governing the evolution of envelopes among different lentiviruses. When comparing different lentiviruses, the envelope framework remains more accessible to our observation for TM, where sequence homologies can still be found, than for SU, where no sequence homology remains. In the latter case, the spatial framework can be distinguished only by indirect reflections such as the relative position of variable and conserved domains (Fig. 2). Second-site suppressions, that is, the possibility of compensating for an inactivating mutation by another mutation located in a distant domain, could be helpful to elucidate the framework structure.

The robustness of the structural framework, combined with the high frequency of mutations and duplications, provides favorable grounds for genetic innovation. This might explain the existence of multiple additional features that makeup the genomic complexity of lentiviruses, for instance, the unusually long signal peptide, the unusually long cytoplasmic domain, and *nef*, a specific "acquisition" of primate lentiviruses. We may speculate that the long CD and *nef* were generated by successive duplication events at the 3′ end of lentiviral genome. As expected, such additional features, absent from the "minimal" replication competent genome typified by murine or avian oncoviruses, are not required for the "basic" replicative cycle. It is thus tempting to think that these additional functions confer significant advantages for long-term persistence in the complex and hostile environment provided by the infected organism in vivo.

References

Alizon M, Wain-Hobson S, Montaggnier L, Sonigo P (1986) Genetic variability of the AIDS virus: nucleotide sequence analysis of two isolates from African patients. Cell 46: 633–74

Armstrong JA, Dawkins RL, Horne R (1985) Retroviral infection of the accessory cells and the immunological paradox in AIDS. Immunol Today 6: 121

Avrameas A, Strosberg AD, Moraillon A, Sonigo P, Pancino G (1993) Serological diagnosis of feline immunodeficiency virus (FIV) infection based on synthetic peptides from Env glycoproteins. Res Virol (in press)

Baillou A, Janvier B, Mayer R, Kanki P, M'Boup S, Denis F, Goudeau A, Essex M, Barin F (1991) Site-directed serology using synthetic oligopeptides representing the C-terminus of the external glycoproteins of HIV-1, HIV-2 or SIVmac may distinguish subtypes among primate lentiviruses. AIDS Res Hum Retroviruses 7: 767–771

Ball JM, Rushlow KE, Issel CJ, Montelaro RC (1992) Detailed mapping of the antigenicity of the surface unit glycoprotein of equine infectious anemia virus by using synthetic peptide strategies. J Virol 66: 732–742

Beisel CE, Edwards JF, Dunn LL, Rice NR (1993) Analysis of multiple mRNAs from pathogenic equine infectious anemia virus (EIAV) in an acutely infected horse reveals a novel protein, Ttm, derived from the carboxy terminus of the EIAV transmembrane protein. J Virol 67: 832–842

Berman PW, Gregory TJ, Riddle L, Nakamura GR, Champe MA, Porter JP, Wurm FM, Hershberg RD, Cobb EK, Eichberg JW (1990) Protection of chimpanzees from infection by HIV-1 after vaccination with recombinant glycoprotein gp120 but not gp160. Nature 345: 622–625

Braun MJ, Clements JE, Gonda MA (1987) The visna virus genome: evidence for a hypervariable site in the env gene and sequence homology among lentivirus envelope proteins. J Virol 61: 4046–4054

Broliden P-A, Gegerfelt AV, Clapham P, Rosen J, Feniö E-M, Wahren B, Broliden K (1992) Identification of human neutralization inducing regions of the human immunodeficiency virus type 1 envelope glycoproteins. Proc Natl Acad Sci USA 89: 461–465

Buchmeier MJ, Zinkernagel RM (1992) Immunodominant T cell epitope from signal sequence. Science 257: 1142

Bugge TH, Lindhardt BO, Hansen LL, Kusk P, Hulgaard E, Holmbäck K, Klasse PJ, Zeuthen J, Ulrich K (1990) Analysis of a highly immunodominant epitope in the human immunodecifiency virus type 1 transmembrane glycoprotein, gp41, defined by a human monoclonal antibody. J Virol 64: 4123–4129

Chakrabarti L, Guyader M, Alizon M, Daniel MD, Desrosiers RC, Tiollais P, Sonigo P (1987) Molecular cloning and nucleotide sequence of simian immunodeficiency virus from macaques. Nature 328: 543–547

Chakrabarti L, Emerman M, Tiollais P, Sonigo P (1989) The C-terminal domain of the transmembrane envelope protein of SIVmac modulates infectivity. J Virol 63: 4395–4403

Chappey C, Hazout S (1992) A method for delineating structurally homogeneous regions in protein sequences. CABIOS 8: 225–260

Chong Y-H, Ball JM, Issel CJ, Montelaro RC, Rushlow KE (1991) Analysis of equine humoral responses to the transmembrane envelope glycoprotein (gp45) of equine infectious anemia virus. J Virol 65: 1013–1018

Cianciolo GJ, Copeland TD, Oroszlan S, Snyderman R (1985) Inhibition of lymphocyte proliferation by a synthetic peptide homologous to retroviral envelope proteins. 230: 453–455

Cianciolo GJ, Bogerd H, Snyderman R (1988) Human retrovirus-related synthetic peptides inhibit T lymphocyte proliferation. Immunol Lett 19: 7–14

Dedera D, Gu R, Ratner L (1992) Conserved cysteine residues in the human immunodeficiency virus type 1 transmembrane envelope protein are essential for precursor envelope cleavage. J Virol 66: 1207–1209

Dubay JW, Roberts SJ, Hahn BH, Hunter E (1992) Truncation of the human immunodeficiency virus type 1 transmembrane glycoprotein cytoplasmic domain blocks virus infectivity. J Virol 66: 6616–6625

Ellerbrok H, D'Auriol L, Vaquero C, Sitbon M (1992) Functional tolerance of the HIV-1 envelope signal peptide to mutations in the amino-terminal and hydrophobic regions. J Virol 66: 5114–5118

Embretson J, Zupancic M, Ribas JL, Burke A, Racz PK, Tenner-Racz K, Haase AT (1993) Massive covert infection of helper T lymphocytes and macrophages by HIV during the incubation period of AIDS. Nature 362: 359–362

Fontenot JD, Hoover EA, Elder JH, Montelaro RC (1992) Evaluation of feline immunodeficiency virus and feline leukemia virus transmembrane peptides for serological analysis. J Clin Microbiol 30: 1885–1890

Fox CH, Tenner-Racz K, Ràcz P, Firpo A, Pizzo PA, Fauci AS (1991) Lymphoid germinal centers are reservoirs of human immunodeficiency virus type 1 RNA. J Infect Dis 164: 1051–1057

Francini G, Gurgo C, Guo H-G, Gallo RC, Collati E, Fargnoli KA, Hall LF, Wong-Staal F, Reitz Jr MS (1987) Sequence of simian immunodeficiency virus and its relationship to human immunodeficiency viruses. Nature 328: 539–543

Gabuzda DH, Lever A, Terwilliger E, Sodroski J (1992) Effects of deletions in the cytoplasmic domain on biological functions of human immunodeficiency virus type 1 envelope glycoprotein. J Virol 66: 3306–3315

Gallaher WR, Ball JM, Barry RF, Griffin MC, Montelaro RC (1989) A general model for the transmembrane proteins of HIV and other retroviruses. AIDS Res Hum Retroviruses 5: 431–440

Gnann JW, McCormick JB, Mitchell S, Nelson JA, Oldstone MBA (1987a) Synthetic peptide immunoassay distinguishes HIV type 1 and HIV type 2 infections. Science 237: 1346–1349

Gnann JW, Nelson JA, Oldstone MBA (1987b) Fine mapping of an immunodominant domain in the transmembrane glycoprotein of human immunodeficiency virus. J Virol 61: 2639–2641

Gnann JW, Schwimmbeck PL, Nelson JA, Truax AB, Oldstone MBA (1987c) Diagnosis of AIDS using a 12 aminoacid peptide representing an immunodominant epitope of human immunodeficiency virus. J Infect Dis 156: 261–267

Guyader M, Emerman M, Sonigo P, Clavel F, Montagnier L, Alizon M (1987) Genetic organization and transactivation of human immunodeficiency virus type 2. Nature 326: 662–669

Haffar OK, Dowbenko DJ, Berman PW (1991) The cytoplasmic tail of HIV-1 gp160 contains regions that associate with cellular membranes. Virology 180: 439–441

Henderson RA, Michel H, Sakaguchi K, Shabanowitz J, Appella E, Hunt DF, Hengelhard VH (1992) HLA-A2.1-associated peptides from a mutant cell line: a second pathway of antigen presentation. Science 255: 1264–1266

Hirsch V, Riedel N, Mullins JI (1987) Genome organization of STLV-3 is similar to that of the AIDS virus except for a truncated transmembrane protein. Cell 49:307–319
Hirsch VM, Edmonson P, Murphey-Corb M, Arbeille B, Johnson PR, Mullins JI (1989) SIV adaption to human cells. Nature 341: 373–374
Homsy J, Meyer M, Tateno M, Clarkson S, Levy JA (1989) The Fc and not CD4 receptor mediates antibody enhancement of HIV infection in human cells. 244: 1357–1360
Homsy J, Tateno M, Levy JA (1988) Antibody-dependent enhancement of HIV infection. Lancet 1: 1285–1286
Homsy J, Meyer M, Levy JA (1990) Serum enhancement of human immunodeficiency virus (HIV) infection correlates with disease in HIV-infected individuals. J Virol 64: 1437–1440
Huet T, Cheynier R, Meyerhans A, Roelants G, Wainhobson S (1990) Genetic organization of a chimpanzee lentivirus related to HIV-1. Nature 345: 356–359
Hunter E, Swanstrom R (1990) Retrovirus envelope glycoproteins. Curr Top Microbiol Immunol 157: 187–253
Issel CJ, Horohov DW, Lea DF, Adams WV, Hagius SD, McManus JM, Allison AC, Montelaro RC (1992) Efficacy of inactivated whole-virus and subunit vaccines in preventing infection and disease caused by equine infectious anemia virus. J Virol 66: 3398–3408
Javaherian K, Langlois AJ, McDanal C, Ross KL, Eckler LI, Jellis CL, Profy AT, Rusche JR, Bolognesi DP, Putney SD, Matthews TJ (1989) Principal neutralizing domain of the human immunodeficiency virus type 1 envelope protein. Proc Natl Acad Sci USA 86; 6768–6772
Jiang S, Lin K, Neurath AR (1991) Enhancement of human immunodeficiency virus type 1 infection by antisera to peptides from the envelope glycoproteins gp120/gp41. J Exp Med 174: 1557–1563
Jouault T, Chapuis F, Olivier R, Parravicini C, Bahraoui E, Gluckman J-C (1989) HIV infection of monocytic cells: role of antibody-mediated virus binding to Fc-gamma receptors. AIDS 3: 125–133
Kalyanaraman VS, Rodriguez V, Veronese F, Rahman R, Lusso P, DeVico AL, Copeland T, Oroszlan S, Gallo RC, Sarngadharan MG (1990) Characterization of the secreted, native gp120 and gp160 of the human immunodeficiency virus type 1. AIDS Res Hum Retroviruses 6: 371–380
Kestler HW, Ringler DJ, Mori K, Panicali DL, Sehgal PK, Daniel MD, Desrosiers RC (1991) Importance of the nef gene for maintenance of high virus loads and for development of AIDS. Cell 65: 651–662
Kirchhoff F, Jentsch KD Stuke A, Mons J, Hunsmann G (1990) Genomic divergence of an HIV-2 from a German AIDS patient probably infected in Mali. AIDS 4, 847–857
Klasse PJ, Pipkorn R, Blomberg J (1988) Presence of antibodies to a putatively immunodeficiency virus (HIV) envelope glycoprotein gp41 is strongly associated with health among HIV-positive subjects. Proc Natl Acad Sci USA 85: 5225–5229
Knowles DP, Cheevers WP, McGuire TC, Brassfield AL, Harwood WG, Stem TA (1991) Structure and genetic variability of envelope glycoproteins of two antigenic variants of caprine arthritis-encephalitis lentivirus. J Virol 65: 5744–5750
Kodama T, Wooley DP, Naidu YM, Kestler HWI, Daniel MD, Li Y, Desrosiers RC (1989) Significance of premature stop codons in env of SIV J Virol 63: 4709–4714
Lange J, Vahine A, Dekker J, Wolf FD, Goudsmit J (1989) Antibodies to an HIV-1 synthetic peptide with partial homology to MULV P15E do not protect against AIDS 3: 402–403
Leonard CK, Spellmann MW, Riddle L, Harris RJ, Thomas JN, Gregory TJ (1990) Assignment of intrachain disulfide bonds and characterization of potential glycosylation sites of the type 1 recombinant human immunodeficiency virus envelope glycoprotein (gp120)expressed in Chinese hamster ovary cells. J Biol Chem 265: 10373–10382
Matthews T, Langlois A, Robey W, Chang N, Gallo R, Fischinger P, Bolognesi D (1986) Restricted neutralization of divergent human-T-lymphotropic virus type III isolates by antibodies to the major envelope glycoprotein. Proc Natt Acad Sci USA 83: 9709–9713
McGuire TC, Adams DS, Johnson GC, Klevjer-Anderson P, Barbee DD, Gorham JR (1986) Acute arthritis in caprine arthritis-encephalitis virus challenge exposure of vaccinated or persistently infected goats. Am J Vet Res 47: 537–540
Mitani M, Cianciolo GJ, Snyderman R, Yasuda M, Good RA, Day NK (1987) Suppressive effect on polyclonal B-cell activation of a synthetic peptide homologous to a transmembrane component of oncogenic retroviruses. Proc Natl Acad Sci USA 84: 237–240
Modrow S, Hahn BH, Shaw GM, Gallo RC, Wong-Staal F, Wolf H (1987) Computer-assisted analysis of envelope protein sequences of seven human immunodeficiency virus isolates: prediction of antigenic epitopes in conserved and variable regions. J Virol 61: 570–578

Montefiori DC, Robinson WE, Hirsch VM, Modliszewski A, Mitchell WM, Johnson PR (1990) Antibody-dependent enhancement of SIV infection—further characterization and cross reactivity between macaque and sooty Mangabey isolates. J Med Primatol 19: 269–278
Müller MC, Saksena NK, Nerrienet E, Chappey C, Hervé VMA, Durand J-P, Legal-Campodonico P, Lang M-C, Digoutte J-P, Georges AJ, Georges-Courbot MC, Sonigo P, Barré-Sinoussi F (1993) Simian immunodeficiency viruses from central and western Africa: evidence for a new species-specific lentivirus in tantalus monkeys. J Virol 67: 1227–1235
Mulligan MJ, Yamshchikov GV Jr, Ritter GD, Gao F, Jin MJ, Nail CD, Spies CP, Hahn BH, Compans RW (1992) Cytoplasmic domain truncation enhances fusion activity by the exterior glycoprotein complex of human immunodeficiency virus type 2 in selected cell types. J Virol 66: 3971–3975
Myers G, Rabson AB, Berzofsky JA, Smith TF, Wong-Staal F (1990) Human retroviruses and AIDS. Los Alamos National Laboratory, Los Alamos
Myers G, MacInnes K, Korber B (1992) The emergence of simian/human immunodeficiency virus. AIDS Res Hum Retrov 8: 373–386
Naidu YM, Kestler HW III, Li Y, Butler CV, Silva DP, Schmidt DK, Troup CD, Sehgal PK, Sonigo P, Daniel MD, Desrosiers RC (1988) Characterization of infectious molecular clones of simian immunodeficiency virus (SIVmac) and human immunodeficiency virus type 2: persistent infection of rhesus monkeys with molecularly cloned SIVmac. J Virol 62: 4691–4696
Neurath AR, Strick N, Taylor P, Rubinstein P, Stevens CE (1990) Search for epitope-specific antibody responses to the human immunodeficiency virus (HIV-1) envelope glycoproteins signifying resistance to disease development. AIDS Res Hum Retroviruses 6: 1183–1192
Norrby E, Biberfeld G, Chiodi F, von Gegerfeldt A, Nauclér A, Parks E, Lerner R (1987) Discrimination between antibodies to HIV and to related retroviruses using site-directed serology Nature 329: 248–250
Norrby E, Chiodi F, Whalley A, Parks E, Naucler A, Costa CM, Torstensson R, Biberfeld G (1989) Site-directed enzyme-linked immunosorbent assay with a synthetic simian immunodeficiency virus SIVmac peptide identifying antibodies against the HIV-1. AIDS 3: 17–20
Nothwehr SF, Gordon JI (1990) Structural features in the NH2-terminal region of a model eukariotic signal peptide influence the site of its cleavage by signal peptidase. J Biol Chem 265: 17202–17208
Oberste MS, Greenwood JD, Gonda MA (1991) Analysis of the transcription pattern and mapping of the putative rev and env splice junctions of bovine immunodeficiency-like virus. J Virol 65: 3932–3937
Oldstone BA (1991) Molecular anatomy of viral persistence. J Virol 65: 6381–6386
Oldstone MBA, Tishon A, Lewicki H, Dyson HJ, Feher VA, Assa-Munt N, Wright PE (1991) Mapping the anatomy of the immunodominant domain of the human immunodeficiency virus gp41 transmembrane protein: peptide conformation analysis using monoclonal antibodies and proton nuclear magnetic resonance spectroscopy. J Virol 65: 1727–1734
Palker TJ, Tanner ME, Scearce RM, Streilein RD, Clark ME, Haynes BF (1989) Mapping of immunogenic regions of human T cell leukemia virus type 1 (HTLV-1) gp46 and gp21 envelope glycoproteins with env-encoded synthetic peptides and a monoclonal antibody to gp46. J Immunol 142: 971–978
Pancino G, Chappey C, Saurin W, Sonigo P (1993a) B epitopes and selection pressures in feline immunodeficiency virus envelope glycoproteins. J Virol 67: 664–672
Pancino G, Fossati I, Chappey C, Castelot S, Hurtrel B, Moraillon A, Klatzmann D, Sonigo P (1993b) Structure and variations of feline immunodeficiency virus envelope glycoproteins. Virology 192: 659–662
Pantaleo G, Graziosi C, Demarest JF, Butini L, Montroni M, Fox CH, Orenstein JM, Kotler DP, Fauci AS (1993) HIV infection is active and progressive in lymphoid tissue during the clinically latent stage of disease. Nature 362: 355–358
Patarca R, Haseltine WA (1984) Similarities among retrovirus proteins. Nature 312: 496
Pedersen NC, Johnson L, Birch D, Theilen GH (1986) Possible immunoenhancement of persistent viremia by feline leukemia virus envelope glycoprotein vaccines in challenge-exposure situations where whole inactivated virus vaccines were protective. Vet Immunol Immunopathol 11: 123–148
Phillips TR, Talbott RL, Lamont C, Muir S, Lovelace K, Elder JH (1990) Comparison of two host cell range variants of feline immunodeficiency virus. J Virol 64: 4605–4613

Porterfield JS (1986) Antibody-dependent enhancement of viral infectivity. Adv Virus Res 31: 335–355

Ratner L, Haseltine W, Patarca R, Livak KJ, Starcich B, Josephs SF, Doran ER, Rafalski JA, Whitehorn EA, Baumeister K, Ivanoff L, Petteway SR, Pearson J, Lautenberger ML, Papas JA, Ghrayeb TS, Chang NT, Gallo RC, Wong-Staal F (1985) Complete nucleotide sequence of the AIDS virus HTLV-III. Nature 313: 277–284

Redfield RR, Birx DL, Ketter N, Tramont E, Polonis V, Davis C, Brundage JF, Smith G, Johnson S, Fowler A, Wierzba T, Shafferman A, Volvovitz F, Oster C, Burke DS (1991) A phase 1 evaluation of the safety and immunogenicity of vaccination with recombinant gp160 in patients with early human immunodeficiency virus infection. N Engl J Med 324: 1677–1684

Rice NR, Henderson LE, Sowder RC, Copeland, Oroszlan S, Edwards JF (1990) Synthesis and processing of the transmembrane protein of equine infectious anemia virus. J Virol 64: 3770–3778

Robinson WE, Montefiori DC, Mitchell WM (1988) Antibody-dependent enhancement of human immunodeficiency virus type 1 infection. Lancet 1: 790–794

Robinson WE, Montefiori DC, Mitchell WM, Prince AM, Alter HJ, Dreesman GR, Eichberg JW (1989) Antibody-dependent enhancement of human immunodeficiency virus type 1 (HIV-1) infection in vitro by serum from HIV-1-infected and passively immunized chimpanzees. Proc Natl Acad Sci USA 86: 4710–4714

Robinson WE, Kawamura T, Lake D, Masuho Y, Mitchell WM, Hersh EM (1990a) Antibodies to the primary immunodominant domain of human immunodeficiency virus type I (HIV-I) glycoprotein 41 enhance HIV-I infection in vitro. J Virol 64: 5301–5305

Robinson WE, Montefiori DC, Mitchell WM (1990b) Complement-mediated antibody-dependent enhancement of HIV-1 infection requires CD4 and complement receptors. Virology 175: 600–604

Robinson WE, Gorny MK, Xy JI, Mitchell WM, Zolla-Pazner S (1991) Two immunodominant domains of gp41 bind antibodies which enhance human immunodeficiency virus type I infection in vitro. J Virol 65: 4169–4176

Ruegg CL, Monell CR, Strand M (1989a) Identification, using synthetic peptides, of the minimum amino acid sequence from the retroviral transmembrane protein p15E required for inhibition of lymphoproliferation and its similarity to gp21 of human T-lymphotropic virus types I and II. J Virol 63: 3250–3256

Ruegg CL, Monell CR, Strand M (1989b) Inhibition of lymphoproliferation by a synthetic peptide with sequence identity to gp41 of human immunodeficiency virus type 1. J Virol 63: 3257–3260

Rushlow K, Olsen K, Stiegler G, Payne SL, Montelaro RC, Issel CJ (1986) Lentivirus genomic organization: complete nucleotide sequence of the env gene of equine infectious anemia vius. Virology 155: 309–321

Saltarelli M, Querat G, Konings DAM, Vigne R, Clements JE (1990) Nucleotide sequence and transcriptional analysis of molecular clones of CAEV which generate infectious virus. Virology 179: 347–364

Sattentau QJ, Moore JP (1991) Conformational changes induced in the human immunodeficiency virus envelope glycoprotein by soluble CD4 binding. J Exp Med 174: 407–415

Schrier RD, Gnann JWJ, Langlois AJ, Shriver K, Nelson JA, Oldstone MBA (1988) B- and T-lymphocyte responses to an immunodominant epitope of human immunodeficiency virus. J Virol 62: 2531–2536

Schulz TF, Jameson BA, Lopalco L, Siccardi AG, Weiss RA, Moore JP (1992) Conserved structural features in the interaction between retroviral surface and transmembrane glycoproteins? AIDS Res Hum Retroviruses 8: 1571–1580

Schwarz DE, Tizard R, Gilbert W (1983) Nucleotide sequence of the Rous sarcoma virus. Cell 32: 853–869

Shimizu H, Morikawa S, Yamaguchi K, Tsuchie H, Hachimori K, Ushijima H, Kitamura T (1990) Shorter size of transmembrane glycoprotein of an HIV-1 isolate. AIDS 4: 575–576

Sonigo P, Alizon M, Staskus K, Klatzmann D, Cole S, Danos O, Tiollais P, Haase A, Wain-Hobson S (1985) Nucleotide sequence of the visna lentivirus: relationship to the AIDS virus. Cell 43: 369–382

Sonigo P, Barker C, Hunter E, Wain-Hobson S (1986) Nucleotide sequence of Mason-Pfizer monkey virus, an immunosuppressive D-type retrovirus. Cell 45: 375–385

Sonigo P, Girard M, Dormont D (1990) Design and trails of AIDS vaccines. Immunol Today 11: 465–471

Stephens EB, Butfiloski EJ, Monck E (1992) Analysis of the amino terminal presequence of the feline immunodeficiency virus glycoprotein: effects of deletions on the intracellular transport of gp95. Virology 190: 569–578

Syu W-J, Lee W-R, Du B, Yu Q-C, Essex M, Lee T-H (1991) Role of conserved gp41 cysteine residues in the processing of human immunodeficiency virus envelope precursor and viral infectivity. J Virol 65: 6349–6352

Takeda A, Tuazon CU, Ennis FA (1988) Antibody-enhanced infection by HIV-1 via FC receptor-mediated entry. Science 242: 580–583

Talbott RL, Sparger EE, Lovelace KM, Fitch WM, Pedersen NC, Luciw PA, Elder JH (1989) Nucleotide sequence and genomic organization of feline immunodeficiency virus. Proc Natl Acad Sci USA 86: 5743–5747

Tenner-Racz K, Racz P, Schmidt H, Dietrich M, Kern P, Louie A, Gartner S, Popovic M (1988) Immunohistochemical, electron microscopic and in situ hybridization evidence for the involvment of lymphatics in the spread of HIV-1. AIDS 2: 299–309

Tristem M, Marshall C, Karpas A, Hill F (1990) Origin of vpx in lentiviruses. Nature 347: 341–342

Tsujimoto H, Cooper RW, Kodama T, Fukasawa M, Miura T, Ohta Y, Ishikawa K-I, Nakai M, Frost E, Roelants G, Roffi J, Hayami M (1988) Isolation and characterisation of simian immunodeficiency virus from mandrills in Africa and its relationship to other human and simian immunodeficiency viruses. J Virol 62: 4044–4050

Venable RM, Pastor RW, Brooks BR, Carson FW (1989) Theoretically determined three dimensional structures for amphipathic segments of the HIV-1 gp41 envelope protein. AIDS Res Hum Retroviruses 5: 7–22

Von Heijne G (1986) A new method for predicting signal sequence cleavage sites. Nucleic Acids Res 14: 4683–4690

Wang JJG, Steel S, Wisniewolski R, Wang CY (1986) Detection of antibodies to human T-lymphotropic virus type III by using a synthetic peptide of 21 amino acid residues corresponding to a highly antigenic segment of gp41 envelope protein. Proc Natl Acad Sci USA 83: 6159–6163

Watson MEE (1984) Compilation of published signal sequences. Nucleic Acids Res 12: 5145–5164

Werner A, Baier M, Cichutek K, Kurth R, Becker W (1990) SIV grows unchanged in human cells. Nature 344: 113

Wilk T, Pfeiffer T, Bosch V (1992) Retained in vitro infectivity and cytopathogenicity of HIV-1 despite truncation of the C-terminal tail of the env gene product. Virology 189: 167–177

Xu J-Y, Gorny MK, Palker T, Karwowska S, Zolla-Pazner S (1991) Epitope mapping of two immunodominant domains of gp41 transmembrane protein of human immunodeficiency virus type 1, using ten human monoclonal antibodies. J Virol 65: 4832–4838

Yu X, Yuan X, McLane MF, Lee T-H, Essex M (1993) Mutations in the cytoplasmic domain of human immunodeficiency virus type 1 transmembrane protein impair the incorporation of Env proteins into mature virions. J Virol 67: 213–221

Mucosal Transmission of Simian Immunodeficiency Virus

C.J. MILLER

1 Introduction

Although human immunodeficiency virus (HIV) is transmitted primarily by heterosexual contact, very little is understood about the biology of heterosexual transmission of HIV. Epidemiological studies indicate that HIV is not efficiently transmitted by sexual contact. In fact, many other sexually transmitted diseases (STDs) are more efficiently transmitted. For example, 22%–25% of individuals are infected by one exposure to *Neisseriia gonorrheae* (HOLMES et al. 1970; HOOPER et al. 1988), and hepatitis B virus is transmitted in 20%–30% of exposures (JUDSON 1981; MOLSEY 1975). In contrast, less than 15% of those monogamous individuals repeatedly exposed to an infected sexual partner become infected with HIV (HOLMBERG et al. 1989). It is estimated that for a single sexual contact, the infectivity of HIV is 0.3% (HEARST and HULLEY 1988; LAWRENCE et al. 1990; PADIAN et al. 1987; PETERMAN et al. 1988). However, some individuals become infected after

California Regional Primate Research Center, and Department of Veterinary Pathology, School of Veterinary Medicine, University of California Davis, Davis, CA 95616, USA

Current Topics in Microbiology and Immunology, Vol. 188

a single or a few sexual contacts (PETERMAN et al. 1988; STASKEWSKI et al. 1987), while others remain uninfected despite hundreds of contacts (LAWRENCE et al. 1990). In the United States, HIV is apparently more efficiently transmitted from men to women than from women to men (PADIAN et al. 1987, 1991). Several cofactors have been identified which increase the risk of an individual acquiring HIV through heterosexual contact. Cervical ectopy, receptive anal intercourse, genital ulcer disease, and infection with other STDs are the most significant factors associated with HIV infection of women (HOLMBERG et al. 1988; HOLMES and KREISS 1988, PLUMMER et al. 1991) while the presence of an intact foreskin and genital ulcer disease of the penis are the risk factors most often associated with HIV infection in men (CAMERON et al. 1989; SIMONSEN et al. 1988).

Three possible explanations exist to explain the variability in the sexual transmission of HIV. First, unique factors in the HIV-infected individual (stage of disease, immune response, presence of other STDs) may influence infectivity. Some HIV-positive persons are highly infectious and others, presumably, are not (PADIAN et al. 1988; PETERMAN 1990; STASKEWSKI et al. 1987). The degree to which an individual is infectious may be related to the stage of disease. During primary HIV infection the levels of antigen and virus in plasma are high (CLARK et al. 1991; COUTINHO et al. 1987; GOUDSMIT et al. 1987), and this initial viremia may be associated with high levels of HIV in genital secretions. HIV is more common in the semen of men with AIDS than in semen of healthy chronically infected individuals (ANDERSON et al. 1992). Second, unique factors in specific HIV strains may influence infectiousness. Some HIV strains may be inherently more likely than other strains to be shed in the secretions of infected persons, or some strains may, by virtue of their cellular tropism, have an increased affinity for target cells in the reproductive tract. Third, unique factors (stage of menstrual cycle, nutritional status, or the presence or absence of the risk factors listed above) may influence the susceptibility of individuals to HIV infection following exposure. Some people may be highly susceptible to HIV infection, while others may be more resistant.

The relative importance of these biological factors can best be assessed using an animal model of heterosexual transmission of HIV. Perhaps even more critical is the need for an animal model to test the ability of candidate vaccines and topical virucidal preparations to prevent the mucosal transmission of HIV. The similarities between human and rhesus monkey reproductive anatomy and physiology (HENDRICKX and CUKIERSKI 1987) and the close genetic and biological homology of HIV and simian immunodeficiency virus (SIV) have led to the development of several nonhuman primate models of mucosal transmission of lentiviruses. This chapter summarizes the nonhuman primate animal models of mucosal transmission of SIV that have been developed.

2 SIV/Rhesus Macaque Model of Genital Transmission

To answer some of the questions regarding the heterosexual transmission of HIV, an animal model of sexual transmission of HIV was developed by inoculating SIV onto the genital mucosa of rhesus macaques (MILLER et al. 1989). Simple application of SIV onto the intact genital mucosa of mature and immature rhesus macaques results in virus transmission. The disease induced by this route of inoculation is indistinguishable from that seen in intravenously inoculated animals (MILLER et al. 1989,1990). This model is being used to understand the interaction between virus and host during the sexual transmission of HIV.

3 Vaginal Transmission of SIV

SIV has been transmitted to female rhesus macaques by infusing a cell-free virus suspension into the vaginal canal through a soft plastic pediatric nasogastric feeding. This technique does not result in gross trauma to the vaginal mucosa, and any microtrauma produced would be less than that which occurs during natural mating. The initial studies using this technique involved multiple inoculations spread over the menstrual cycle of the animals. The cell-free SIV was suspended in 1 ml RPMI-1640 for the inoculation procedure. Following inoculation the animals were monitored for viremia and seroconversion using standard virological and serological techniques (MILLER et al. 1989). Thus, two animals were infected after eight inoculations (twice a week for 4 weeks) with 10^4 intravenous animal infectious doses (IVAID) of cell-free SIV (MILLER et al. 1989). One animal was infected but a second animal remained uninfected following eight inoculations (twice a week for 4 weeks) with 10^3 IVAID of cell-free SIV (MILLER et al. 1989).

In an effort to determine the efficiency with which a single intravaginal inoculation transmits SIV, 13 animals were inoculated intravaginally once with 10^4 IVAID of cell-free SIV. Of the 13 animals inoculated, 9 animals became systemically infected with SIV (MILLER et al. 1990). A single inoculation with this relatively large amount of SIV was not capable of producing infection in all the animals inoculated. In addition, 16 mature female rhesus monkeys were inoculated intravaginally with 0.5 ml of the virus stock (5×10^3 IVAID), and eight became infected. Thus it seems that 0.5 ml of the undiluted cell-free virus stock is a 50% genital animal infectious dose (MARTHAS et al. 1992; C. Miller, unpublished data).

The animals used in the single inoculation studies described above were in all stages of the menstrual cycle at the time of virus inoculation. There did not appear to be any correlation between any particular point in the menstrual cycle and the susceptibility of the animal to infection. This is a significant consideration because the thickness of the vaginal mucosa changes dramatically during the menstrual cycle. In women and rhesus macaques (MILLER et al. 1992b), the vaginal mucosa is composed of a stratified squamous epithelium that is relatively thick during the follicular phase of the cycle (under the influence of estrogen) and relatively thin during the luteal phase of the cycle (under the influence of progesterone). Thus, a transmission study was designed to determine whether the stage of the menstrual cycle at the time of inoculation is a factor in the susceptibility to vaginal transmission of SIV. Six animals were intravaginally inoculated once with 10^4 50% tissue culture infective doses ($TCID_{50}$; approximately 10^4 IVAID) of cell-free SIV between days 10 and 12 of the menstrual cycle. This is the follicular phase of the cycle and is the stage at which the epithelium is the thickest. Four of the six animals exposed at this stage became infected with SIV. Six animals were inoculated intravaginally once with 10^4 $TCID_{50}$ (approximateiy 10^4 IVAID) of cell-free SIV between days 20 and 22 of the menstrual cycle. This is the luteal phase of the cycle and is the stage at which the vaginal epithelium is the thinnest. Five of the six animals exposed at this stage became infected with SIV. Thus, there is not an obvious difference in the susceptibility of animals vaginally exposed to SIV at opposite points of the menstrual cycle. It is possible that the relatively large doses of SIV used in the inoculation obscured subtle differences in the relative susceptibility of the animals in the two groups.

In order to locate the site of the SIV target cells in the female reproductive tract, four hysterectomized female macaques were inoculated intravaginally with cell-free SIV_{mac} (MILLER et al. 1992a). The cervix and uterus of the animals were surgically removed, and the vagina was formed into a blind pouch. The animals were allowed to heal completely prior to virus inoculation, and culposcopic examination revealed that the vaginal epithelium completely lined the vaginal pouch. The animals were inoculated with 0.5 ml of the virus stock (5×10^3 IVAID) and two of the four animals were infected. Thus, the lack of the cervix and uterus did not affect the dose of cell-free virus required for the genital transmission of SIV. This finding strongly suggests that target cells for SIV are present in the vaginal epithelium.

Both cell-free and cell-associated HIV are present in the genital secretions of infected individuals, and it is not clear whether either one or both forms of the virus are involved in sexual transmission. The ability of cell-associated SIV to infect vaginally inoculated animals was assessed with a series of transmission studies. A stock of cell-associated SIV was produced to determine the number of SIV-infected cells required for the genital transmission of SIV. The stock consisted of peripheral blood mononuclear cells (PBMC) obtained from ten uninfected rhesus macaques, pooled and infected with SIV in vitro. The cells were washed, resuspended in cryoprotectant, frozen, and stored in liquid nitrogen. After 2 weeks an aliquot of the cells was thawed, and using end-point dilution culture techniques it was determined that the stock contained 2 SIV-infected cells/100 viable cells. To titrate the cell-associated stock in vivo two animals were inoculated intravenously with 100 SIV-infected cells, and two more were intravenously inoculated with 10 SIV-infected cells. The two animals inoculated with 100 infected cells became viremic, while one of the two animals inoculated with 10 infected cells became infected. To determine the number of infected cells required for genital transmission two mature female rhesus macaques were inoculated intravaginally with 100 infected cells, and four were inoculated intravaginally with 10 infected cells. None of the intravaginally inoculated animals became infected. Next, a higher dose of cell-associated SIV was used for intravaginal inoculation of the same animals; four animals were inoculated with 10^3 infected cells, and two were inoculated with 10^4 infected cells. Again, none of the intravaginally inoculated animals became infected despite being exposed to this high dose of cell-associated virus (MILLER 1992).

The unsuccessful attempts to transmit SIV by genital inoculation of cell-associated virus suggest that cell-associated HIV may not play a significant role in sexual transmission. The difficulty in transmitting cell-associated SIV may be related to the pH in the vagina. The vaginal pH in women (MOLLER and KASPERSEN 1991) and monkeys (MILLER 1993) is low (4.0–6.5). Tissue culture cells (uninfected CEM $\times$ 174 cells) suspended in RPMI-1640 survive less than 1 h after infusion into the vagina of rhesus monkeys (MILLER 1992). Of autologous lymphocytes suspended in Hank's medium, 90% are killed within 30 min after infusion into a women's vagina (D. Anderson, personal communication). Thus, it seems likely that the only cells that contain infectious SIV particles at the time of inoculation can contribute to genital transmission. The end-point culture technique used to titrate the cell-associated stock overestimates the amount of infectious SIV present in the cells at the time of genital inoculation. By maintaining the infected cells in culture, integrated SIV genome can produce infectious virions that result in positive cultures. It is unlikely that lymphocytes in the vagina can survive long enough to produce infectious virions from integrated genome. Based on these considerations it is likely that studies attempting to quantify the amount of HIV (or SIV) in genital secretions based on the presence of virus nucleic acid or virus isolation probably overestimate the amount of infectious virus present in secretions at the time of transmission.

In an ejaculate, lymphocytes may be able to survive somewhat longer in the vagina. The pH of seminal plasma ranges from 7 to 8, and lymphocytes and cell-free virus can survive for extended periods in these conditions. After ejaculation the pH of the vagina rises to 7.0 or higher (FOX and MELDRUM 1973). Of autologous lymphocytes suspended in seminal plasma, 90% are alive 30 min after infusion into a women's vagina (D. Anderson, personal communication). Thus, cells infused into the vagina are much more stable if they are in seminal plasma. However, these observations do not conclusively establish the importance of cell-associated virus in sexual transmission. The life-span of foreign lymphocytes in the vaginal canal is still limited due to the presence of bacteria and the fact that cells are washed from the vagina by the action of vaginal secretions. Thus it is not clear that there is sufficient time for cells that contain integrated virus genome in the ejaculate to produce infectious virions.

It was recently reported that HIV-1 can be transmitted to chimpanzees by placing infected cells into the cervical canal (GIRARD et al. 1992). In this study, one animal became viremic and seropositive following endocervical inoculation with 300 HIV-1 infected PBMC. A second animal inoculated with 100 infected cells remained uninfected while a third animal inoculated with 30 infected cells was transiently viremic but seronegative. Thus the anatomic site into which infected cells are placed affects the ability of the cells to transmit infection. The pH of the endocervical canal is much closer to neutral than the pH of the vagina, and it is possible that cells could survive for extended periods in the lumen of the endocervical glands. The significance of this report is not clear because it is not known whether lymphocytes in an ejaculate are capable of reaching the endocervical canal.

To assess the importance of seminal plasma in the genital transmission of SIV (and HIV), a series of transmission studies based on intravaginal inoculation of a mixture of seminal plasma and SIV-infected cells or cell-free SIV were conducted. Six animals were inoculated with 10^3 $TCID_{50}$ (approximately 10^3 IVAID) of cell-free SIV immediately after placing 0.25 ml human seminal plasma into the vaginal canal. The inoculations were performed between days 10 and 12 of the menstrual cycle. All six of the animals became infected (Miller, unpublished results). As reported above, only four of six animals were infected after inoculation with ten fold more cell-free SIV without the presence of seminal plasma. Thus, the presence of seminal plasma increases the efficiency of vaginal SIV transmission. A large study is currently underway to determine the extent of this seminal plasma effect. It is clear, however, that a single dose of cell-free SIV can produce infection in 100% of intravaginally inoculated animals. This finding offers a simple and efficient challenge system to test the ability of vaccines and chemical agents to prevent the genital transmission of SIV.

4 Genital Transmission of SIV in Male Rhesus Macaques

The initial transmission studies in male monkeys (MILLER et al. 1989) utilized another stock of cell-free SIV_{mac}. This stock was also grown in human PBMC and titrated in vivo. One milliliter of a 10^{-3} dilution of the virus stock inoculated intravenously infected two of two animals and 1 ml of a 10^{-4}

dilution inoculated intravenously infected two of two animals. One milliliter of a 10^{-5} dilution of the virus stock failed to infect four intravenously inoculated animals. Based on these results 1 ml of 10^{-4} diluation is between 1 and 10 IVAID of this stock of cell-free SIV_{mac}.

The initial genital transmission studies in male monkeys were conducted by infusing the virus preparations through a 2.5 mm outer-diameter (8 F), soft, plastic, pediatric nasogastric feeding tube (American Pharmaseal, Valencia, CA) into the urethra of male rhesus macaques immobilized with ketamine HCl. As with the female animals, the initial studies in the males (MILLER et al. 1989) involved multiple inoculations. Two mature male macaques were inoculated intraurethrally with 1 ml of the undiluted virus stock (10^4 IVAID) once a week for 3 weeks. One animal was viremic at 16 days postinoculation (after two inoculations), and one was not viremic until 31 days postinoculation (after three inoculations). Four juvenile male monkeys were inoculated intraurethrally with 1 ml (10^4 IVAID) of the cell-free SIV_{mac} virus stock used for the female transmission studies described above. The animals were inoculated twice a week over 2 weeks. All were viremic by 14 days postinoculation (after four inoculations).

To simulate more closely the natural exposure of men to HIV, four mature male macaques were inoculated genitally by placing 1 ml cell-free SIV (10^4 IVAID) onto the foreskin and glans of the penis (MILLER et al. 1990). The inoculum was allowed to drip freely off the penis, and a portion of the inoculum came into contact with the urethral os. The inoculations were repeated twice over a 1-week period. After two inoculations two of the four animals became infected. These results demonstrate that SIV can be transmitted across the intact mucosa (stratified squamous epithelium) of the foreskin and glans, and that target cells for SIV are present in the mucosa. Further support for these conclusions is available from virus localization studies in chronically infected rhesus macaques (MILLER et al. 1991). In one mature male macaque SIV-infected cells were found within the stratified squamous epithelium of the foreskin (Fig. 1). Immunohistochemial studies demonstrate that $CD4^+$ Langerhans' cells are abundant in the foreskin mucosa, and that lesser numbers of macrophages are also present at this site (Miller, unpublished results). Thus, antigen-presenting cells (Langerhans' cells or macrophages) in the mucosa of the foreskin and glans may be the target cell of SIV (or HIV).

A recent report provides evidence that SIV can be transmitted to male macaques through mating with infected female macaques (KULLER et al. 1992 a or b). In this study two uninfected adult male pigtailed macaques were housed with SIV-infected females. Mating behavior between the animals was observed. Both males became viremic and seropositive following different periods of exposure to the infected females. Although the possibility that the infection was transmitted by biting or scratching cannot be ruled out, the results of this study suggest that SIV can be transmitted from female to male by heterosexual contact.

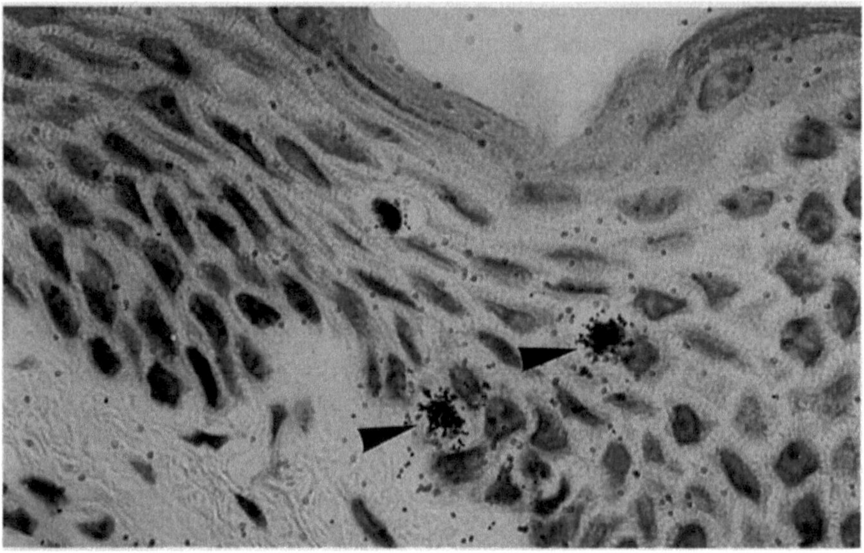

Fig. 1. In situ hybridization demonstrating two SIV-infected cells (*arrowheads*) in the foreskin mucosa (stratified squamous epithelium) of a chronically infected rhesus macaque. Immunohistochemical studies (not shown) demonstrate that CD4$^+$ Langerhans' cells and macrophages are present in this location. Hematoxylin counterstain $\times 400$

5 SIV/Macaque Model of Rectal Transmission

Several groups have infected macaques by intrarectal inoculation with SIV. In one study three of four juvenile male rhesus macaques inoculated intrarectally with 10^2 ro 10^3 IVAID of cell-free SIV_{mac} became seropositive and viremic (PAUZA et al. 1992). In another report (CRANAGE et al. 1992), four of four monkeys became infected following intrarectal inoculated with 1500-fold more cell-free SIV_{mac} than required for intravenous infection. This paper also reported that four animals vaccinated intramuscularly with a whole killed SIV vaccine were protected from intrarectal challenge with the same dose of virus. In a series of experiments conducted by another group of investigators, six pig-tailed macaques were inoculated intrarectally with cell-free SIV_{mne} (MORTON et al. 1989). Two animals were inoculated once with 1×10^3 TCID, and one of these animals became viremic and seropositive. Two animals were inoculated with 10^2 TCID, and two were inoculated with 10 TCID; all of thse animals remained virus negative and seronegative. However, one of the animals inoculated with 10^2 TCID was reported to have died of an AIDS-like disease. A more recent report from the same group described the rectal transmission of a single cell clone of SIV_{mne}; SIV_{mne} CL E11S (KULLER et al. 1992). Animals inoculated with 10^5–10^4 TCID became viremic and seropositive. The virus dose required for seroconversion was 10^4 TCID, approximately 10^3 more virus than the minimum needed for seroconversion after intravenous inoculation (KULLER et al. 1992).

6 SIV Transmission at Other Mucosal Sites

It was recently reported that placing 500 $IVAID_{50}$ of cell-free SIV into the conjunctival sac infected one of two animals inoculated, while one monkey inoculated conjunctivally with 100 $IVAID_{50}$ of cell-free SIV did not become infected (CONWAY et al. 1991). Another group reported that a neonatal rhesus macaque became infected following placement of cell-free SIV_{mac} on the oral and conjunctival mucosa (JENNINGS et al. 1989). Unpublished studies at this primate center also demonstrate that SIV_{mac} can be transmitted to neonatal macaques by conjunctival inoculation. Three neonatal macaques were inoculated by placing 10^3 IVAID of cell-free SIV_{mac} in one conjunctival sac. One of the three animals became seropositive and viremic, while the other two remained uninfected.

Several reports provide support for the possibility of oral transmission of SIV. Fifteen infants born to rhesus monkey dams experimentally infected with SIV_{smm} during pregnancy were all virus negative at birth, and all the infants remained SIV negative until 9 months after birth. Two of the infants seroconverted at 9 and 12 months postpartum, and SIV was isolated from the PBMC of one of these animals (MCCLURE et al. 1991). SIV was isolated from the milk of the mothers of these two infants at 9 months postpartum. These observations suggest that SIV (as HIV) can be transmitted by ingestion of milk that contains virus. Apparently biting and being bitten are modes of horizontal transmission of SIV in group housed juvenile male rhesus macaques (LOWENSTINE et al. 1991). In this study four animals experimentally infected with SIV_{stm} were housed with five uninfected animals (contact animals). Behavioral observations documented the interactions between the animals. Viremia was detected in two of the contant animals 4 weeks after the group was formed. All the contact animals were bitten during the first week, but the two animals that became infected were the only contact animals observed biting the experimentally infected animals. This raises the possibility that SIV was transmitted due to infected blood contacting the oral epithelium. It is probably significant that the oral and conjunctival mucosa, like the vaginal mucosa (see below), is a stratified squamous epithelium that contains Langerhans cells.

7 Isolation of SIV from Genital Secretions

Both cell-free and cell-associated SIV can be isolated from the vaginal secretions of chronically infected female rhesus macaques (MILLER et al. 1990). Vaginal secretions from five SIV-infected female rhesus macaques were collected three times (once every 2 weeks for 6 weeks) and cultured for virus isolation. Samples contaminated with menstrual blood were discarded. SIV was isolated from the cells and cell-free filtrates of vaginal secretions from two of five females on the first collection, from the cells and filtrate from one of four animals on the next collection, and from the cells of one female and the cells and filtrates of two females on a third collection. The positive samples were collected at all stages of the menstrual cycle, and all the animals were healthy and viremic at the time the samples were collected. These findings suggest that SIV can be shed at any stage of the menstrual cycle, and the fact that virus was never recovered from the vaginal secretions of two animals suggests that individual factors may be important in determining if SIV is present in vaginal secretions.

SIV can be isolated from the semen of chronically infected male macaques (MILLER et al. 1989). Paired semen and blood samples were obtained from each of two SIV-infected male rhesus. Ten paired semen and blood specimens were taken over 8 months from one intravenously inoculated animal. Two of ten semen specimens were positive while ten of ten blood specimens were SIV positive. One of the virus-positive semen samples was collected a week before the animal was killed due its deteriorating clinical condition. SIV was also isolated from tissue homogenates made from the epididymis of this animal. The total number of mononuclear cells recovered from the ejaculates of this male ranged from 0 to 2×10^6 cells/ml. The samples from which virus was cultured contained approximately 1×10^6 cells/ml. The degree to which an HIV-infected individual is infectious may be related to the stage of disease. During primary HIV infection the levels of antigen and virus in plasma are high (CLARK et al. 1991; COUTINHO et al. 1987; GOUDSMIT et al. 1987), and this initial viremia may be associated with high levels of HIV· in genital secretions. As yet studies have not been undertaken in humans to test this hypothesis, but in the SIV/rhesus macaque model SIV can be isolated from the semen of intravenously inoculated macaques as early as 10 days postinoculation (Miller, unpublished results).

8 Localization of SIV in the Reproductive Tract of Chronically Infected Rhesus Macaques

The identity and location of the cellular targets of SIV during genital transmission are yet to be determined. SIV was efficiently transmitted to hysterectomized female macaques by inoculation of cell-free virus into blind

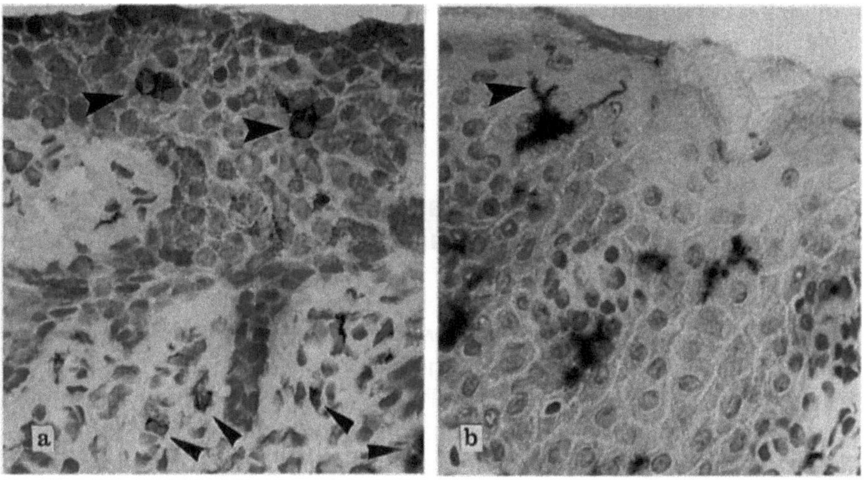

Fig. 2a, b. Immunohistochemical staining of frozen sections of vagina from mature rhesus macaque. **a** $CD68^+$ monocyte/macrophages are abundant in the superficial submucosa (*small arrowheads*) and can also be detected occasionally in the mucosa (stratified squamous epithelium) of the vagina (*large arrowheads*). **b** $CD1a^+$ Langerhans' cells in the vaginal mucosa. Note the dendritic processes extending from the cell bodies. In one case the process can be seen to branch (*arrowhead*). Hematoxylin counterstain $\times$ 300

vaginal pouches (MILLER et al. 1992a). The 50% infectious dose of cell-free virus was the same in the intact and hysterectomized animals, indicating that sufficient target cells are present in the intact vaginal mucosa for transmission of the virus. Since only a few $CD4^+$ T cells are present in the submucosa of the vagina (MILLER et al. 1992b), the most likely target cells in the vaginal mucosa are macrophages or Langerhans' cells (Fig. 2; MILLER et al. 1992a, c). In chronically infected female rhesus macaques, SIV-infected cells are present in the uterus, cervix, and vagina (MILLER et al. 1992c). SIV-infected cells are most numerous in the submucosa of the ectocervix and vagina, and the morphology of the infected cells is consistent with lymphocytes and monocyte/macrophages. SIV-infected cells are also found within the stratified squamous epithelium of the vagina. Some of the infected cells in this location have a dendritic morphology consistent with Langerhans' cells (MILLER et al. 1992c, 1993) Langerhans' cells of the skin have been reported to be infected with HIV in AIDS patients (GIANNETTI et al. 1993; TSCHACHLER et al. 1987; ZAMBRUNO et al. 1991). The finding that cells in the vaginal epithelium are infected with SIV is the first indication that mucosal Langerhans' cells might be infected with a lentivirus (MILLER et al. 1992c).

In cervical biopsy material obtained from HIV-infected women, T cells and macrophages were determined to be the cell types infected with HIV (POMERANTZ et al. 1988). These findings are consistent with our results in SIV-infected monkeys. All the cells that were found to be infected in the genital tract of both human and rhesus monkey presumably express the CD4 molecule. There was no evidence that non-$CD4^+$ cell types (epithelial cells) were infected in the reproductive tracts of either the monkeys or humans.

9 Route of Virus Dissemination Following Mucosal Inoculation of SIV

The detection of SIV-infected cells with dendritic morphology within the vaginal epithelium suggests that Langerhans' cells may have a role as target cells in the sexual transmission of HIV and SIV. These antigen-presenting cells are potentially efficient disseminators of these viruses from the genital mucosa to draining lymph nodes. These cells function to transport antigen from the stratified squamous epithelium to the draining lymph node (HOEFSMIT et al. 1982; KRAAL et al. 1986; KRIPKE et al. 1990; KUPIEC–WEGLINSKI et al. 1988; SHELLEY and JUHLIN 1976, 1977; SILBERBERG–SINAKIN et al. 1976; STEINMAN 1991). Blood dendritic cells (of which Langerhans' cell precursors are a subset) can be infected with HIV (LANGHOFF et al. 1991; MACATONIA et al. 1990; PATTERSON et al. 1991; PATTERSON and KNIGHT 1987). Furthermore, when infected in vitro, these cells produce much higher levels of virus than T cells, but they do not exhibit the usual cytopathic effects associated with HIV infection (LANGHOFF et al. 1991). Thus, Langerhans' cells in the vagina and ectocervix (BJERCKE et al. 1983; EDWARDS and MORRIS 1985; HACKEMANN et al. 1968; MORRIS et al. 1983; YOUNES et al. 1968) may be especially well suited as target cells for the sexual transmission of HIV and SIV (MILLER et al. 1992a, c).

Based on these observations, the following hypothesis has been proposed to explain the cellular events in the mucosal transmission of HIV and SIV (MILLER et al. 1992a, 1993). The virus infects antigen-presenting cells (macrophages and Langerhans' cells) in the vaginal mucosa and is carried by these cells to the iliac and obturator lymph nodes draining the genital tract. In rhesus macaques, india ink inoculated into the vaginal submucosa localizes in these lymph nodes within 72 h (Miller, unpublished results). It is possible that the virus initially replicates in the target cells (macrophages and Langerhans' cells) and undergoes subsequent rounds of replication in the draining lymph nodes prior to spreading to more proximal lymphoid nodes and finally to the bloodstream and distant lymphoid tissue.

There has been speculation that trauma to tissues during intercourse might be necessary for heterosexual transmission of HIV, and that virus can gain direct access to the bloodstream through these wounds; however, this is very unlikely. The hemodynamic pressures in the peripheral vasculature do not permit the movement of particles (cells or viruses) from tissues directly into blood vessels. Blood cells (including $CD4^+$ T cells) that escape from the peripheral vasculature do not reenter the bloodstream but travel through lymphatics to reach the draining lymph nodes. This is the same pathway that vaginal macrophages and Langerhans' cells would take following infection with HIV (MILLER et al. 1992a, 1993). It is reasonable to assume that trauma would increase the amount of virus that crosses the mucosa, and that the hemorrhage associated with the trauma would increase the number of $CD4^+$ target cells in the submucosa of the vagina. The increased number of virions and target cells might increase the efficiency of HIV transmission but probably would not affect the route of virus dissemination from the genital tract to systemic lymphoid tissue (MILLER et al. 1992a, 1993) This route of virus dissemination has important implications for generating vaccine-mediated protection against the heterosexual transmission of HIV. In heterosexual transmission the initial virus-host cell interaction occurs in the genital mucosa or submucosa, and the early events in the dissemination of HIV occur in the draining lymph node. Thus, to prevent dissemination of the virus to the systemic lymphoid tissue the protective immune response (both antibody and cell-mediated immunity) must be present in the genital mucosa, submucosa, and in the draining lymph node. A strong systemic immune response may provide another level of protection from infection after sexual contact with an infected partner, but a genital mucosal immune response would concentrate the protective response at the initial sites of infection.

References

Anderson DJ, O'Brien TR, Martinez A, Seage GR, Padian N, Horsburgh RC, Mayer KH (1992) Effects of disease stage and zidovudine therapy on the detection of human immunodeficiency virus type-1 in semen. JAMA 267: 2769–2774

Bjercke S, Scott H, Braathen LR, Thorsby E (1983) HLA-DR-expressing langerhans-like cells in vaginal and cervical epithelium. Acta Obstet Gynecol Scand 62: 585–589

Cameron DW, Simonsen JN, D'Costa LJ, Ronald AR, Maitha GM, Gakinya MN, Cheang M, Ndinya AJ, Piot P, Brunham RC (1989) Female to male transmission of human immunodeficiency virus type 1: risk factors for seroconversion in men [see comments]. Lancet 2: 403–407.

Clark SJ, Saag MS, Decker WD, Campbell-Hill S, Roberson JL, Velderkamp PJ, Kappes JC, Hahn BH, Shaw GM (1991) High titers of cytopathic virus in the plasma of patients with symptomatic primary HIV-1 infection. N Engl J Med 324: 954–960

Conway MD, Davison-Fairburn B, Martin LN, Insler MS, Murphey-Corb M (1991) Infection of rhesus monkeys with topical instillation of simian immunodeficiency virus $(SIV)_{B670}$ into the conjunctival sac. J Med Primatol 20: 152–155

Coutinho RA, Goudsmit J, Paul DA, de Wolf F, lange J, van der Noordaa J (1987) The natural history of HIV infection in homosexual men. Ann Inst Pasteur Virol 138: 67–74

Cranage MP, Baskerville A, Ashworth LAE, Dennis M, Cook N, Sharpe S, Farrar G, Rose J, Kitchin PA, Greenway PJ (1992) Intrarectal challenge of macaques vaccinated with formalin-activated simian immunodeficiency virus. Lancet 339: 273–274

Edwards JNT, Morris HB (1985) Langerhans cells and lymphocyte subsets in the female genital tract. Br J Obstet Gynaecol 92: 974–982

Fox C, Meldrum S (1973) Continuous measurement by radio-telemetry of vaginal pH during human coitus. J Reprod Fertil 33: 69–75

Giannetti A, Zambruno G, Cimarelli A, Marconi A, Negroni M, Girolomoni G, Bertazzoni U (1993) Direct detection of HIV-1 RNA in epidermal Langerhans cells of infected patients. J Acquir Immune Defic Syndr 6: 329–333
Girard M, Mahoney J, Rimskey L, Barre-Sinoussi F, Muchmore E, Weinhold K, Fultz PN (1992) HIV-1 genital infection: a chimpanzee model (Abstract). 7th Colloque Des "Cent Gardes". Fondation Merieux, Paris, pp. 75–79
Goudsmit J, Lange JMA, Paul DA, Dawson GJ (1987) Antigenemia and antibody titers to core and envelope antigens in AIDS, AIDS-related complex and subclinical human immunodeficiency virus infection. J Infect Dis 155: 558–560
Hackemann M, Grubb C, Hill KR (1968) The ultrastructure of normal squamous epithelium of the human cervix uteri. J Ultrastruct Res 22: 443–457
Hearst N, Hulley SB (1988) Preventing the heterosexual spread of AIDS. Are we giving our patients the best advice? JAMA 259: 2428
Hendrickx AG, Cukierski MA (1987) Reproductive and developmental toxicology in nonhuman primates. In: Graham CR (ed) Preclinical safety of biotechnology products intended for human use. Liss New York, pp 73–88
Hoefsmit ECM, Duijvestijn AM, Kamperdijk EWA (1982) Relation between Langerhans cells, veiled cells, and interdigitating cells. Immunobiology 161: 255–265
Holmberg SD, Stewart JA, Gerber AR, Byers RH, Byers RH, Lee FK, O'Malley PM, Nahmias AJ (1988) Prior herpes simplex virus type 2 infection as a risk factor for HIV infection JAMA 259: 1048–1050
Holmberg SD, Horsburgh CRJ, Ward JW, Jaffe HW (1989) Biologic factors in the sexual transmission of human immunodeficiency virus. J Infect Dis 160: 116–125
Holmes KK, Kreiss J (1988) Heterosexual transmission of human immunodeficiency virus: overview of a neglected aspect of the AIDS epidemic. J Acquir Immune Defic Syndr 1: 602–610
Holmes KK, Johnson DW, Trostle H (1970) An estimate of the risk of men acquiring gonorrhea by sexual contact with infected females. Am J Epidemiol 91: 170–174
Hooper RR, Reynolds GH, Jones OG, Zaidi A, Weisner PJ, Latimer KP, Lester A, Campbell AF, Harrison WO, Karney WW, Holmes KK (1988) Cohort study of venereal disease. I. The risk of gonorrhea transmission from infected women to men. Am J Epidemiol 108: 136–144
Jennings M, Lerche N, Tillery S, Yee J, Antipa L, Chesney P (1989) Infection with simian immunodeficiency virus (SIV) via conjunctival and oral mucosae in a newborn rhesus macaque (Abstract). 5th International Conference on AIDS, Montreal
Judson FN (1981) Epidemiology of sexually transmitted hepatitis B infection in heterosexuals: a review. Sex Transm Dis 8 Suppl 336–343
Kraal G, Breel M, Janse M, Bruin G (1986) Langerhans cells, veiled cells and interdigitating cells in the mouse recognized by a monoclonal antibody. J Exp Med 163: 981–997
Kripke M, Munn C, Jeevan A, Tang J-M, Bucana C (1990) Evidence that cutaneous antigen-presenting cells migrate to regional lymph nodes during contact sensitization. J Immunol 145: 2833–2837
Kuller L, Beneviste RE, Clark EA, Katze MG, Morton WR (1992a) Successful intrarectal inoculation of two macaque species with SIVmne CL E11S (Abstract). 10 Annual Symposium on Nonhuman Primate Models for AIDS, San Juan, Puerto Rico
Kuller L, Beneviste RE, Watanabe R, Tsai C-C, Morton WR (1992b) Transmission of SIVmne from female to male Macaca nemestrina. J Med Primatol 21: 299–307
Kupiec-Weglinski J, Austyn J, Morris P (1988) Migration patterns of dendritic cells in the mouse. Traffic from the blood, and T-cell-dependent and independent entry into lymphoid tissue. J Exp Med 167: 632–645
Langhoff EF, Terwilliger EF, Bos JH, Kalland KH, Pozansky MC, Bacon OML, Haseltine WA (1991) Replication of human immunodeficiency virus type 1 in primary dendritic cell cultures. Proc Natl Acad Sci USA 88: 7998–8002
Lawrence DN, Jason JM, Holman RC, Murphy JJ (1990) HIV transmission from hemophiliac men to their heterosexual partners. In: Alexander NJ, Gablenick HL, Spieler JM (eds) Heterosexual transmission of AIDS. Liss, New York, pp 35–53
Lowenstine LJ, Jennings MB, Yee JL, Vogel AP, Scott L, Lerche NW, Hendrickx AG (1991) Horizontal transmission of SIVstm among group-housed juvenile male rhesus macaques (Abstract). 9th Annual Symposium on Nonhuman Primate Models for AIDS, Seattle
Macatonia S, Lau R, Patterson S, Pinching A, Knight S (1990) Dendritic cell infection, depletion and dysfunction in HIV-infected individuals. Immunology 71: 38–45

Marthas ML, Miller CJ, Sutjipto S, Higgins J, Torten J, Unger RE, Kiyono H, McGhee JR, Marx PA, Pedersen NC (1992) Efficacy of live-attenuated and whole-inactivated simian immunodeficiency virus vaccines against vaginal challenge with virulent SIV. J Med Primatol 21: 99–107

McClure HM, Anderson DC, Fultz PN, Ansari AA, Jehuda-Cohen T, Villinger F, Klumpp SA, Switzer W, Lockwood E, Brodie A, Keyserling H (1991) Maternal transmission of SIVsmm in rhesus macaques. J Med Primatol 20: 182–187

Miller CJ (1992) Use of the SIV/rhesus macaque model of the heterosexual transmission of HIV in vaccine research. Vacc Res 1: 295–301

Miller CJ (1993) Use of the SIV/rhesus macaque system to test virucides designed to prevent the sexual transmission of HIV. In: Gablenick H, Speiler J (eds) Barrier contraception. Wiley-Liss, New York (in press)

Miller CJ, Alexander NJ, Sutjipto S, Lackner AA, Hendrickx AG, Gettie A, Lowenstine LJ, Jennings M, Marx PA (1989) Genital mucosal transmission of simian immunodeficiency virus: animal model for heterosexual transmission of human immunodeficiency virus. J Virol 63: 4277–4284

Miller CJ, Alexander NJ, Sutjipto S, Hendrickx AG, Jennings M, Marx PA (1990) Effect of virus dose and nonoxynol-9 on the genital transmission of SIV in rhesus macaques. J Med Primatol 19: 401–409

Miller C, Vogel P, Dandekar S, Hendrickx A, Alexander N, Marx P (1991) Localization of SIV in the reproductive tract of rhesus macaques: relationship to the distribution of T cells and macrophages (Abstract). 7 International Conference on AIDS, Florence

Miller CJ, Alexander NJ, Vogel P, Anderson J, Marx PA (1992a) Mechanism of genital transmission of SIV: a hypothesis based on transmission studies and the location of SIV in the genital tract of chronically infected female rhesus macaques. J Med Primatol 21: 64–68

Miller CJ, McChesney M, Moore PF (1992b) Langerhans cells, macrophages and lymphocyte subsets in the vervix and vagina of rhesus macaques. Lab Invest 67: 628–634

Miller CJ, Vogel P, Alexander NJ, Sutjipto S, Hendrickx AG, Marx PA (1992c) Localization of SIV in the genital tract of chronically infected female rhesus macaques. Am J Pathol 141: 655–660

Miller CJ, McGhee JR, Gardner MB (1993) Mucosal immunity, HIV transmission and AIDS. Lab Invest 68: 129–145

Moller BR, Kaspersen P (1991) The acidity of the vagina. In: Horowitz B, Marc PA (eds) Vaginitis and vaginosis. Wiley-Liss New York, pp 63–67

Molsey JW (1975) The epidemiology of viral hepatitis: an overview. Am J Med Sci 270: 253–270

Morris HHB, Gatter KC, Stein H, Mason DY (1983) Langerhans cells in human cervical epithelium: an immunohistologic study. Br J Obstet Gynaecol 90: 400–411

Morton WR, Kuller L, Benveniste RE, Clark EA, Tsai C-C, Gale MJ, Thouless ME, Overbaugh J, Katze MG (1989) Transmission of the simian immunodeficiency virus SIV_{mne} in macaques and baboons. J Med Primatol 18: 237–245

Padian N, Wiley J, Winkelstein W (1987) Male-to-female transmission of human immunodeficiency (HIV): current results, infectivity rates, and San Francisco population seroprevalence estimates (Abstract). 3rd International Conference on AIDS Washington

Padian N, Wiley J, Glass S, Marquis L, Winkelstein W (1988) Anomalies of infectivity in the heterosexual transmission of HIV (Abstract). 4th International Conference on AIDS, Stockholm

Padian NS, Shiboski SC, Jewell NP (1991) Female to male transmission of human immunodeficiency virus. JAMA 266: 1664–1667

Patterson S, Knight SC (1987) Susceptibility of human peripheral blood dendritic cells to infection by human immunodeficiency virus. J Gen Virol 68: 1177–1181

Patterson S, Gross J, Bedford P, Knight SC (1991) Morphology and phenotype of dendritic cells from peripheral blood and their productive and non-productive infection with human immunodeficiency virus type-1. Immunology 72: 361–367

Pauza CD, Emau P, Salvato MS, Trivedi P, MacKenzie D, Malkovsky M, Schultz KT (1992) Pathogenesis of SIVmac251 after atraumatic inoculation of the rectal mucosa in rhesus macaques (Abstr). 10th Annual Symposium on Nonhuman Primate Models for AIDS, San Juan, Puerto Rico

Peterman TA (1990) Facilitators of HIV transmission during sexual contact. In: Alexander NJ, Gablenick HL, Spieler JM (eds) Heterosexual transmission of AIDS. Liss, New York, pp 55–68

Peterman TA, Stoneburner RL, Allen JR, Jaffe HW, Curran JW (1988) Risk of human immunodeficiency virus from heterosexual adults with transfusion associated infections. JAMA 259: 55–58
Plummer FA, Simonsen JN, Cameron DW, Ndinya AJ, Kreiss JK, Gakinya MN, Waiyaki P, Cheang M, Piot P, Ronald AR (1991) Cofactors in male-female sexual transmission of human immunodeficiency virus type 1. J Infect Dis 163: 233–239
Pomerantz RJ, de la Monte SM, Donegan CE, Rota TR, Vogt MW, Craven DE, Hirsch MS (1988) Human immunodeficiency virus (HIV) infection of the uterine cervix. Ann Intern Med 108: 321–327
Shelley W, Juhlin L (1976) Langerhans cells form a reticuloepithelial trap for external contact allergens. Nature 1977, 261: 46–47
Shelley W, Juhlin L (1977) Selective uptake of contact allergens by the Langerhans cell. Arch Dermatol 113: 187–192
Silberberg-Sinakin I, Thorbecke GJ, Baer RL, Rosenthal SA, Berezowsky V (1976) Antigen-bearing Langerhans cells in skin, dermal lymphatics and in lymph nodes. Cell Immunol 25: 137–151
Simonsen JN, Cameron DW, Gakinya MN, Ndinya AJ, D'Costa LJ, Karasira P, Cheang M, Ronald AR, Piot P, Plummer FA (1988) Human immunodeficiency virus infection among men with sexually transmitted diseases. Experience from a center in Africa. N Engl J Med 319: 274–278
Staskewski S, Schieck E, Rehmet S, Helm EB, Stille W (1987) HIV transmission from a male after only two sexual contacts. Lancet 2: 628
Steinman RM (1991) The dendritic cell system and its role in immunogenicity. Annu Rev Immunol 9: 271–296
Tschachler E, Groh V, Popvic M, Mann DL, Konrad K, Safai B, Eron L, diMarzo-Veronese F, Wolff K, Stingl G (1987) Epidermal Langerhans cells: a target for HTLV-III/LAV infection. J Invest Dermatol 88: 233–237
Younes MS, Robertson EM, Bencosme SA (1968) Electron microscope observations on Langerhans cells in the cervix. Am J Obstet Gynecol 102: 397–403
Zambruno G, Mori L, Marconi A, Mongiardo N, De Rienzo B, Bertazzoni U, Giannetti A (1991) Detection of HIV-1 in epidermal Langerhans cells of HIV-infected patients using polymerase chain reaction. J Invest Dermatol 96: 979–982

Comparison of Regulatory Features Among Primate Lentiviruses

K.T. JEANG and A. GATIGNOL

1 Introduction

It has been more than a decade since the description of primate lentiviruses as the etiological agents for immunodeficiencies (BARRE–SINOUSSI et al. 1983; GALLO et al. 1984; LEVY et al. 1984; CLAVEL et al. 1986). Since that time, besides the human immunodeficiency viruses (HIV-1 and HIV-2), molecular isolates of simian immunodeficiency viruses (SIVs; DESROSIERS 1990), as well as related viruses in cattle, cats, sheep, and goats, have been obtained. For HIVs and SIVs, extensive nucleotide sequencing information is available; the sequences from 30 or more individual HIV isolates and 20 or more SIV isolates are available in one recent compilation (MYERS et al. 1992a). Based on this information, the phylogenetic relationship between (and among) HIVs and SIVs has been proposed (reviewed in DESROSIERS 1990; MYERS et al. 1992b). From this perspective, we can operationally categorize the primate lentiviruses into four groups, represented by (a) HIV-1 and SIV_{cpz}; (b) SIV_{smm}, SIV_{mac}, and HIV-2; (c) SIV_{agm}; and (d) SIV_{mnd} (DESROSIERS 1990; MYERS et al. 1992b).

From a separate perspective, clues about the relatedness between viruses can be derived from similar/distinct usage of regulatory means. For example,

Laboratory of Molecular Microbiology, National Institute of Allergy and Infectious Diseases, National Institutes of Health, Bethesda, MD 20892, USA

Current Topics in Microbiology and Immunology, Vol. 188

herpesviruses that share a thymidine kinase function can be presumptively classified from those that do not. Similarly, RNA viruses that are obligatory for reverse transcription are in an obvious category separate from those that replicate otherwise. It is in this vein that we compare the regulatory features between HIV and SIV. We discuss here the mechanisms used by HIV and SIV to regulate transcriptional and posttranscriptional expression. Wherever possible a comparison is made between the two groups of viruses; however, because disproportionate data are available about HIV-1, some aspects are reviewed within this singular context while we await parallel information about the other viruses.

2 Transcriptional Regulation

Upon infection of cells the genomes of lentiviruses, as those of other retroviruses, are reverse-transcribed into DNA copies and integrate into host chromosomes. Although unintegrated viral DNAs are prevalent inside the cell (PANG et al. 1990; ZACK et al. 1990; BUKRINSKY et al. 1991), evidence is persuasive that only the integrated provirus leads to productive transcription (SAKAI et al. 1993). This implies that cellular mechanisms can distinguish between integrated and unintegrated proviral DNAs and suggests that an understanding of viral expression should consider the contribution of chromatin. It is noteworthy that the regulation of integrated chromatin-associated genes can be markedly different from that of naked unintegrated DNA (FELSENFELD 1992; ADAMS and WORKMAN 1993). In the study of lentiviruses, this point often has been overlooked in approaches that utilize assays based on transient transfections or cell-free transcription.

2.1 Basal Transcription of HIV/SIV Long Terminal Repeats

The basic organization of the lentivirus promoter is similar to that of other eucaryotic genes (see reviews, see DYNAN 1989; JOHNSON and MCKNIGHT 1989) in containing conserved 'enhancers' and TATA. Sequence components necessary for RNA polymerase II directed transcription are present in the U3 of lentivirus long terminal repeats (LTRs; reviewed in JONES 1989; PAVLAKIS and FELBER 1990). This portion of the LTR (454 nucleotides in length for HIV-1; 555 nucleotides for HIV-2) contains binding motifs for many cellular factors (see Fig. 1). Here there is approximately 54% (aligning LAI to ROD) identity in nucleotide sequence between HIV-1 and HIV-2/SIV. High conservation is seen in the distal and proximal (towards R) ends of U3 (MYERS et al. 1992a). The middle part of U3 has gaps of insertions in HIV-2

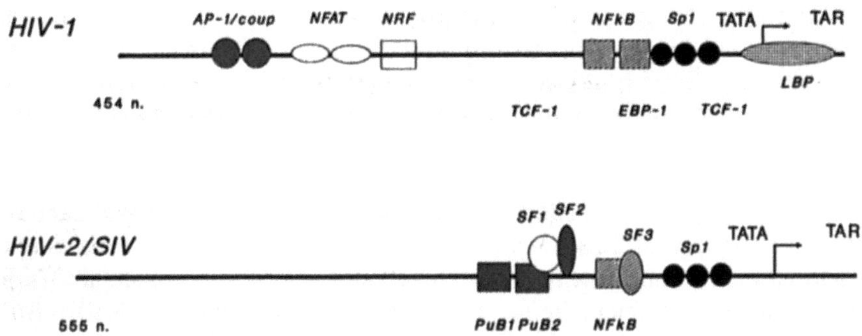

Fig. 1. DNA-binding proteins found in the LTRs of HIV-1 and HIV-2/SIV. Schematic examples of some of the binding sites by cellular transcription factors for the HIV-1 (*above*) or the HIV- 2/SIV LTRs (*below*). For sake of simplicity not all described proteins have been diagrammed. Most of the factors are discussed in the text; TCF-1 is a factor described by WATERMAN et al. (1991)

not found in HIV-1. Perhaps most noteworthy and best studied of the sequence motifs in U3 are the highly conserved TATAA box (BUCHER and TRIFONOV 1986), the triply reiterated Sp1 motifs (JONES et al. 1986), and the NF-κB site(s) (LENARDO and BALTIMORE 1989; two copies in HIV-1, one copy in HIV-2/SIV).

Detailed dissections of the HIV-1 LTR using DNAse I footprinting have delineated many binding sites for cellular proteins both upstream and downstream of the transcriptional start (GARCIA et al. 1987; WU et al. 1988). A few prototypic (but non-exhaustive) examples of binding factors are diagrammed in Fig. 1. For instance, between nucleotides -330 and -350 are two binding sites for the AP-1 (FRANZA et al. 1988) and/or coup (COONEY et al. 1991) proteins. More proximally, an authentic and a related motif for NFAT (SHAW et al. 1988; CRABTREE 1989) are seen. Binding of a negative regulatory factor (NRF; GARCIA et al. 1987) to nucleotide positions -173 to -159 has also been described.

Separate from binding assays, functional analyses have shown that deletion of the AP-1/coup motifs perturb little the LTR transcriptional activity (VAISHNAV and WONG-STAAL 1991). The NFAT sites, however, direct response in T cells to stimulation of antigen receptor (SHAW et al. 1988). The mechanistic role of NRF is unclear. In virus studies, if sequences in the HIV-1 LTR which span -156 to -420 (defined previously as a negative regulatory element, NRE; ROSEN et al. 1985; and encompass the above binding proteins) are removed from an infectious provirus, the resulting HIV replicates better than an otherwise isogenic NRE-containing virus (Y. LU et al. 1990).

Elements that immediately flank the HIV-1 TATAA box have been characterized in great detail. Specifically, the upstream NF-κB and Sp1 motifs (NABEL and BALTIMORE 1987; KAWAKAMI et al. 1988; JONES et al. 1986) have been studied in both subgenomic and viral settings. The NF-κB binding sites are recognized by the EBP-1 protein in resting epithelial cells (WU et al. 1988) and by NF-κB and Rel-related proteins (NABEL and BALTIMORE 1987; FRANZA et al. 1987) in activated T-cells. In transfections of subgenomic fragments these motifs are the primary determinants of the magnitude of basal transcription (BERKHOUT and JEANG 1992). Surprisingly, removal of NF-κB does not render HIV replication incompetent; the NF-κB sites in HIV-1 are not needed for viral infectivity in human T-cells such as MT4 (LEONARD et al. 1989). In other studies, it was shown further that there is a functional redundancy between NF-κB and Sp1 for viral replication. Thus NF-κB becomes crucial when an HIV-1 proviral LTR is compromised in Sp1 function (PARROT et al. 1991; ROSS et al. 1991); and additional reiterations of Sp1 binding-sites can, in themselves also, influence HIV-1 replication (KOKEN et al. 1992). A "take-home" message emerges that, so long as transcription is maintained in the HIV LTR (how it is maintained seems not to be absolutely critical), productive virus life-cycle ensues. In agreement with this, CHANG et al. (1993) have created chimeric replication-competent HIVs in which both NF-κB and Sp1 sites have been substituted completely with unrelated enhancers. However, there is conflicting information suggesting that, depending on cell-type, NF-κB and Sp1 and their relative spatial orientation to each other are critical for HIV-1 replication (KIM et al. 1993).

Based on sequence relatedness between the U3s of HIV-2/SIV and HIV-1, many sites for DNA-binding proteins are expected to be cross-conserved. Direct experimental verification of this has been limited. There are specific instances in which binding proteins (e.g., LBP-1) for HIV-1 sequences have been shown to interact with the corresponding sequences in HIV-2 (JONES et al. 1988). However, because HIV-2 pathogenesis is not identical to HIV-1 and has a longer clinical latency (CLAVEL et al. 1987; ANCELLE et al. 1987; DUFOORT et al. 1988), it would not be unexpected if some cellular factors uniquely interact with the HIV-2/SIV LTR (see Fig. 1, bottom).

Indeed, it has been shown that HIV-2 and HIV-1 do not respond in the same manner to T-cell activation (TONG-STARKSEN et al. 1990; ARYA 1990; MARKOVITZ et al. 1990). In particular, while HIV-1 utilizes the doubly reiterated NF-κB motifs in signal transduction, HIV-2 (with only one copy of NF-κB site and a divergent second site) appears to recruit additional cellular factors. There is no consensus as to what these additional factors are. TONG-STARKSEN et al. (1990) have suggested that AP-3 binding to the divergent NF-κB site (position -79 to -95) is related to activation of the HIV-2 LTR. Alternatively, ARYA (1990) has proposed that additional sequences at -178 to -186 and -220 to -228 function as a second responsive enhancer. A third finding emphasizes the importance of two purine-rich (PuB1 and PuB2; Fig. 1) motifs (MARKOVITZ et al. 1992) positioned between nucleotides -130 and -180. This region is recognized by a novel Ets-related transcription factor, Elf-1 (LEIDEN et al. 1992). In SIV_{mac}, three novel binding proteins have been described (SF1, SF2, SF3; WINANDY et al. 1992). Simian factors 1 and 2 (SF1, SF2; Fig. 1), in phorbol ester stimulated T-cells, bind to a region between −116 to −140, while SF3 (Fig. 1), in unstimulated cells, binds to a region overlapping the NF-κB motif (WINANDY et al. 1992). It has been speculated that the SF2 protein may be related to the PuB1/PuB2 (LEIDEN et al. 1992) factor, and that SF3 may serve a related function as AP-3 (TONG-STARKSEN et al. 1990) in HIV-2 (WINANDY et al. 1992).

As discussed for HIV-1, binding studies are useful as clues leading to further analyses in the virus. Results from chimeric HIV-1/SIV_{agm} constructions suggest (but do not prove formally) that overlapping mechanisms may govern HIV-1 and SIV infection of lymphocytes in vitro (SHIBATA et al. 1990c). Thus some of the observations described for HIV-1 may apply generally to SIV (and vice versa). Consistent, in part, with this was the recent report (BELLAS et al. 1993) that the NF-κB sites are not necessary for SIV_{mac} infection of T-cells in vitro (but are used in macrophages). However, an in vivo example from $SIV_{smm\text{-}PBJ14}$ (DEWHURST et al. 1990) suggests that NF-κB does play a role in SIV pathogenesis (MARTIN 1990).

2.2 Tat-Activated Transcription

Lentivirus transcription can be viewed on two levels. Immediately after integration, proviral transcription is directed by cellular factors that recognize binding motifs in the LTR (see Sect. 2.1). Shortly thereafter, synthesis of viral Tat (*trans*-activator of transcription) protein switches the transcriptional regimen from basal to activated expression (see reviews, VAISHNAV and WONG-STAAL 1991; JEANG et al. 1991; CULLEN and GARRETT 1992). Below, we discuss the role of Tat in lentivirus transcription.

2.2.1 Structure of the Tat Protein

Tat (ARYA et al. 1985; SODROSKI et al. 1985) is a regulatory protein whose function is highly conserved among all lentiviruses (MYERS et al. 1992b). An intact Tat function is essential for viability of HIV-1 (DAYTON et al. 1986; FISHER et al. 1986), HIV-2 (SHIBATA et al. 1990a), and SIV_{agm} (SHIBATA et al. 1990b). Tat mRNA is encoded by two separate exons. Depending on the virus, the protein is approximately 101 amino acids (most isolates of HIV-1, see MYERS et al. 1992a; the commonly discussed 86 amino acid Tat is an exception found for LAI/Bru derivatives), 130 amino acids (HIV-2), or 119 amino acids (SIV_{agm}; Fig. 2a). The size differences between HIV-1 and HIV-2 Tat can be roughly accounted for by two unique "insertions," in the latter protein, of 24 amino acids in the N-terminus and 14 amino acids in the C-terminus (BERKHOUT et al. 1990b). Deletion of these two "insertions," however, do not affect function (BERKHOUT et al. 1990b).

There is 30%–40% relatedness in the amino acids of the three (HIV-1, HIV-2, SIV_{agm}) Tat proteins, which can be aligned functionally in a linear

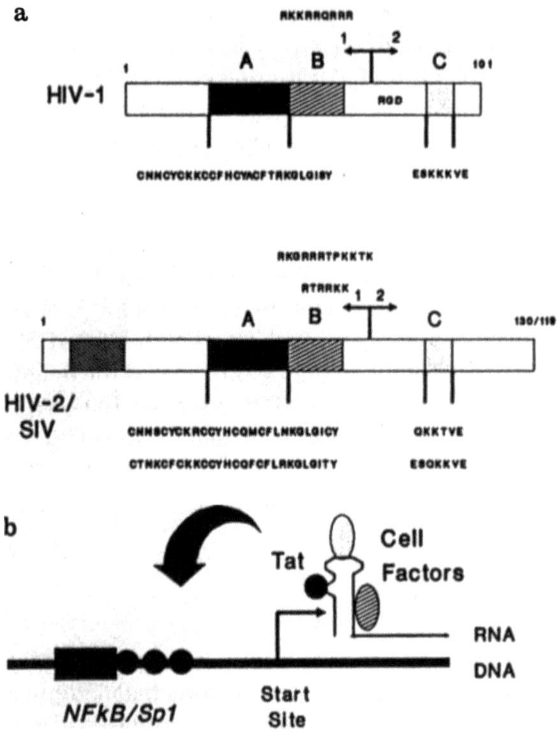

Fig. 2a, b. Structure of Tat protein and one model of Tat interaction with TAR RNA and promoter DNA. **a** Schematic representations of the HIV-1 (*above*) and HIV-2/SIV Tat proteins. The conserved A, B, and C regions represent the activation domain, the basic nuclear/RNA binding domain, and a domain, of unknown function in the second exon. *Below*, the amino acid sequences are shown for HIV-2rod (*upper*) and SIV_{agm} (*lower*). The C homology often has been overlooked because most reviews have regarded HIV-1 Tat as 86 amino acids in length (e.g., see VAISHNAV and WONG-STALL 1991). However, with the exception of LAI/Bru derived Tat proteins, Tat from most HIV-1 isolates is actually around 101 or 102 amino acids. The RGD motif is present only in HIV-1 Tat. **b** A model illustrating that Tat contacts nascent TAR RNA and then "bends backward" to interact with factors bound at the promoter DNA. Details are discussed in the text

manner (Fig. 2a). In a comparison between HIV-2/SIV and HIV-1, three regions in Tat are seen easily to be conserved. Toward the N-terminus, in the first coding exon of Tat, there is a cysteine-rich region (region A, Fig. 2a) encompassing 26 amino acids (VIGLIANTI and MULLINS 1988) that has been shown to be a minimal activation domain for lentivirus Tat proteins (DERSE et al. 1991). Immediately flanking region A (Fig. 2a), toward the C-terminus, is a conserved stretch of basic amino acids (region B, Fig. 2a; KUPPUSWAMY et al. 1989). This region (WEEKS et al. 1990; WEEKS and CROTHERS 1991; CALNAN et al. 1991) with a few flanking amino acids (CHURCHER et al. 1993) is widely considered as the binding domain of Tat for TAR RNA (see below).

In functional studies, it has been determined that this apparently well-conserved region cannot be identically exchanged between HIV-1 and HIV-2 (ELANGOVAN et al. 1992; CHANG and JEANG 1992; RHIM and RICE 1993). Domain swaps indicate that the strain-specific amino acid composition of this region determines a preferential activation by HIV-2 Tat of its cognate LTR over that of HIV-1 (EMERMAN et al. 1987; BERKHOUT et al. 1990b; FENRICK et al. 1989).

The second coding exon of HIV-1 (amino acids 73–101) and HIV-2 (amino acids 99–130) Tat has been shown to be dispensible for transcriptional function (SEIGEL et al. 1986; KUPPUSWAMY et al. 1989; EMERMAN et al. 1987). However, similar experiments with SIV_{mac} Tat indicated that this part of the protein may contribute to optimal *trans*-activation (VIGLIANTI and MULLINS 1988). An explanation for this apparent discrepancy may lie in differences in the assays since we have recently found that the second exon of HIV-1 Tat appears to assist in activation of integrated (but not unintegrated) LTRs (Jeang, unpublished). Another difference between the second exons of HIV-1 Tat and HIV-2/SIV Tat is the presence of an RGD cell adhesion signal (BRAKE et al. 1990) in the former (Fig. 2a) but not seen in the latter. This amino acid motif has been suggested to be important (BRAKE et al. 1990) for the cellular uptake of diffusible Tat protein (FRANKEL and PABO 1988). Whether this is important in vivo remains to be confirmed.

2.2.2 Functional Interaction of Tat, TAR, and Promoter

Lentivirus Tat protein activates transcription from its cognate LTR promoter using a (yet) novel route (see reviews, SHARP and MARCINIAK 1989; JEANG et al. 1991). Tat represents an example of a *trans*-activator that functions mechanistically through a bipartite target composed of both RNA and DNA (Fig. 2b). Our current understanding suggests that Tat is an RNA-binding protein (DINGWALL et al. 1989; ROY et al. 1990; CORDINGLEY et al. 1990) that physically attaches to a leader RNA (TAR, Fig. 3; FENG and HOLLAND 1988; MUESING et al. 1987) during nascent transcription (BERKHOUT et al. 1989) and then interacts with promoter DNA elements (SOUTHGATE et al. 1990, 1991; BERKHOUT et al. 1990a; SELBY and PETERLIN, 1990; KAMINE et al. 1991) in a "bend backward" manner (Fig 2b). TAR RNA serves as a physical tether bringing Tat to the promoter where it influences initiation (LASPIA et al. 1989; JEANG and BERKHOUT 1992) or processivity (reviewed in CULLEN 1990) events.

The above model is based upon contributions from many laboratories (reviewed in JEANG et al. 1991). It emerged from the initial observation that the Tat-responsive element (TAR) mapped to a sequence downstream of the +1 RNA start site (ROSEN et al. 1985; HAUBER and CULLEN 1988; JAKOBOVITS et al. 1988; MUESING et al. 1987). Although this sequence could interact with DNA-binding proteins (GARCIA et al. 1987; JONES et al. 1988; KATO et al.

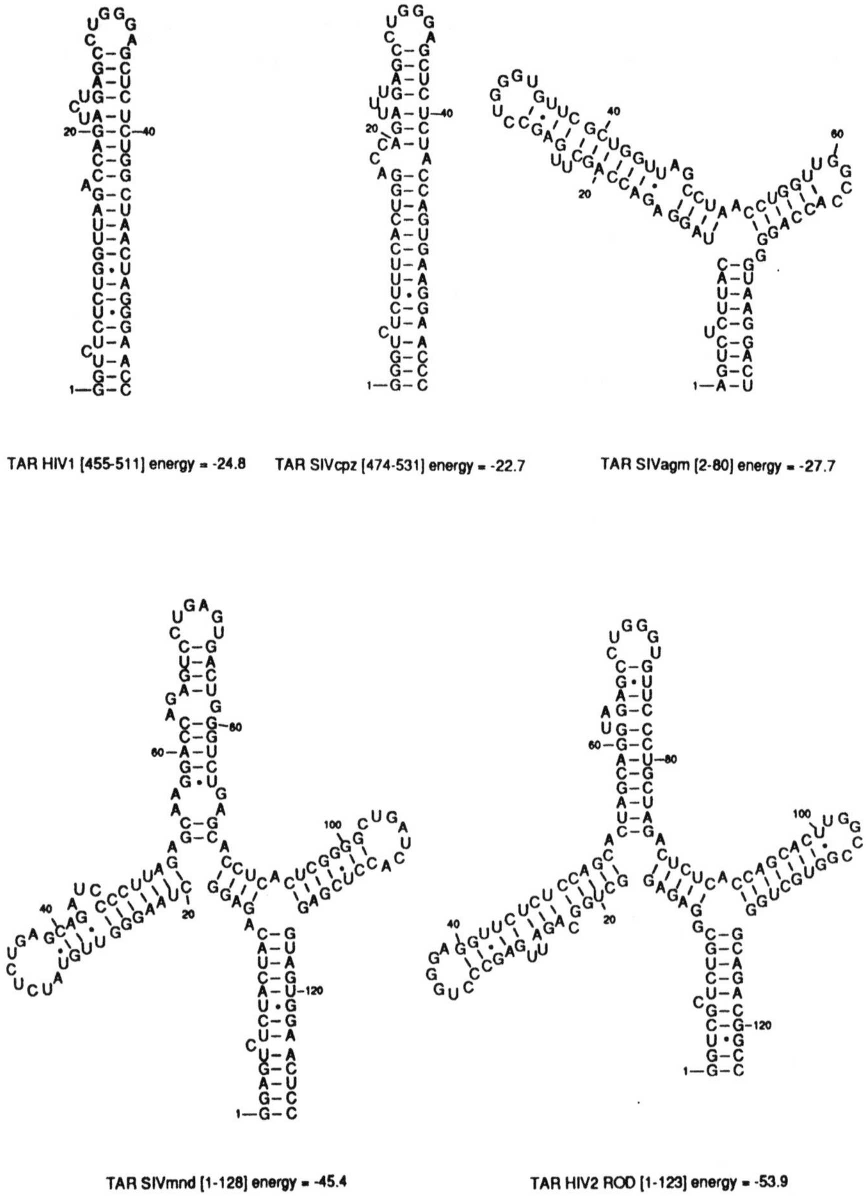

Fig. 3. Predicted secondary structures of HIV and SIV TAR RNAs. Folding was performed using the Zuker program (ZUKER 1989; JAEGER et al. 1989). The different viral strains are as indicated, the nucleotide numbers in each structure are *bracketed*, and the free energy is expressed in kcal/mol. The sequences are from HIV1NL43 (ADACHI et al. 1986; revised by A. Buckler-White, personal communications; GenBank accession number M19921), SIV_{cpz} (HUET et al. 1990; accession X52154), HIV2 ROD (GUYADER et al. 1987; accession M15390), SIV AGMTYO (FUKASAWA et al. 1988; accession X07805), SIV MNDGB1 (TSUJIMOTO et al. 1989; accession X15781). Each folding is shown as one secondary structure except HIV2 in which two alternative forms are shown. Note that each of the alternate forms of HIV2 TAR reveals relatedness to either SIV_{mnd} or SIV_{agm}

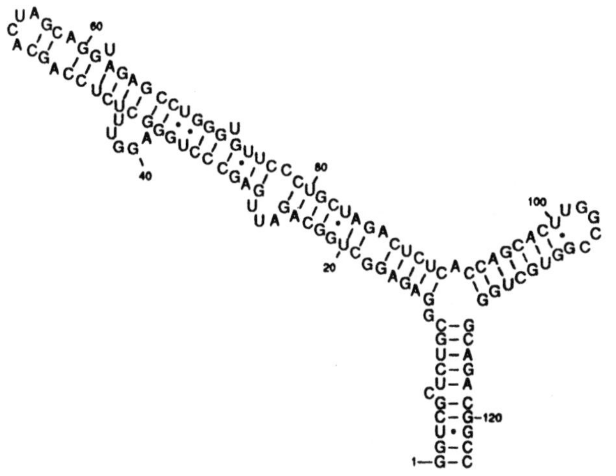

TAR HIV2 ROD [1-123] energy = -53.4

Fig. 3. *(Continued)*

1991), it was recognized quickly that this region in its RNA form is a structured stem-bulge-loop hairpin (Fig. 3; MUESING et al. 1987; FENG and HOLLAND 1988). Subsequent functional and biochemical analyses confirmed that TAR RNA (not DNA) is important for function; in particular the intactness of the bulge and loop elements (BERKHOUT and JEANG 1989; ROY et al. 1990) and the nucleotide pairs in the stem that immediately flank the bulge (BERKHOUT and JEANG 1991) are critical.

Tat mechanism of *trans*-activation is conserved between HIV-1 and HIV-2/SIV. Even widely diverged primate lentiviruses (i.e., HIV-1 and SIV_{mnd}) conserve a TAR RNA structure (Fig. 3; SAKURAGI et al. 1992; BERKHOUT 1992) and Tat function (SAKURAGI et al. 1992). As with HIV-1, the secondary structure of HIV-2 TAR is important for *trans*-activation (EMERMAN et al. 1987; BERKHOUT et al. 1990b; FENRICK et al. 1989). Recent information indicates that HIV-2 Tat, as its HIV-1 counterpart, also uses TAR-2 RNA as a simple tether (ELANGOVAN et al. 1992; CHANG and JEANG 1992; RHIM and RICE 1993). Furthermore, the ultimate interaction of HIV-2/SIV_{mac} Tat at the DNA promoter is likely to be with promoter-bound Sp1 (ANDERSON and CLEMENTS 1992) in a manner similar to that proposed for HIV-1 (HARRICH et al. 1989; BERKHOUT et al. 1990a; SOUTHGATE et al. 1991; KAMINE et al. 1991; ZIMMERMANN et al. 1991; BERKHOUT and JEANG 1992). Finally, one shared observation between HIV-1 and SIV is that small changes in Tat that do not obviously affect transcription can influence dramatically viral viability and tropism (CHENG-MAYER et al. 1991; SAKAI et al. 1992). This suggests an additionally important function for Tat in the infectivity of the virus, separate from its role in transcription.

3 Posttranscriptional Regulation

One crucial aspect of retrovirus replication is the balance between spliced and unspliced RNAs. Because full-length genomic RNA is itself a mRNA (for Gag, Pol), a pre-RNA for spliced mRNAs, and an RNA substrate for packaging into virions, the regulated equilibrium between spliced and unspliced RNA is very important. While other retroviruses have evolved *cis*-inhibitory RNA elements to evade splicing (McNALLY et al. 1991; STOLFUS and FOGARTY 1989), lentiviruses utilize Rev to subserve this function. This posttranscriptional function is especially important for primate lentiviruses because of the efficiency and complexity of splicing in both HIVs (SCHWARTZ et al. 1990) and SIVs (VIGLIANTI et al. 1990). Despite the large number of splice donors/acceptors in the primate lentivirus genomes most splicing signals are conserved strictly between HIVs and SIVs, with few exceptions (e.g., an extra 5′ splice donor found in the TAR sequence in HIV-2/SIV; VIGLIANTI et al. 1990).

Rev is a small nuclear phosphoprotein that is conserved highly between all primate lentiviruses. Intact Rev function has been verified to be critical for the viability of HIV-1 (SODROSKI et al. 1986; TERWILLIGER et al. 1988), HIV-2 (SHIBATA et al. 1990a), and SIV_{agm} (SHIBATA et al. 1990b). The Rev open reading frame is encoded by two separate exons, and in HIV-1 and HIV-2 there is approximately 60% homology in amino acids (MYERS et al. 1992b).

A prototypic representation of primate lentivirus Rev is shown in Fig. 4. HIV-1 Rev is 116 amino acids in length while SIV Rev is approximately 100 amino acids (e.g., 89 amino acids for SIV_{agm155}; 107 amino acids for SIV_{mm239}). Mutagenesis studies on HIV-1 (MALIM et al. 1989a, 1991; HOPE et al. 1990a,b;) and HIV-2 (DILLON et al. 1991) Rev proteins are in general agreement with each other on the regions necessary for function. There are three regions in Rev that can be viewed as important (Fig. 4). Between amino acids 35 and 50 is an arginine-rich stretch (MALIM et al. 1989b; OLSEN et al. 1990; MALIM and CULLEN 1991; COCHRANE et al. 1990b; VENKATESH et al. 1990; HOPE et al. 1990a) that is important in nuclear localization and RNA binding (DALY et al. 1989; ZAPP and GREEN 1989; HEAPHY et al. 1990;

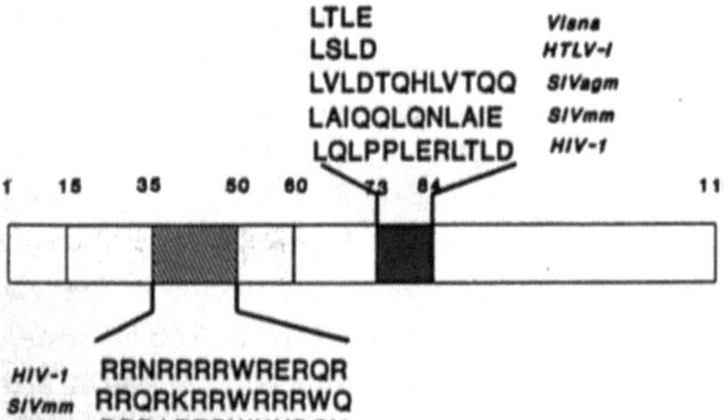

Fig. 4. The prototypic structure of primate lentivirus Rev. The linear representation highlights domains for protein multimerization (amino-acids 15–35, 50–60), for nuclear localization and RNA binding, and for activation. (Diagram is after JEANG et al. 1991; data on the leucine-rich motif from HOPE et al. 1991)

COCHRANE et al. 1990a; MALIM et al. 1990; HOLLAND et al. 1990; DAEFLER et al. 1990; DILLON et al. 1990). A second region rich in leucines (amino acids 73–83; Fig. 4) has been implicated in "activation function" (HOPE et al. 1991; MALIM et al. 1991). A four amino acid core motif within this effector domain is conserved between Rev equivalents from HIVs, SIVs, HTLVs, and visna (HOPE et al. 1991). Thus while specific Rev equivalents may recognize different RNA targets in varying contexts, this finding suggests that all Revs interact with the same cellular machinery in similar fashions. A third segment(s) of Rev (amino acids 15–35 and 50–60; Fig. 4), necessary for function (HOPE et al. 1990a; MALIM et al. 1991; MALIM and CULLEN 1991), likely participates in protein multimerization. It has been suggested variously that active Rev, inside cells, is a protein multimer (OLSEN et al. 1990; ZAPP et al. 1991; MALIM and CULLEN 1991; HEAPHY et al. 1991).

The role of Rev is to facilitate the cytoplasmic expression of unspliced (or singly spliced) viral mRNAs. As such one expects the Rev-responsive element (RRE) to be present in unspliced (or singly spliced) but not multiply spliced RNAs. RRE indeed fulfills this criterion; it is a highly secondary-structured RNA element (LE et al. 1990; approximately 250 nucleotides; see Fig. 5) located in *env*, that is removed as intronic sequence in multiply spliced mRNAs (MALIM et al. 1989c). Surprisingly similar RRE structures can be derived by free-energy computations for even widely disparate primate lentiviruses (e.g., compare HIV-1 RRE with SIV_{mnd} RRE; Fig. 5; DAYTON et al. 1989; HEAPHY et al. 1990; SAKAI et al. 1991); however, there are suggestions that in vivo RRE structures need not reflect in vitro foldings (BARTEL et al. 1991).

Given the remarkable similarities between HIV-1 and SIV in "effector" domains (Fig. 4; HOPE et al. 1991) and in RRE structures (Fig. 5), one would expect facile interchangeability between the Rev varieties of primate lentiviruses. This is, however, not the case. A nonreciprocity in function has been documented between HIV-1 and HIV-2/SIV. Thus, while HIV-2/SIV Rev varieties work well with their cognate RRE forms, they are poorly functional with HIV-1 RRE (MALIM et al. 1989a; DILLON et al. 1990; CHENG et al. 1990; LEWIS et al. 1990; SAKAI et al. 1990a, b). HIV-1 Rev, however, can activate HIV-1, as well as HIV-2/SIV, RRE-containing RNAs in a promiscuous manner. The finding that the Rev effector domain functions when targeted to (any) RNA via a heterologous RNA-binding domain (MCDONALD et al. 1992) suggests that the above non-reciprocity can be explained as flexibility (inflexibility) in recognition of RRE varieties by HIV-1 (HIV- 2/SIV) Rev.

Three mechanisms have been proposed to explain Rev function. Rev has been suggested to effect the selective transport of unspliced RNAs (FELBER et al. 1989; reviewed in PAVLAKIS and FELBER 1990), to directly affect formation of splicing complexes (CHANG and SHARP 1989; X. LU et al. 1990; KJEMS et al. 1991b), and to influence the efficiency of translation of unspliced mRNAs (ARRIGO and CHEN 1991; D'AGOSTINO et al. 1992) in the cytoplasm. It should be noted that, unlike in a recent study of RRE function

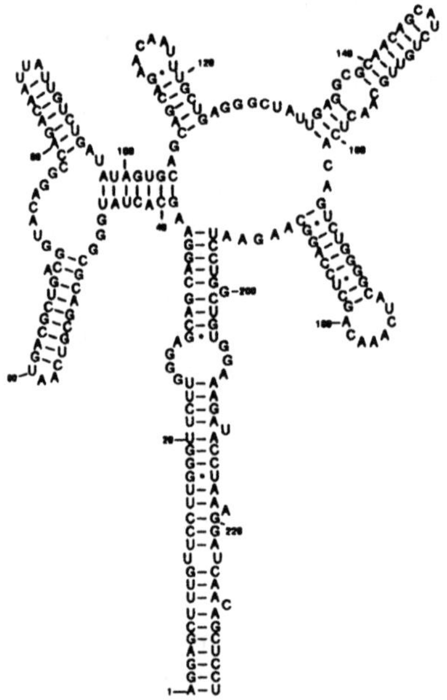

RRE HIV1 [7759-7992] energy = -84.6

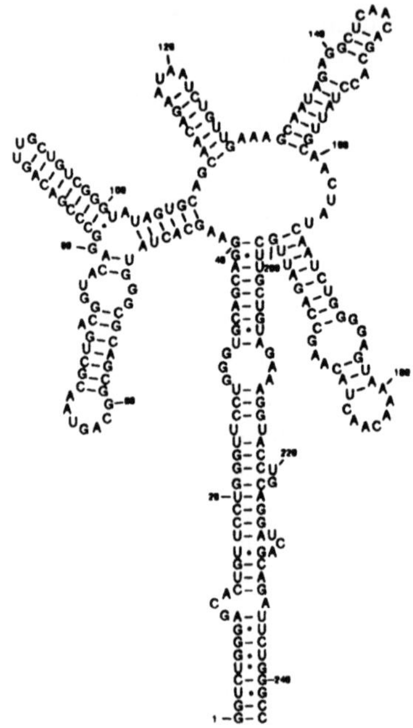

RRE SIVcpz [7801-8044] energy = -76.2

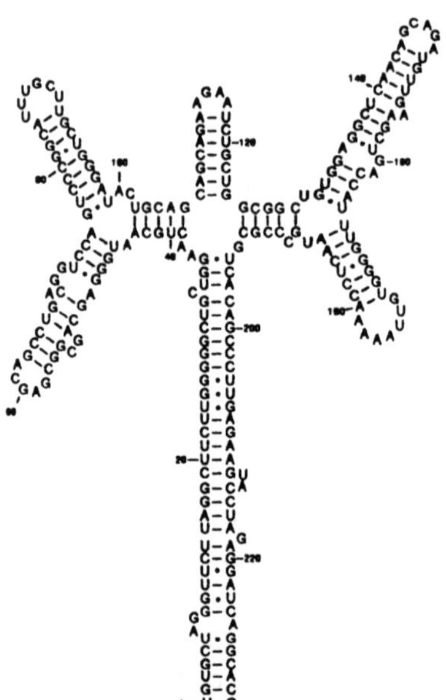

RRE SIVagm [7386-7616] energy = -86.4

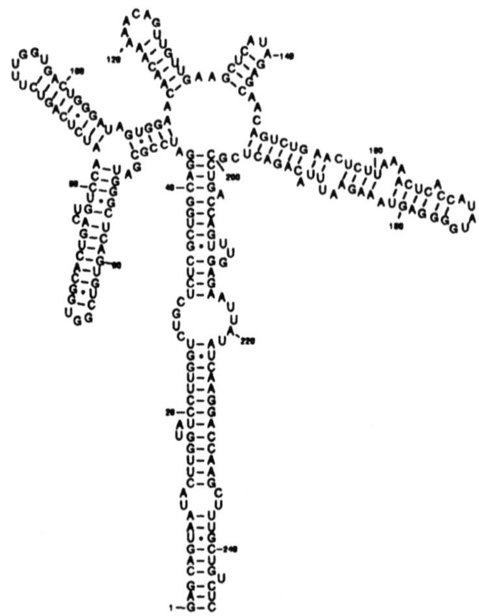

RRE SIVmnd [7369-7614] energy = -79.0

Fig. 5. *(Continued)*

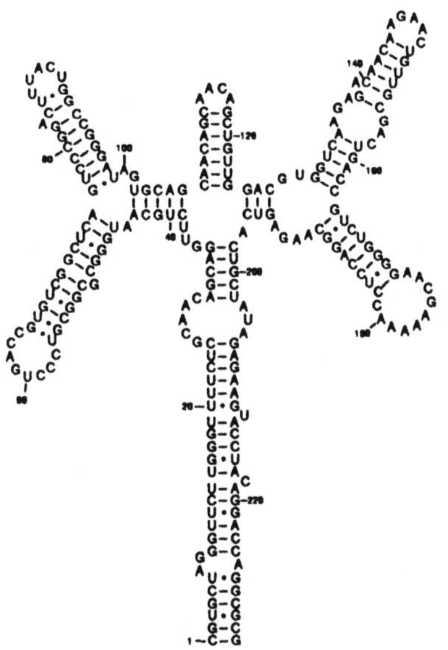

RRE HIV2 ROD [7662-7892] energy = -77.4

Fig. 5. Predicted secondary structures of HIV and SIV RRE RNAs. Folding program and sequence references are as in Fig. 3. In each case, except for SIV_{mnd}, several alternative structures can be derived. For HIV-1, we illustrate the structure most closely related to that verified experimentally based on nuclease and mutagenesis analyses (MALIM et al. 1990; HOLLAND et al. 1990; KJEMS et al. 1991a; BARTEL et al. 1991). The structure of SIV_{cpz} is chosen from all alternatives based upon the highest thermodynamic stability. Based upon this, we chose to illustrate the structures for HIV2 and SIV_{agm} that are most similar. They are the second and third most stable forms for SIV_{agm} and HIV2, respectively. The presented secondary structures are likely the biologically relevant forms of RRE since this configuration is the only common form among these isolates

in authentic HIV infections (DAYTON et al. 1993), all of the above conclusions have been derived from subviral contexts. Since, in general, use of selected reductionist approaches can potently bias answers (SAVAGEAU 1991), further studies using HIV/SIV infections should help in resolving mechanistic contentions.

4 Other Regulators

While this discussion has focused primarily upon the roles of Tat and Rev in the primate lentivirus life cycle, one should not overlook differential contributions to HIV-1 versus SIV/HIV-2 from other regulatory factors. This topic is diverse and complex and, in itself, merits a complete discourse. However, for the sake of brevity we simply raise as examples the (nonexhaustive) consideration of three areas: (a) other lentiviral proteins (e.g., vpr, vpu), (b) other nonlentiviral factors (i.e., coinfection with other viruses), and (c) host cell factors.

Both vpr and vpu have regulatory functions for lentiviruses (reviewed in VAISHNAV and WONG-STAAL 1990). While vpr appears to play comparable roles for human and simian lentiviruses (COHEN et al. 1990; YU et al. 1990; HATTORI et al. 1990; LANG et al. 1993), vpu function is restricted necessarily to HIV-1 (-like) lentiviruses, since with the exception of an isolate of SIV_{cpz} (HUET et al. 1990) *vpu* is not conserved in HIV-2/SIVs. Thus, as with some aspects of Tat/Rev functions, this represents another instance of non-identity between the primate lentiviruses. Another area of likely difference is in the role(s) of coinfecting agents. One would not expect SIV-infected monkeys and HIV-1 infected humans to be identically coinfected with other viruses. In studies of HIV-1, co-infecting viruses contribute to activation (e.g., HTLV: SIEKEVITZ et al. 1987; ZACK et al. 1988; herpes; GENDELMAN et al. 1986; MOSCA et al. 1987; NELSON et al. 1988). Analogous (though not necessarily identical) coinfections in monkeys are likely and may account for species-specific regulation/pathology.

A third regulatory difference between HIV-1 and HIV-2/SIV is in host cell factors. Above we discuss some differences in DNA-binding proteins for the respective LTRs (see Sect. 2.1), More recent studies indicate that RNA-binding proteins are likely to be important for lentivirus gene regulation. A number of HIV-1 "specific" RNA-binding proteins have been characterized (GATIGNOL et al. 1991, 1993; SHELINE et al. 1991; WU et al. 1991; VAISHNAV and WONG-STAAL 1991; CONSTANTOULAKIS et al. 1993). Characterization of HIV-2/SIV "specific" factors will provide bases for further comparisons and should yield clues as to similarities/differences in regulation.

5 Concluding Remarks

This review highlights the general similarities and some specific differences in the regulatory mechanisms used by the primate lentiviruses. From the perspective of the regulatory proteins Tat and Rev and their respective RNA targets, TAR and RRE, one can conclude that much is conserved between HIV-1 and HIV-2/SIV. The fact that non-reciprocal effects have been described for both *trans*-activation axes reinforces the idea that there remain unique properties individualized to the different species. The clear absence of the accessory *vpu* gene in the HIV-2/SIV subgroup is consistent with this. Having stated the above, it is clear that SIV, with its ability to infect $CD4^+$ cells and to cause AIDS in monkeys, represents the best animal model for in vivo HIV-1 pathogenesis. The recent elucidation of some of the pathogenic determinants for SIV (DEWHURST et al. 1990; KESTLER et al. 1991; DANIEL et al. 1992; LANG et al. 1993) should go a long way toward our developing a comparable understanding of HIV- 1.

Acknowledgments. We thank Riri Shibata, Eric Freed, and Li-Min Huang for readings of manuscript and for discussions, and Chuck Buckler for assistance with computer homology matches. K.T.J. would like to thank M. Martin and the many colleagues inside and outside of the Laboratory of Molecular Microbiology who over the years have stimulated thoughts of HIV-1 transcription, and M.-L. Hammarskjold and D. Rekosh for discussions on Rev. Work in our group is supported by the Council for Tobacco Research, USA Inc., and by the intramural AIDS targeted anti-viral program from the Office of the Director of the NIH.

References

Adachi A, Gendelman HE, Koenig S, Folks T, Willey RL, Rabson A, Martin MA (1986) Production of acquired immunodeficiency syndrome-associated retrovirus in human and nonhuman cells transfected with an infectious molecular clone. J Virol 59: 284–291

Adams C, Workman J (1993) Nucleosome displacement in transcription. Cell 72: 305–308

Ancelle R, Bredberg U, Chiodi F, Bottiger B, Fenyo E, Norrby E, Biberfeld G (1987) Long incubation period for HIV-2 infection. Lancet 1: 688–689

Anderson MG, Clements JE (1992) Two strains of SIVmac show differential transactivation mediated by sequences in the promoter. Virology 191: 559–568

Arrigo S, Chen ISY (1991) Rev is necessary for translation but not cytoplasmic accumulation of HIV-1 vif, vpr, and env/vpu RNAs. Genes Dev 5: 808–819

Arya SK (1990) Human immunodeficiency virus type-2 gene expression: two enhancers and their activation by T-cell activators. New Biol 2: 57–65

Arya SK, Guo C, Josephs SF, Wong-Staal F (1985) Trans-activator gene of human T-lymphotropic virus type III (HTLV-III). Science 229: 69–73

Barre-Sinoussi F, Chermann JC, Rey F, Nugeyre MT, Chamaret S, Gruest J, Dauguet C, Axler-Blin C, Vezinet-Brun E, Rouzious C, Rozenbaum W, Montagnier L (1983) Isolation of a T-lymphotropic retrovirus from a patient at risk for acquired immune deficiency syndrome (AIDS). Science 220: 868–871

Bartel DP, Zapp ML, Green MR, Szostak JW (1991) HIV-1 Rev regulation involves recognition of non-Watson-Crick base pairs in viral RNA. Cell 67: 529–536

Bellas R, Hopkins N, Li Y (1993) The NF-kB bindng site is necessary for efficient replication of simian immunodeficiency virus of macaques in primary macrophages but not in T cells in vitro. J Virol 67: 2908–2913

Berkhout B (1992) Structural features in TAR RNA of human and simian immunodeficiency viruses: a phylogenetic analysis. Nucleic Acids Res 20: 27–31

Berkhout B, Jeang K-T (1989) Trans Activation of human immunodeficiency virus type 1 is sequence specific for both the single-stranded bulge and loop of the trans-acting-responsive hairpin: a quantitative analysis. J Virol 63: 5501–5504

Berkhout B, Jeang K-T (1991) A detailed mutational analysis of TAR RNA; critical spacing between the bulge and loop recognition domains. Nucleic Acids Res 19: 6169–6176

Berkhout B, Jeang K-T (1992) Functional roles for TATA/promoter and enhancers in basal and Tat-induced expression of the HIV-1 LTR. J Virol 66: 139–149

Berkhout B, Silverman RH, Jeang K-T (1989) Tat trans-activates the human immunodeficiency virus through a nascent RNA target. Cell 59: 273–282

Berkhout B, Gatignol A, Rabson AB, Jeang K-T (1990a) TAR-independent activation of the HIV-1 LTR: evidence that Tat requires specific regions of the promoter. Cell 62: 757–767

Berkhout B, Gatignol A, Silver J, Jeang KT (1990b) Efficient trans-activation by the HIV-2 Tat protein requires a duplicated TAR RNA structure. Nucleic Acids Res 18: 1839–1846

Brake DA, Debouk C, Biesecker G (1990) Identification of an Arg-Gly-Asp (RGD) cell adhesion site in human immunodeficiency virus type 1 transactivation protein, tat. J Cell Biol 111: 1275–1281

Bucher P, Trifonov E (1986) Compilation and analysis of eukaryotic POL II promoter sequences. Nucleic Acids Res 14: 10013–10017

Bukrinsky M, Stanwick T, Dempsey M, Stevenson M (1991) Quiescent T lymphocytes as an inducible virus reservoir in HIV-1 infection. Science 254: 423–427

Calnan BJ, Biancalana S, Hudson D, Frankel AD, (1991) Analysis of arginine-rich peptides from the HIV Tat protein reveals unusual features of RNA-protein recognition. Genes Dev 5: 201–210

Chang Y-N, Jeang K-T (1992) The basic RNA-binding domain of HIV-2 Tat contributes to preferential trans-activation of a TAR2-containing LTR. Nucleic Acids Res 20: 5465–5472
Chang D, Sharp P (1989) Regulation by HIV Rev depends upon recognition of splice sites. Cell 59: 789–795
Cheng-Mayer C, Shioda T, Levy J (1991) Host range, replicative, and cytopathic properties of human immunodeficiency virus type 1 are determined by very few amino acid changes in tat and gp 120. J Virol 65: 6931–6941
Chang L-J, McNulty E, Martin M (1993) Human immunodeficiency viruses containing heterologous enhancer/promoters are replication competent and exhibit different lymphocyte tropisms. J Virol 67: 743–752
Cheng S-M, Blume M, Lee S-G, Hung P, Hirsch V, Johnson P (1990) The simian immunodeficiency virus (SIV) rev gene regulates env expression. J Med Primatol 19: 167–176
Churcher M, Lamont C, Hamy F, Dingwall C, Green SM, Lowe AD, Butler JG, Gait MJ, Karn J (1993) High affinity binding of TAR RNA by the human immunodeficiency virus type-1 tat protein requires base pairs in the RNA stem and amino acid residues flanking the basic region. J Mol Biol 230: 90–110
Clavel F, Guetard D, Brun-Vezinet F, Chamaret S, Rey M, Santos-Ferreira M, Laurent A, Dauguet C, Katlama C, Rouzioux C, Klatzmann D, Champalimaud J, Montagnier L (1986) Isolation of a new human retrovirus from West African patients with AIDS. Science 233: 343–346
Clavel F, Mansinho K, Chamaret S, Guetard D, Favier V, Nina J, Santos-Ferreira MO, Champalimaud J-L, Montagnier L (1987) Human immunodeficiency virus type 2 infection associated with AIDS in West Africa. N Engl J Med 316: 1180–1185
Cochrane A, Chen C, Rosen C (1990a) Specific interaction of the human immunodeficiency virus Rev protein with a structured region in env mRNA. Proc Natl Acad Sci USA 87: 1198–1202
Cochrane A, Perkins A, Rosen C (1990b) Identification of sequences important in the nucleolar localization of human immunodeficiency virus Rev: relevance of nucleolar localization to function. J Virol 64: 881–885
Cohen E, Dehni G, Sodroski J, Haseltine W (1990) Human immunodeficiency virus vpr product is a virion-associated regulatory protein. J Virol 64: 3097–3099
Constantoulakis P, Campbell M, Felber JB, Nasioulas G, Afonina E, Pavlakis G, (1993) Inhibition of Rev-mediated HIV-1 expression by an RNA binding protein encoded by the interferon-inducible 9-27 gene. Science 259: 1314–1318
Cooney AJ, Tsai SY, O'Malley B, Tsai MJ (1991) Chicken ovalbumin upstream promoter transcription factor binds to a negative regulatory region in the human immunodeficiency virus type 1 long terminal repeat. J Virol 65: 2853–2860
Cordingley MG, LaFemina RL, Callahan PL, Condra JH, Sardana VV, Graham DJ, Nguyen TM, LeGrow K, Gotlib L, Schlabach AJ, Colonno RJ (1990) Sequence-specific interaction of Tat protein and Tat peptides with the transactivation responsive sequence element of human immunodeficiency virus type 1 in vitro. Proc Natl Acad Sci USA 87: 8985–8989
Crabtree G. (1989) Contingent genetic regulatory events in T-lymphocyte activation. Science 243: 355–361
Cullen BR (1990) The HIV-1 Tat protein: an RNA sequence-specific processivity factor? Cell 63: 655–657
Cullen BR, Garrett ED (1992) A comparison of regulatory features in primate lentiviruses. AIDS Res Hum Retroviruses 8:387–393
Daefler S, Klotmena M, Wong-Staal F (1990) Trans-activating Rev protein of the human immunodeficiency virus 1 interacts directly and specifically with its target RNA. Proc Natl Acad Sci USA 87: 4571–4575
D'Agostino D, Felber B, Harrison J, Pavlakis G (1992) The Rev protein of human immunodeficiency virus type 1 promotes polysomal association and translation of gag/pol and vpu/env mRNAs. Mol Cell Biol 12: 1375–1386
Daly T, Cook K, Gray G, Malone T, Rusche J. (1989) Specific binding of HIV-1 recombinant Rev protein to the Rev-responsive element in vitro. Nature 343: 816–819
Daniel M, Kirchhoff F, Czajak S, Sehgal P, Desrosiers RC, (1992) Protective effects of a live attenuated SIV vaccine with a deletion in the nef gene. Science 258: 1938–1941
Dayton AI, Sodroski JG, Rosen CA, Goh WC, Haseltine WA (1986) The transactivator gene of the human T Cell lymphotropic virus type III is required for replication. Cell 44: 941–947

Dayton E, Powell D, Dayton A (1989) Functional analysis of CAR, the target sequence for the Rev protein of HIV-1. Science 246: 1625–1629
Dayton E, Konings D, Lim S, Hsu R, Butini L, Pantaleo, G, Dayton A (1993) The RRE of human immunodeficiency virus type 1 contributes to cell-type-specific viral tropism. J Virol 67: 287–2878
Derse D, Carvalho M, Carroll R, Peterlin BM (1991) A minimal lentivirus Tat. J Virol 65: 7012–7015
Desrosiers, RC (1990) The simian immunodeficiency viruses. Annu Rev Immunol 8: 557–578
Dewhurst S, Embretson JE, Anderson DC, Mullins JI, Fultz P (1990) Sequence analysis and acute pathogenicity of molecularly cloned SIVsmm-pbj14. Nature 345: 636–640
Dillon P, Nelbock P, Perkins A, Rosen C (1990) Function of the human immunodeficiency virus types 1 and 2 Rev proteins is dependent on their ability to interact with a structured region present in env gene mRNA. J Virol 64: 4428–4437
Dillon P, Nelbock P, Perkins A, Rosen C (1991) Structural and functional analysis of the human immunodeficiency virus type 2 Rev protein. J Virol 65: 445–449
Dingwall C, Ernberg I, Gait MJ, Green SM, Heaphy S, Karn J, Lowe AD, Singh M, Skinner MA, Vallerio R (1989) Human immunodeficiency virus 1 tat protein binds trans-activation responsive region (TAR) RNA in vitro. Proc Natl Acad Sci USA 86: 6925–6929
Dufoort G, Courouce AM, Ancelle-Park R, Bletry O (1988) No clinical signs 14 year after HIV-2 transmission via bolld transfection. Lancet 2: 510
Dynan WS (1989) Modularity in promoters and enhancers. Cell 58: 1–4
Elangovan B, Subramanian T, Chinnadurai G (1992) Functional comparison of the basic domains of the Tat proteins of human immunodeficiency virus types 1 and 2 in trans activation. J Virol 66: 2031–2036
Emerman M, Guyader M, Montagnier L, Baltimore D, Muesing MA (1987) The specificity of the human immunodeficiency virus type 2 trans-activator is different from that of human immunodeficiency virus type 1. EMBO J 6: 3755–3760
Felber B, Hadzopoulou-Cladras M, Cladaras C, Copeland T, Pavlakis G (1989) Rev protein of human immunodeficiency virus type 1 affects the stability and transport of the viral mRNA. Proc Natl Acad Sci USA 86: 1495–1499
Felsenfeld G (1992) Chromatin as an essential part of the transcriptional mechanism. Nature 355: 219–224
Feng S, Holland EC (1988) HIV-1 tat trans-activation requires the loop sequence within TAR. Nature 334: 165–167
Fenrick R, Malim M, Hauber J, Le SY, Maizel J, Cullen B (1989) Functional analysis of the tat trans-activator of human immunodeficiency virus type II. J Virol 63: 5005–5012
Fisher AG, Feinberg SF, Josephs SF, Harper ME, Marselle LM, Reyes G, Gonda MA, Aldovini A, Debouk C, Gallo RC, Wong-Staal F (1986) The trans-activator gene of HTLV-III is essential for virus replication. Nature 320: 367–371
Frankel A, Pabo C (1988) Cellular uptake of the Tat protein from human immunodeficiency virus. Cell 55: 1189–1193
Franza BR, Josephs SF, Gilman M, Ryan W, Clarkson B (1987) Characterization of cellular proteins recognizing the HIV enhancer using a microscale DNA-affinity precipitation assay. Nature 330: 391–395
Fukasawa M, Miura T, Hasegawa A, Morikawa S, Tsujimoto H, Miki K, Kitamura T, Hayami M (1988) Sequence of simian immunodeficiency virus from African green monkey, a new member of HIV/SIV group. Nature 333: 457–461
Gallo R, Salahuddin S, Popovic M, Shearere G, Kaplan M, Haynes B, Palker T, Redfield R, Oleske J, Safai B, White G, Foster P, Markham P (1984) Frequent detection and isolation of cytopathic retroviruses (HTLV-III) from patients with AIDS and at risk for AIDS. Science 224: 500–503
Garcia JA, Wu FK, Mitsuyasu R, Gaynor RB (1987) Interactions of cellular proteins involved in the transcriptional regulation of the human immunodeficiency virus. EMBO J 6: 3761–3770
Gatignol A, Buckler-White A, Berkhout B, Jeang K-T (1991) Characterization of a human TAR RNA-binding protein that activates the HIV-1 LTR. Science 251: 1597–1600
Gatignol A, Buckler C, Jeang K-T (1993) Relatedness of an RNA-binding motif in human immunodeficiency virus type 1 TAR RNA-binding protein TRBP to human P1/dsl kinase and Drosophila staufen Mol Cell Biol 13: 2193–2202

Gendelman HE, Phelps W, Feigenbaum L, Ostrove JM, Adachi A, Howley PM, Khoury G, Ginsberg HS, Martin M (1986) Trans-activation of the human immunodeficiency virus long terminal repeat sequence by DNA viruses. Proc Natl Acad Sci USA 83: 9759–9763
Guyader M, Emerman M, Sonigo P, Clavel F, Gluckman J-C, Alizon M (1987) Genome organization and transactivation of the human immuno-deficiency virus type 2. Nature 326: 662–669
Harrich D, Garcia J, Wu F, Mitsuyasu R, Gonzalez J, Gaynor RB (1989) Role of Sp1-binding domains in in vivo transcriptional regulation of the human immunodeficiency virus type 1 long terminal repeat. J Virol 63: 2585–2591
Hattori N, Michaels F, Fargnoli K, Marcon L, Gallo R, Franchini G (1990) The human immunodeficiency virus type 2 vpr gene is essential for productive infection of human macrophages. Proc Natl Acad Sci USA 87: 8080–8084
Hauber J, Cullen BR (1988) Mutational analysis of the trans-activation-responsive region of the human immunodeficiency virus type I long terminal repeat. J Virol 62: 673–679
Heaphy S, Dingwall C, Erngerg I, Gait M, Green S, Karn J, Lowe A, Singh M, Skinner M (1990) HIV-1 regulator of virion expression (Rev) protein binds to an RNA stem-loop structure located within the Rev response element region. Cell 60: 685–693
Heaphy S, Finch J, Gait M, Karn J, Singh M (1991) Human immunodeficiency virus type 1 regulator of virion expression rev, forms nucleoprotein filaments after binding to a purine rich "bubble" located within the rev-responsive region of viral mRNAs. Proc Natl Acad Sci USA 88: 7366–7370
Holland S, Ahmad N, Maitra R, Wingfield P, Venkatassan S (1990) Human immunodeficiency virus Rev protein recognizes a target sequence in rev-responsive element RNA within the context of RNA secondary structure. J Virol 64: 5966–5975
Hope T, Huang X, McDonald D, Parslow T (1990a) Steroid-receptor fusion of the human immunodeficiency virus type 1 Rev transactivator: mapping cryptic functions of the arginine-rich motif. Proc natl Acad Sci USA 88: 7787–7791
Hope T, McDonald D, Huang X, Low J, Parslow T (1990b) Mutational analysis of the human immunodeficiency virus type 1 Rev transactivator: essential residues near the amino terminus. J Virol 64: 5360–5366
Hope T, Bond BL, McDonald D, Klein N, Parslow T (1991) Effector domains of human immunodeficiency virus type 1 Rev and human T-cell leukemia virus type I Rex are functionally interchangeable and share an essential peptide motif. J Virol 65: 6001–6007
Huet T, Cheynier R, meyerhans A, Roelants G, Wain-Hobson S (1990) Genetic organization of a chimpanzee lentivirus related to HIV-1. Nature 345: 356–359
Jaeger JA, Turner DH, Zuker M (1989) Predicting optimal and suboptimal secondary structure for RNA. Methods Enzymol 183: 281–306
Jakobovits A, Smith DH, jakobovits EB, Capon DJ (1988) A discrete element 3′ of human immunodeficiency virus (HIV-1) and HIV-2 mRNA initiation sites mediates transcriptional activation by an HIV trans-activator. Mol Cell Biol 8: 2555–2561
Jeang K-T, Berkhout B (1992) Kinetics of HIV-1 long terminal repeat transactivation. Use of intragenic ribozyme to assess rate-limiting steps. J Biol Chem 267: 17891–17899
Jeang K-T, Chang YN, Berkhout B, Hammarskjold M-L, Rekosh D (1991) Regulation of HIV expression: mechanisms of action of Tat and Rev. AIDS 5: S3–S14
Johnson PF, McKnight SL (1989) Eucaryotic transcriptional regulatory proteins. Annu Rev Biochem 58: 799–839
Jones KA (1989) HIV trans-activation and transcription control mechanisms. New Biol 1: 127–135
Jones KA, Kadonaga JT, Luciw PA, Tjian R (1986) Activation of the AIDS retrovirus promoter by the cellular transcription factor, Spl. Science 232: 755–759
Jones KA, Luciw PA, Duchange N (1988) Sturctural arrangements of transcription control domains within the 5′-untranslated leader regions of the HIV-1 and HIV-2 promoters. Genes Dev 2: 1101–1114
Kamine J, Subramanian T, Chinnadurai G (1991) Sp1-dependent activation of a synthetic promoter by human immunodeficiency virus type I Tat protein. Proc Natl Acad Sci USA 88: 8510–8514
Kato H, Hirikoshi H, Roeder R (1991) Repression of HIV-1 transcription by a cellular protein. Science 251: 1476–1479
Kawakami K, Schneidereit C, Roeder RG (1988) Identification and purification of a human immunoglobulin enhancer-binding protein (NF-κB) that activates transcription from a

human immunodeficiency virus type 1 promoter in vitro. Proc Natl Acad Sci USA 82: 488–492
Kestler H, Ringler D, Mori K, Panicali D, Desrosiers R (1991) Importance of the nef gene for maintenance of high virus load and for development of AIDS. Cell 65: 651–662
Kim J, Gonzalez-Scarano F, Zeichner S, Alwine J (1993) Replication of type 1 human immunodeficiency viruses containing linker substitution mutations in the -201 to -130 region of the long terminal repeat. J Virol 67: 1658–1662
Kjems J, Brown M, Chang DD, Sharp PA (1991a). Structural analysis of the interaction between the human immunodeficiency virus rev protein and the rev response element. Proc Natl Acad Sci USA 88: 683–687
Kjems J, Frankel A, Sharp PA (1991b) Specific regulation of mRNA in vitro by a peptide from HIV-1. Rev Cell 67: 169–178
Koken SE, van Wamel J, Goudsmit J, Berkhout B, Geelen JL (1992) Natural variants of the HIV-1 long terminal repeat: analysis of promoters with duplicated DNA regulatory motifs. Virology 191: 968–972
Kuppuswamy M, Subramanian T, Srinivasan A, Chinnadurai G (1989) Multiple functional domains of Tat, the transactivator of HIV-1, defined by mutational analysis. Nucleic Acids Res 17: 3551–3561
Lang S, Weeger M, Stahl-Hennig C, Coulibaly C, Hunsmann G, Muller J, Muller-Hermelink H, Fuchs D, Wachter H, Daniel M, Desrosiers R, Fleckenstein B (1993) Importance of vpr for infection of rhesus monkeys with simian immunodeficiency virus. J Virol 67: 902–912
Laspia MF, Rice AP, Matthews MB (1989) HIV-1 Tat protein increases transcriptional initiation and stabilizes elongation. Cell 59: 283–292
Le SY, Malim M, Cullen B, Maizel JV (1990) A highly conserved RNA folding region coincident with the Rev-response element of primate immunodeficiency viruses. Nucleic Acids Res 18: 1613–1623
Leiden J, Wang C, Petryniak B, Markovitz D, Nabel G, Thompson C (1992) A novel Ets-related transcription factor, Elf-1, binds to human immunodeficiency virus type 2 regulatory elements that are required for inducible trans-activation in T cells. J Virol 66: 5890–5897
Lenardo M, Baltimore D (1989) NF-κB; a pleiotropic mediator of inducible and tissue-specific gene control. Cell 58: 227–229
Leonard J, Parrott C, Buckler-White A, Turner W, Ross E, Martin M, Rabson A (1989) The NF-κB binding sites in the HIV-1 long terminal repeats are not required for virus infectivity. J Virol 63: 4919–4924
Levy J, Hoffman A, Kramer S, Landis JA, Shimabukuro J, Oshiro L (1984) Isolation of lymphocytopathic retroviruses from San Francisco patients with AIDS. Science 225: 840–842
Lewis N, Williams J, Rekosh D, Hammarskjold M-L (1990) Identification of a cis-acting element in human immunodeficiency virus type 2 (HIV-2) that is responsive to the HIV-1 rev and human T-cell leukemia virus types I and II rex proteins. J Virol 64: 1690–1697
Lu X, Heimer J, Rekosh D, Hammarskjold M-L (1990) U1 small nuclear RNA plays a direct role in the formation of a rev-regulated human immunodeficiency virus env mRNA that remains unspliced. Proc Natl Acad Sci USA 87: 7598–7602
Lu Y, Touzhan N, Stenzel M, Dorfman T, Sodroski J, Haseltine W (1990) Identification of cis-acting repressive sequences within the negative regulatory element of human immunodeficiency virus type 1. J Virol 64: 5226–5229
Malim MH, Cullen BR (1991) HIV-1 structural gene expression requires the binding of multiple Rev monomers to the viral RRE: implications for HIV-1 latency. Cell: 241–248
Malim MH, Bohnlein S, Fenrick R, Le S-Y, Maizel J, Cullen BR (1989a) Functional comparison of the Rev trans-activators encoded by different primate immunodeficiency virus species. Proc Natl Acad Sci USA 86: 8222–8226
Malim MH, Bohnlein S, Hauber J, Cullen BR (1989b) Functional dissection of the HIV-1 Rev trans-activator-derivation of a trans-dominant repressor of Rev function. Cell 58: 205–214
Malim MH, Hauber J, Le S-Y, Maizel J, Cullen BR (1989c) The HIV-1 rev trans-activator acts through a structured target sequence to activate nuclear export of unspliced viral mRNA. Nature 338: 254–257
Malim MH, Tiley LS, McCarn DF, Rusche JR, Hauber J, Cullen BR (1990) HIV-1 Structural gene expression requires binding of the Rev transactivator to its RNA target sequence. Cell 60: 675–683

Malim MH, McCarn DF, Tiley L, Cullen BR, (1991) Mutational definition of the human immunodeficiency virus type 1 Rev activation domain. J Virol 65: 4248–4254
Markovitz D, Hannibal M, Perez V, Gauntt C, Folks T, Nabel G (1990) Differential regulation of human immunodeficiency viruses (HIVs): a specific regulatory element in HIV-2 responds to stimulation of the T-cell antigen receptor. Proc Natl Acad Sci USA 87: 9098–9102
Markovitz D, Smith M, Hilfinger J, Hannibal M, Petryniak B, Nabel G (1992) Activation of the human immunodeficiency virus type 2 enhancer is dependent on purine box and kB regulatory elements. J Virol 66: 5479–5484
Martin MA (1990) Fast-acting slow virus. Nature 345: 572–573
McDonald D, Hope T, Parslow T (1992) Posttranscriptional regulation by the human immunodeficiency virus type 1 Rev and human T-cell leukemia virus type I Rex protein through a heterologous RNA binding site. J Virol 66: 7232–7238
McNally MT, Gotarek R, Beemon K (1991) Characterization of Rous sarcoma virus intronic sequences that negatively regulate splicing. Virology 185: 99–108
Mosca J, bednarik D, Raj N, Rosen C, Sodroski J, Haseltine W, Hayward G, Pitha P (1987) Activation of human immunodeficiency virus by herpesvirus infection: identification of a region within the long terminal repeat that responds to a trans-acting factor encoded by herpes simplex virus 1. Proc Natl Acad Sci USA 84: 7408–7412
Muesing MA, Smith DH, Capon DJ (1987) Regulation of mRNA accumulation by a human immunodeficiency virus trans-activator protein. Cell 48: 691–701
Myers G, Berzofsky J, Korber B, Smith R, Pavlakis GN (1992a) Human retroviruses and AIDS. A compilation and analysis of nucleic acid and amino acid sequences. Los Alamos Laboratories, Los Alamos
Myers G, MacInnes K, Korber B (1992b) The emergence of simian/human immunodeficiency viruses. AIDS Res Hum Retroviruses 8: 373–386
Nabel G, Baltimore D (1987) An inducible transcription factor activates expression of human immunodeficiency virus in T cells. Nature 326: 711–713
Nelson J, Reynolds-Kholer C, Oldstone M, Wiley CA (1988) HIV and HCMV coinfect brain cells in patients with AIDS. Virology 165: 286–290
Olsen H, Cochrane A, Dillon P, Nalin C, Rosen C (1990) Interaction of the human immunodeficiency virus type 1 Rev protein with a structured region in env mRNA is dependent on multimer formation mediated through a basic stretch of amino acids. Genes Dev 4: 1357–1364
Pang S, Koyanagi Y, Miles S, Wiley C, Vinters HV, Chen ISY (1990) High levels of unintegrated HIV-1 DNA in brain tissue of AIDS dementia patients. Nature: 85–89
Parrot C, Seidner T, Duh E, Leonard J, Theodore TS, Buckler-White A, Martin MA, Rabson AB (1991) Variable role of the long terminal repeat Sp1-binding sites in human immunodeficiency virus replication in T lymphocytes. J Virol 65: 1414–1419
Pavlakis GN, Felber BK (1990) Regulation of expression of human immunodeficiency virus. New Biol 2: 20–31
Rhim H, Rice AP (1993) TAR RNA binding properties and relative transactivation activities of human immunodeficiency virus type 1 and type 2 Tat proteins. J Virol 67: 1110–1121
Rosen CA, Sodroski JG, Haseltine WA (1985) The location of cis-acting regulatory sequences in the human T cell lymphotropic virus type III (HTLV-III/LAV) long terminal repeat. Cell 41: 813–823
Ross E, Buckler-White A, Rabson AB, Englund G, Martin MA (1991) Contribution of NF-κB and Spl binding motifs to the replicative capacity of human immunodeficiency virus type 1: distinct patterns of viral growth are determined by T-cell types. J Virol 65: 4350–4358
Roy S, Parkin NT, Rosen C, Itovitch J, Sonenberg N (1990) Structural requirements for trans-activation of human immunodeficiency virus type 1 long terminal repeat-directed gene expression by Tat: importance of base pairing, loop sequence, and bulges in the Tat-responsive sequence. J Virol 64: 1402–1406
Sakai H, Shibata R, Miura T, Hayami M, Ogawa K, Kiyomasu T, Ishimoto A, and Adachi A (1990a) Complementation of the rev gene mutation among human and simian lentiviruses. J Virol 64: 2202–2207
Sakai H, Siomi H, Shida H, Shibata R, Kiyomusu T, Adachi A (1990b) Functional comparison of transactivation by human retrovirus rev and rex genes. J Virol 64: 5833–5839
Sakai H, Shibata R, Sakuragi J, Kiyomasu T, Kawamura M, Hayami M, Ishimota A, Adachi A (1991) Compatibility of rev gene activity in the four groups of primate lentiviruses. Virology 184: 513–520

Sakai H, Sakuragi J, Sayuri S, Shibata R, Adachi A (1992) Functional analysis of biologically distinct genetic variants of simian immunodeficiency virus isolated from a mandrill. Virology 189: 161–166
Sakai H, Kawamura M, Sakuragi J, Sakuragi S, Shibata R, Ishimoto A, Ono N, Ueda S, Adachi A (1993) Integration is essential for efficient gene expression of human immunodeficiency virus type 1. J Virol 67: 1169–1174
Sakuragi J, Sakai H, Sakuragi S, Shibata R, Wain-Hobson S, Hayami M, Adachi A (1992) Functional classification of simian immunodeficiency virus isolates from a chimpanzee by transactivators. Virology 189: 354–358
Savageau MA (1991) Reconstructionist molecular biology. New Biol 3: 190–197
Schwartz S, Felber B, Benko D, Fenyo E, Pavlakis G (1990) Cloning and functional analysis of multiply spliced mRNA species of human immunodeficiency virus type 1. J Virol 64: 5448–5456
Selby MJ, Peterlin BM (1990) Trans-activation by HIV-1 Tat via a heterologous RNA binding protein. Cell 62: 769–776
Seigel LJ, Ratner L, Josephs SF, Derse D, Feinberg MB, Reyes GA, O'Brien, SJ, Wong-Staal F (1986) Transactivation induced by human T-lymphotropic virus type III (HTLV-III) maps to a viral sequence encoding 58 amino acids and lacks tissue specificity. Virology 148: 226–231
Sharp PA, Marciniak RA (1989) HIV TAR: an RNA enhancer? Cell 59: 229–230
Shaw J-P, Utz PJ, Durand DB, Toole JJ, Emmel E, Crabtree G (1988) Identification of a putative regulatory of early T cell activation genes. Science 241: 202–205
Sheline C, Milocco L, Jones KA (1991) Two distinct nuclear transcription factors recognize loop and bulge residues of the HIV-1 TAR RNA hairpin. Genes Dev 5: 2508–2520
Shibata R, Miura T, Hayami M, Ogawa K, Sakai H, Kiyomasu T, Ishimoto A, Adachi A (1990a) Mutational analysis of the human immunodeficiency virus type 2 (HIV-2) genome in relation to HIV-1 and simian immunodeficiency virus SIVagm. J Virol 64: 742–747
Shibata R, Miura T, Hayami M, Sakai H, Ogawa K, Kiyomasu T, Ishimotao A, Adachi A (1990b) Construction and characterization of an infectious DNA clone and of mutants of simian immunodeficiency virus isolated from the African green monkey. J Virol 64: 307–312
Shibata R, Sakai H, Kiyomusu T, Ishimoto A, Hayami M, Adachi A (1990c) Generation and characterization of infectious chimeric clones between human immunodeficiency virus type 1 and simian immunodeficiency virus from an African green monkey. J Virol 64: 5861–5868
Siekevitz M, Josephs S, Dukovich M, Peffer N, Wong-Staal F, Greene W (1987) Activation of the HIV-1 LTR by T cell mitogens and the trans-activator protein of HTLV-I. Science 238: 1575–1579
Sodroski JG, Patarca R, Rosen CA, Wong-Staal F, Haseltine WA (1985) Location of the trans-activating region on the genome of human T-cell lymphotropic virus type III. Science 229: 74–77
Sodroski J, Goh W, Rosen C, Dayton A, Terwilliger E, Haseltine W (1986) A second post-transcriptional trans-activator gene required for HTLV-III replication. Nature 321: 74–77
Southgate CD, Green MR (1991) The HIV-1 tat protein activates transcription from an upstream DNA-binding site: implicatons for tat function. Genes Dev 5: 2496–2507
Southgate C, Zapp ML Green MR (1990) Activation of transcription by HIV-1 Tat protein tethered to nascent RNA through another protein. Nature 345: 640–642
Stolfus M, Fogarty S (1989) Multiple regions in the Rous sarcoma virus src gene intron acts in cis to affect the accumulation of unspliced RNA. J Virol 63: 1669–1676
Terwilliger E, Burghoff R, Sia R, Sodroski J, Haseltine W, Rosen C (1988) The art gene product of human immunodeficiency virus is required for replication. J Virol 62: 655–658
Tong-Starksen SE, Welsh TM, Peterlin BM (1990) Differences in transcriptional enhancers of HIV-1 and HIV-2. Response to T cell activation signals. J Immunol 145: 4348–4354
Tsujimoto H, Hasegawa A, Maki N, Fukasawa M, Miura T, Speidel S, Cooper RW, Moriyama EN, Gojobori T, Hayami M (1989) Sequence of a novel simian immunodeficiency virus from a wild-caught African mandrill. Nature 341: 539–541
Vaishnav YN, Wong-Staal F (1991) The biochemistry of AIDS. Annu Rev Biochem 60: 577–630
Vaishnav YN, Vaishnav M, Wong-Staal F (1991) Identification and characterization of a nuclear factor that specifically binds to the Rev response element (RRE) of human immunodeficiency virus type 1 (HIV-1). New Biol 3: 142–150

Venkatesh L, Mohammed S, Chinnadurai G (1990) Functional domains of the HIV-1 rev gene required for trans-regulation and subcellular localization. Virology 176: 39–47
Viglianti GA, Mullins JI (1988) Functional comparison of transactivation by simian immunodeficiency virus from rhesus macaques and human immunodeficiency virus type 1. J Virol 62: 4523–4532
Viglianti GA, Sharma P, Mullins JI (1990) Simian immunodeficiency virus displays complex patterns of RNA splicing. J Virol 64: 4207–4216
Waterman M, Fischer W, Jones KA (1991) A thymus-specific member of the HMG protein family regulates the human T cell receptor Ca enhancer. Genes Dev 5: 656–669
Weeks KM, Crothers DM (1991) RNA recognition by Tat-derivied peptides: interaction in the major groove? Cell 66: 577–588
Weeks KM, Ampe C, Schultz SC, Steitz TA, Crothers DM (1990) Fragments of the HIV-1 Tat protein specifically bind TAR RNA. Science 249: 1281–1285
Winandy S, Renjifo B, Li Y, Hopkins N (1992) Nuclear factors that bind two regions important to transcriptional activity of the simian immunodeficiency virus long terminal repeat. J Virol 66: 5216–5223
Wu FK, Garcia JA, Harrich D, Gaynor RA (1988) Purification of the human immunodeficiency virus type 1 enhancer and TAR binding proteins EBP-1 and UBP-1. EMBO J 7: 2117–2129
Wu FK, Garcia J, Sigman D, Gaynor R (1991) tat regulates the binding of the human immunodeficiency virus trans-activating region RNA loop-binding protein TRP-185. Genes Dev 5: 2128–2140
Yu X-F, Matsuda M, Essex M, Lee T-H (1990) Open reading frame vpr of simian immunodeficiency virus encodes a virion-associated protein. J Virol 64: 5688–5693
Zack J, Cann A, Lugo J, Chen ISY (1988) HIV-1 production from infected peripheral blood T cells after HTLV-I induced mitogenic stimulation. Science 240: 1026–1029
Zack J, Arrigo S, Weitsman S, Go A, Haislip A, Chen IS (1990) HIV-1 entry into quiescent primary lymphocytes: molecular analysis reveals a labile latent viral structure. Cell 61: 213–222
Zapp M, Green MR (1989) Sequence-specific RNA binding by the HIV-1 Rev protein. Nature 342: 714–716
Zapp M, Hope T, Parslow T, Green MR (1991) Oligomerization and RNA binding domains of the type 1 human immunodeficiency virus Rev protein: a dual function for an arginine-rich binding motif. Proc Natl Acad Sci USA 88: 7734–7738
Zimmermann K, Dobrovnik M, Ballaun C, Bevec D, Hauber J, Bohnlein E (1991) trans-Activation of the HIV-1 LTR by the HIV-1 Tat and HTLV-I Tax proteins is mediated by different cis-acting sequences. Virology 182: 874–878
Zuker M (1989) On finding all suboptimal foldings of an RNA molecule. Science 244: 48–52

MHC of Nonhuman Primates

D.I. WATKINS

1 Introduction

1.1 The Role of MHC Molecules in the Response to HIV

Our understanding of AIDS has been expanded by studies in nonhuman primates. Several vaccine studies in macaques have suggested that a strong immune response to SIV can be generated in appropriately immunized monkeys (MURPHY-CORB et al. 1989; DESROSIERS et al. 1989). Furthermore, these studies provided the first demonstration that it may be possible to induce a protective immune response against HIV. The cell-mediated immune response to SIV may prove crucial for this vaccine protection. However, to

---

Wisconsin Regional Primate Research Center and the Department of Pathology, University of Wisconsin, Madison, WI 53715-1299, USA

Current Topics in Microbiology and Immunology, Vol. 188

understand the nature of the cellular response to SIV in both immunized and infected monkeys, knowledge of the MHC products involved in the recognition of the virus is crucial. Indeed, in outbred populations, definition of the cellular immune response to a pathogen cannot be carried out without the knowledge of the MHC products involved.

The clinical course following HIV infection in humans usually is characterized by a prolonged period of time between virus infection and the onset of symptomatic disease. This apparent clinical latency can vary considerably from individual to individual. MHC haplotype has been shown in numerous settings to affect profoundly the nature of an immune response generated to a pathogen. Survival following an AIDS virus infection may be related to the host's ability to mount a strong immune response to the virus. It is possible that humans with certain MHC haplotypes may be more or less successful in generating an immune response to HIV. It has recently been demonstrated that several MHC haplotypes (including the *HLA-A1-Cw7-B8-DR3* haplotype) correlate with a rapid decline in $CD4^+$ cell count and increased likelihood of early deterioration after AIDS virus infection (KASLOW et al. 1990; MALLAL et al. 1990; CAMERON et al. 1990; DONALD et al. 1992). Thus, the products of the central or MHC linked genes may play a significant role in defense against HIV.

The use of nonhuman primates in AIDS research has increased the need for an accurate description of the various MHC alleles in several species of non-human primates. Homologues of *HLA-A, -B, -C, -E, -F, DRA, DRB, DQA, DQB*, and *DPB* are found in most species of great apes and Old World primates studied. Unfortunately, however, the development of methods for detecting specific MHC alleles in nonhuman primates lags far behind that in humans. This presents a significant problem for those investigators wishing to use nonhuman primate species as a model for AIDS.

1.2 MHC Class I Molecules

The products of the classical human MHC class I loci (*HLA-A, -B, -C*) are highly polymorphic molecules which can bind peptides in a groove (the antigen recognition site) in the top of the molecule (BJORKMAN et al. 1987a,b; GARRETT et al. 1989). The T cell receptor of the $CD8^+$ T cell then binds to the peptide/MHC complex, triggering destruction of the target cell. β_2-Microglobulin deficient mice that do not express MHC class I molecules show delayed clearance of viruses and are susceptible to a variety of other pathogens (ZIJLSTRA et al. 1990; BENDER et al. 1992; HOU et al. 1992; MULLER et al. 1992; TARELTON et al. 1992;FLYNN et al. 1992). There appears to have been selective pressures for diversity at the amino acid residues that line the antigen recognition site of MHC class I molecules (HUGHES and NEI 1988). This diversity is thought to increase the number of possible different viral or tumor antigens which can be bound and presented to cytotoxic T lymphocytes (CTLs). Thus, these polymorphic classical MHC class I molecules are critical to the surveillance function of an intact immune system.

In contrast to the functional, highly polymorphic MHC class I molecules, the nonclassical human MHC class I gene products (*HLA-E, -F, -G*) are nonpolymorphic, and no function has yet been ascribed to them. Nonclassical MHC class I molecules, however, have been implicated as possible ligand for γ/δ expressing T cells (STROMINGER 1989; PORCELLI et al. 1989; ITO et al. 1990).

1.3 MHC Class II Molecules

Like MHC class I molecules, MHC class II molecules bind peptides and present them to T cells. Despite several similarities between MHC class I and class II molecules, there are also some important differences. MHC class II molecules present peptides to helper $CD4^+$ T lymphocytes whereas MHC class I molecules present peptides to $CD8^+$ CTLs. Unlike MHC class I molecules which are expressed on nearly every cell type, MHC class II molecules are expressed largely by lymphocytes and macrophages. Cellular processing of peptide fragments is also different for MHC class II molecules. Peptides bound by MHC class II molecules are derived largely from the extracellular medium and are processed through acidic endocytic vesicles (NEEFJES and PLOEGH 1992), whereas peptides bound to MHC class I molecules are endogenously derived.

1.4 Central or MHC-Linked Molecules

The MHC haplotypes of an individual clearly play a role in that individual's response to vaccines, infectious disease, and malignancies. Susceptibility to nasopharyngeal carcinoma in Chinese (LU et al. 1990), insulin-dependent diabetes (TODD et al. 1987), and Lyme disease associated arthritis (STEERE et al. 1989) appears to be MHC linked. In addition to the products of the MHC class I and class II loci, there are other MHC-linked gene loci which can also play a critical role in containing an infection. Indeed, correlations between susceptibility to disease and particular MHC class I or class II alleles have often proven incomplete except in a few well-defined cases (*HLA-B27* and ankylosing spondylitis, for example; SCHLOSSTEIN et al. 1973; HAMMER et al. 1990). It has recently been suggested that many of the disease associations involving single MHC alleles may actually be due to the fact that the single MHC allele marks a particular extended MHC haplotype (ALPER 1991). The MHC region contains several loci important to the immune response. In addition to the class I and class II genes, tumor necrosis factor, proteasome (GOLDBERG and ROCK 1992), peptide transporter (DEMARS and SPEIS 1992), and the complement genes are located in the MHC. Thus an entire MHC haplotype (*B8, SC01, DR3*) may mark an individual who is unable to mount an immune response to the hepatitis B vaccine. Individuals homozygous for this haplotype appear to be incapable of responding to this vaccine (ALPER et al. 1989). It has been hypothesized that the inability to mount an effective immune response to the hepatitis B vaccine is recessive and may be due to an as yet undiscovered variant gene present in the *B8, SC01, DR3* haplotype.

2 MHC Class I Molecules and Genes in Nonhuman Primates

In humans more than 100 alleles of the *HLA-A, -B*, and *-C* loci have been cloned and sequenced (ZEMMOUR and PARHAM 1993). In nonhuman primates, however, only 48 sequences from six species have thus far been reported (WATKINS et al. 1992). From these sequences it has been shown that the homologues of the *HLA-A, -B, -E*, and *-F* loci are well preserved in Old World primates whereas New World primates appear to be using MHC class I loci derived from the ancestral homologue of *HLA-G* (WATKINS et al. 1990).

2.1 MHC Class I Molecules and Genes of Chimpanzees

The pioneering studies of Balner at the TNO Primate Center in Holland laid the groundwork for the description of MHC class I alleles in chimpanzees. Using classical serological methods, he was able to generate alloantisera that described 12 alleles at the *A* locus and 20 alleles at the *B* locus (BALNER et al. 1978). Dorf and Metzgar confirmed these studies utilizing the cross-reactivity of human alloantisera with chimpanzee lymphocytes (DORF and METZGAR 1970). Rather curiously, it was shown that alloantisera against the human cross-reactive group *HLA-A1, -A3, -A11* showed a high frequency of cross-reactivity with chimpanzee lymphocytes.

Molecular analysis of chimpanzee class I alleles later confirmed this *HLA-A1, -A3, -A11* cross-reactivity. *HLA-A* genes are comprised of five distinct allele families: the *A1/A3/A11, A9, A10, A19*, and *A2/28* families (KATO et al. 1989; LAWLOR et al. 1990). The six MHC class I cDNAs of the chimpanzee which have been sequenced to date are all related to the *HLA-A1, -A3, -A11* family (LAWLOR et al. 1988, 1990; MAYER et al. 1988; CHEN et al. 1993). Indeed, to date all chimpanzee and pygmy chimpanzee *HLA-A* homologues are related to *HLA-A1, -A3, -A11* (data not shown). This suggests that chimpanzees and pygmy chimpanzees have been through a genetic bottleneck since homologues of only one of the five families of human *HLA-A* alleles is present.

It is difficult to draw any conclusions as to the nature of the *HLA-B* homologues in chimpanzees since only four sequences exist. However, two of these sequences show strong similarity to *HLA-B57*, *-B58* (*B17*) in the α_1-domain. The other two sequences are similar to the newly described *HLA-B48* allele, found almost exclusively in Oriental and New World populations (BELICH et al. 1992; WATKINS et al. 1992). Comparison of the α_2-domain of the chimpanzee *HLA-B* homologues to their human counterparts, however, reveals no particular sequence similarity to either *HLA-B57*, *-B58*, or *-B48*. Whether this simply reflects the fact that these chimpanzee alleles represent homologues of as yet unsequenced *HLA-B* alleles, or whether it means that they are derived from ancestral homologues of the *HLA-B57*, *B58*, *B48* lineages is uncertain. The analysis of more chimpanzee *HLA-B* homologues should provide the answer to this question.

An *HLA-C* and an *HLA-F* homologue were also cloned from the chimpanzee (LAWLOR et al. 1988, 1990). These alleles showed high levels of sequence similarity to their human counterparts, with the *Patr-F* allele only 14 amino acids different from *HLA-F*. We have also cloned chimpanzee genomic sequences from *HLA-E*, *-G*, and *-H* homologues, suggesting that nearly all of the human MHC class I loci are present in the chimpanzee (data not shown). Thus, human and chimpanzees have very similar MHC class I loci.

2.2 MHC Class I Genes of Gorillas

Homologues of *HLA-A*, *-B*, and *-C* have been described in the lowland gorilla (*Gorilla gorilla*; WATKINS et al. 1991; LAWLOR et al. 1991). Like the chimpanzee, none of the sequences of these gorilla alleles are the same as any human class I allele. The *Gogo-A* alleles appear to be more polymorphic than their *-B* locus counterparts. Additionally, these *Gogo-A* alleles show striking similarities to *HLA-A2*, and gene tree construction confirmed that the gorilla *HLA-A* homologues are related to the *HLA-A2* family of human *HLA-A* alleles. Unlike the chimpanzee, therefore, the *Gogo-A* alleles clustered with the other main group of *HLA-A* alleles. Interestingly, an unusual *HLA-A* homologue has been described in the gorilla. This allele appears to be a recombinant between the ancestral homologues of *HLA-A* and *HLA-H* creating an entirely new allele (WATKINS et al. 1991; LAWLOR et al. 1991). Whether this allele is a product of the *Gogo-A* locus, or whether it represents a new gorilla locus is uncertain. Only two *Gogo-B* allelic lineages were present in the seven animals analyzed (LAWLOR et al. 1991). We have now examined an additional two unrelated animals and have still described only these two *Gogo-B* allelic lineages. Given that one of these alleles (*Gogo-B*01*) shows some sequence homology to *HLA-B27*, it is interesting to note that gorillas have a high incidence of reactive arthritis and two animals have been described with clinical signs of ankylosing spondylitis. The five

HLA-C homologues in the gorilla fall into two distinct groups. The alleles of one of these groups is similar to *HLA-Cw*07* whereas the other does not group with any of the human *HLA-C* alleles.

2.3 MHC Class I Genes of Orangutans and Gibbons

Nine homologues of the human *HLA-A*, *-B*, and *-E* genes have been cloned and sequenced from orangutans and gibbons (CHEN et al. 1992; LAWLOR et al. 1990). This demonstrates that the *HLA-A*, *-B*, and *-E* loci must be at least 30 million years old, the last time humans and gibbons had a common ancestor. Surprisingly, we could find no evidence for *HLA-C* related MHC class I cDNAs in these two species, and the orangutan appears to have duplicated its *HLA-B* related locus. Finally, since the *HLA-A* genes of gibbons and orangutans are the product of a recombinant between the ancestors of the *HLA-A* and *HLA-E*, this recombination event (HUGHES and NEI 1989) must have occurred before the ancestors of the gibbon and human diverged.

2.4 MHC Class I Molecules of Macaques

MHC class I gene products play a crucial role in the immune response to pathogens. Thus, an analysis of the T cell immune response to the AIDS virus in macaques must be preceded by a definition of the MHC class I molecules in these species. Unfortunately, there are very few useful serological reagents available for such an analysis in nonhuman primates. Alloantisera have been generated for MHC characterization in cynomolgus monkeys and pigtailed macaques (HEISE et al. 1987; KEEVER and HEISE 1985a, b; GAUR et al. 1989). However, these sera do not define all of the MHC class I molecules of these species. Moreover, these reagents are not widely available for all the investigators who must work with these animals. We have therefore employed another technique of MHC class I typing, a single technique that can be used to type any primate species. Lymphocytes are biosynthetically labeled, and MHC class I molecules are precipitated with an antibody that recognizes phylogenetically conserved determinants on primate MHC class I molecules. These precipitated molecules are then separated on the basis of their charge using an isoelectric focusing gel. Thus, each focused band represents a specific MHC class I allele. We have applied one-dimensional isoelectric focusing (1-D IEF) in the analysis of MHC class I molecules of the macaque. Lymphocytes from each monkey express at least three MHC class I molecules, and considerable polymorphism is present at the macaque MHC class I loci.

1-D IEF has proven useful in defining the MHC restriction of virus-specific CTL in both HTLV-I and SIV infected rhesus monkeys (WATKINS et al. 1988; MILLER et al. 1990, 1991; YAMAMOTO et al. 1990a,b). The definition of the CTL response to the AIDS virus in the rhesus monkey underlines the value of characterizing MHC class I genes and gene products of the macaque to facilitate studying SIV immune responses. SIV_{mac} infected rhesus monkeys generate a $CD8^+$ cell response that suppresses virus replication in vitro. We were able to demonstrate that these SIV_{mac} infected monkeys generate an MHC-restricted cytotoxic response to the gag protein. Indeed, monkeys which share an MHC class I gene (*Mamu-A1*) develop CTLs which recognize only a single fragment of the *gag* protein (peptide 11). This gag epitope is presented to CTL by that shared MHC class I molecule. Thus, transfection of a null cell line with the cDNA encoding the Mamu-A1 allel and pulsing of these targets with peptide 11 sensitizes that cell line for lysis by both clones and bulk populations of SIV_{mac} specific CTL.

2.5 Characterization of MHC Class I cDNAs in Macaques

Using the polymerase chain reaction we have now cloned and sequenced more than ten full-length MHC class I cDNAs in the rhesus and cynomolgus monkeys. Sequence analysis and comparison to the products of the human MHC class I loci have revealed that these macaques express the products of one *HLA-A*-like locus and at least *HLA-B*-like locus. Additionally, we and others have cloned homologues of *HLA-E* and *HLA-F* from macaque peripheral blood lymphocytes (R. Bontrop, personal communication). This suggests that the MHC class I loci of the macaque are very similar to those of humans.

2.6 An HLA-B27/14-like MHC Class I Molecule in Macaques

We have also cloned and sequenced more than 20 partial-length MHC class I cDNAs from three rhesus monkeys with post-*Shigella* reactive arthritis and from one unaffected animal. All three of the affected animals expressed MHC class I molecules that are similar to *HLA-B27/14*. Interestingly, humans with *HLA-B27* are susceptible to reactive arthritis after infection with enteric pathogens (Keat 1983). Glutamic acid at position 45 and cysteine at 67 were present in the majority of these rhesus monkey MHC class I molecules. Several other substitutions important in the conservation of the *HLA-B27*-like B pocket were also present, suggesting that these macaque molecules, like HLA-B27, could perhaps bind to peptides containing a bulky arginine side chain at position 2. Elution and sequencing of radiolabeled peptides bound by the rhesus HLA-B27/14 molecule confirmed that arginine can be bound at the second position.

2.7 Evolution of MHC Class I Genes in Nonhuman Primates

The human *HLA-A* and *-B* loci are more than 30 million years old. That they have been maintained as separate gene loci during evolutionary history suggests that recombinational mechanisms between loci are not a major driving force in the evolution of primate MHC class I molecules. Furthermore, the *HLA-G* and *HLA-F* gene loci have been well preserved in New World primates (WATKINS et al. 1990). Thus, while interlocus recombination events do occur at a low frequency in both humans and great apes, such segmental exchanges must be either very rare or selected against. This stands in contrast to the situation in rodents where gene conversion may play an important role in the evolution of the murine MHC class I loci (RADA et al. 1990). However, there are other explanations for the apparently high rate of gene conversion in mouse and rat MHC class I genes (HUGHES 1991).

3 MHC Class II Molecules and Genes in Nonhuman Primates

MHC class II genes and allelic lineages are remarkably well preserved during evolution of the great apes and Old World primates (KASAHARA et al. 1990). Although there is evidence that diversity at the *DRB* loci in primates has been generated by recombination, *HLA-DRB* and *Patr-DRB* alleles of the same lineage cluster together in phylogenetic trees (KENTER et al. 1992a; GYLLENSTEN et al. 1991). Indeed, Patr-DRβ 4*0104 and HLA-DRβ 4*0104 differ at only three residues in the α_1-domain and the α_2-domains of *DRB* alleles are conserved between chimpanzees and humans (FAN et al. 1989). This sharing of trans-species *DRB* sequence lineages observed even between humans and rhesus macaques (SLIERENDREGT et al. 1992). Furthermore, a rhesus monkey Mamu-DRβ 1*03 allele can present a peptide from *Mycobacterum tuberculosis* to a human T cell clone restricted by HLA-DR-17 (GELUCK et al. 1993). These data therefore imply that the homologues of *HLA-DRA* (FAN et al. 1989), *-DRB*, (KENTER et al. 1992a; GYLLENSTEN et al. 1990), *-DQA* (KENTER et al. 1992b), and *-DQB* (GYLLENSTEN et al. 1990; OTTING et al. 1992) alleles in great apes are stable and evolve in a trans-specific fashion (KLEIN 1987).

3.1 MHC Class II Genes of Chimpanzees

MHC class II *DRB* allelic lineages and loci are very similar between humans and chimpanzees (KENTER et al. 1992b; BRAENDLE et al. 1992). More than 50 chimpanzee *DRB* genes have now been cloned and sequenced. Comparison to human *DRB* genes reveals that the *HLA-DRB1*02, -DRB, 1*03, -DRB 1*07* allelic lineages are present in chimpanzees whereas the chimpanzee homologues of *HLA-DRB1*01, -DRB1*04*, and *-DRB1*09* allelic lineages were not found in chimpanzees. Additionally, there are two chimpanzee allelic lineages that are not present in humans. However, chimpanzee homologues of the *HLA-DRB3, -DRB4, -DRB5, -DRB6*, and *-DRB7* loci were found. Interestingly, the sequence similarities between chimpanzees and humans allows even the presentation of peptides by chimpanzee MHC class II molecules to human T cell proliferative clones (BONTROP et al. 1990). The most polymorphic of the chimpanzee MHC class II loci, the *HLA-DRB1* locus, exhibited considerably less polymorphism than seen in humans. This suggests that perhaps chimpanzees have been through a genetic bottleneck.

DQA and *DQB* genes are also very similar between chimpanzees and humans (OTTING et al. 1992; KENTER et al. 1992b; GAUR et al. 1992; GYLLENSTEN et al. 1990). Like the condensation seen in *DRB1* loci, only five *DQB1* alleles have been described in the chimpanzee. These alleles are derived from only two ancient lineages. Again, therefore, evidence for a founder effect is present in the MHC class II genes of the chimpanzee. Taken together with the finding that all of the *Patr-A* alleles in the chimpanzee are related to only one of the five human *HLA-A* allelic lineages, this suggests that chimpanzees have been through a genetic bottleneck.

3.2 MHC Class II Genes of Gorillas

Cloning and sequencing of exon 2 of 25 MHC class II *DRB* genes from six gorillas shows that all of the human *DRB* lineages are found in nonhuman primates (KUPFERMANN et al. 1992). Four different gorilla *DRB* loci were defined and 13 of the alleles at these loci could be assigned to previously cloned human and great ape lineages. Two new allelic lineages were also found. These studies reiterate the finding of conservation among the MHC class II genes in nonhuman primates and humans.

3.3 MHC Class II Molecules and Genes of Macaques

Definition of rhesus monkey MHC class II genes and gene products will be essential to the delineation of $CD4^+$ T lymphocyte mediated responses to HIV. Macaques mount strong proliferative responses to the env and gag glycoproteins after immunization with either inactivated SIV or recombinant SIV env proteins (MILLS et al. 1991; JONES et al. 1992). Additionally, macaques recognize the HIV core protein p24 in an MHC-restricted fashion after immunization with hybrid HIV p24 Ty virus-like particles (MILLS et al. 1990). Further definition of the MHC class II genes and gene products will therefore be crucial to the planning of additional vaccine strategies in the macaque.

Fortunately, macaque MHC class II genes are very similar to their human and chimpanzee counterparts. Early serological, biochemical, and restriction fragment length polymorphism analysis demonstrated the presence of at least ten rhesus monkey MHC class II haplotypes (GIPHART et al. 1983; OTTING et al. 1988; BONTROP et al. 1989; SCHEPART et al. 1990; MANNING and HEISE 1991; SLIERENDREGT et al. 1991). Sequence analysis of rhesus monkey *DRB* genes revealed that the *DRB1*03*, -*DRB1*04*, and -*DRB1*10* allelic lineages have been preserved in the rhesus monkey (SLIERENDREGT et al. 1992). Surprisingly, rhesus monkey cell lines expressing Mamu-DRβ1*03 alleles can present the p3-13 peptide of the 65-kDa heat-shock protein of *M. tuberculosis* to a human T cell clone restricted by HLA-DR17 (GELUCK et al. 1993). Additionally, homologues of the human *HLA-DRB3*, -*DRB4*, -*DRB5*, and -*DRB6* and polymorphic *HLA-DQA* and *DQB* loci are also found in the rhesus monkey (GYLLENSTEN et al. 1990; KENTER et al. 1992b; OTTING et al. 1992). Cloning and sequencing of five *DRB* cDNAs from another species of macaque, the pigtailed macaque (*Macaca nemestrina*) showed that *DRB1* alleles of the *DRw52* group are probably at least 35 million years old (ZHU et al. 1991).

4 Central or MHC-Linked Genes in Nonhuman Primates

With the exception of the complement genes C3 and C4 and the cytochrome P-450 21-hydroxylase gene (CYP21), little is known about the central or MHC-linked genes of nonhuman primates (ALPER et al. 1972). A tandem duplication of the C4-CYP21 genes is found in the chimpanzee (GRANADOS et al. 1987; KAWAGUCHI et al. 1990; CHRISTIANSEN et al. 1992). Although several different C4 and CYP21 restriction fragment length polymorphisms are seen in the human, orangutan, and rhesus monkey, only one such pattern was observed in chimpanzees and gorillas (BONTROP et al. 1991). Indeed, analysis with *Taq* I, *Sst* I, and *Hind* III failed to detect any differences among chimpanzees at the C4 and CYP21 genes. Again this would support the notion that the chimpanzee has been through a genetic bottleneck soon after its ancestors split from the ancestral lineages leading to humans.

5 Conclusions

MHC genes are well conserved during the evolution of great ape and Old World primates. This conservation is especially evident at the MHC class II loci with chimpanzee and rhesus monkey lymphocytes able to present peptides to human T cell clones. MHC class I loci in nonhuman primates are not as well conserved as their human counterparts. Homologues of the *HLA-C* locus are found only in gorillas and chimpanzees whereas in orangutans and rhesus monkeys the *HLA-B* homologue may have been duplicated. Additionally, MHC class I alleles appear to be undergoing more rapid changes than MHC class II alleles, making them less similar to their counterparts in humans. Despite this divergence, however, certain rhesus monkey MHC class I alleles can bind SIV-derived peptides which are very similar to HIV-derived peptides bound by human MHC class I alleles. Moreover, these rhesus monkey and human MHC class I alleles can show similarities in their α_1-domain. This suggests, therefore, that the peptide binding abilities of both the MHC class I and MHC class II molecules have been conserved between humans and nonhuman primates. These similarities between the MHC class I and class II molecules of human and rhesus monkeys makes SIV infection of macaques an ideal animal model for AIDS.

Acknowledgments. I would like to thank Lori Geurts for preparation of this manuscript. This work was supported by RR00167 and A132426 from the NIH.

References

Alper CA (1991) Major histocompatibility complex allele-disease association in the light of extended haplotypes. In: Farid NR, (ed) Immunogenetics of autoimmune disease, vol 1. CRC, Boca Raton

Alper CA, Robin NI, Refetoff S (1972) Genetic polymorphism in rhesus C3 and Gc globulin. J Immunol 107: 96–98

Alper CA, Kruskall MS, Marcus-Bagley D, Craven DE, Katx AJ, Brink SJ, Dienstag JL, Awdeh Z, Yunis EJ (1989) Genetic prediction of nonresponse to hepatitus B vaccine. N Engl J Med 321: 708–712

Balner H, van Vreeswijk W, Roger JH, D'Amaro J (1978) The major histocompatibility complex of chimpanzees: identification of several new antigens controlled by the A and B loci of ChLA. Tissue Antigens 12: 1–18

Belich MP, Madrigal JA, Hildebrand WH, Zemmour J, Williams RC, Luz R, Petzl-Erler ML, Parham P (1992) Unusual *HLA-B* alleles in two tribes of Brazilian Indians. Nature 357: 326–329

Bender BS, Croghan T, Zhang L, Small PA Jr (1992) Transgenic mice lacking class I major histocompatibility complex-restricted T cells have delayed viral clearance and increased mortality after influenza virus challenge. J Exp Med 175: 1143–1145

Bjorkman PJ, Saper MA, Samraoui B, Bennet WS, Strominger JL, Wiley DC (1987a) Structure of the human class I histocompatibility antigen, HLA-A2. Nature 329: 506–512

Bjorkman PJ, Saper MA, Samraoui B, Bennet WS, Strominger JL, Wiley DC (1987b) The foreign antigen binding site and T cell recognition regions of class I histocompatibility antigens. Nature 329: 512–518

Bontrop RE, Otting N, Broos LAM, Noort MC, Kenter M, Jonker M (1989) RELP analysis of the HLA-, ChLA-, and RhLA-DQ alpha chain gene regions: conservation of restriction sites during evolution. Immunogenetics 30(6): 432–439

Bontrop RE, Elferink DG, Otting N, Jonker M, de Vries RRP (1990) Major histocompatibility complex class II restricted antigen presentation across a species barrier: conservation of restriction determinants in evolution. J Exp Med 172: 53

Bontrop RE, LAM Broos, Otting N, Jonker MJ (1991) Polymorphism of C4 and CYP21 genes in various primate species. Tissue Antigens 37(4): 145–151

Braendle U, Ono H, Vincek V, Klein D, Golubic M, Grahovac B, Klein J (1992) Trans-species evolution of Mhc-DRB haplotype polymorphism in primates: organization of DRB genes in the chimpanzee. Immunogenetics 36(1): 39–48

Cameron PU, Mallal SA, French MAH, Dawkins RL (1990) Major histocompatibility complex genes influence the outcome of HIV infection ancestral haplotypes with C4 null alleles explain diverse HLA associations. Human immunol 29: 282–295

Chen ZW, McAdam SN, Hughes AL, Dogon AL, Letvin NL, Watkins DI (1992) Molecular cloning of orangutan and gibbon MHC class I cDNA: the HLA-A and -B loci diverged over 30 million years ago. J Immunol 148(8): 2547–2554

Chen ZW, Hughes AL, Ghim SH, Letvin NL, Watkins DI (1993) Two more chimpanzee Patr-A locus alleles related to the HLA-A1/A3/A11 family. Immunogenetics (in press)

Christiansen FT, Bontrop RE, Giphart M, Cameron PU, Zhang WJ, Townend D, Jonker M, Dawkins RL (1991) Major histocompatibility complex ancestral haplotypes in the chimpanzee: identification using C4 allotyping. Hum Immunol 31(1): 34–39

DeMars R, Spies T (1992) New genes in the MHC that encode proteins for antigen processing. Trends Cell Biol 2: 81–86

Desrosiers RC, Wyand MS, Kodama T, Ringler DJ, Arthur LO, Sehgal PK, Letvin NL, Daniel MD (1989) Vaccine protection against simian immunodeficiency virus infection. Proc Natl Acad Sci USA 86: 6353–6357

Donald JA, Rudman K, Cooper DW, Baumgart KW, Garsia RJ, Gatenby PA, Rickard KA (1992) Progression of HIV-related disease is associated with HLA DQ and DR alleles defined by restriction fragment length polymorphisms. Tissue Antigens 39: 241–248

Dorf ME, Metzgar RS (1970) The distribution of human HL-A antigens in chimpanzees and gorillas In: Goldsmith EI, Moor-Janhowsk J (eds) Medical Primatology. Proceedings of the second conference on experimental medicine and surgery in primates. Karger, Farmington CT

Fan W, Kasahara M, Gutknecht J, Klein D, Mayer WE, Jonker M, Klein J (1989) Shared class II MHC polymorphisms between humans and chimpanzees. Hum Immunol 26(2): 107–121

Flynn JL, Goldstein MM, Triebold KJ, Koller B, Bloon BR (1992) Major histocompatibility complex class I-restricted T cells are required for resistance to *Mycobacterium tuberculosis* infection. Proc Natl Acad Sci USA 89: 12013–12017

Garrett TPJ, Saper MA, Bjorkman PJ, Strominger JL, Wiley DC (1989) Specificity pockets for the side chains of peptide antigens in HLA-Aw68. Nature 342: 692–696

Gaur LK, Bowden DM, Tsai C-C, Davis A, Clark EA (1989) The major histocompatibility complex, MnLA, of pigtailed macaques: definition of fifteen specificities. Hum Immunol 24(4): 277–294

Gaur LK, Heise ER, Thurtle PS, Nepom GT (1992) Conservation of the HLA-DQB2 locus in nonhuman primates. J Immunol 148(3): 943–948

Geluck A, Elferink DG, Slierendregt BL, van Meijgaarden KE, de Vries RRP, Ottenhoff THM, Bontrop RE (1993) Evolutionary conservation of *Mhc*-DR/peptide/T cell interactions in primates. J Exp Med (in press)

Giphart MJ et al. (1983) The major histocompatibility complex of rhesus monkeys, RhLA. XV. Chemical characterizations of DR locus antigens and antigen 48. Immunogenetics 10: 45–53

Goldberg AL, Rock KL (1992) Proteolvsis, proteasomes and antigen presentation. Nature 357: 375–379

Granados J, Awdeh ZL, Chen J-H, Giles CM, Balner H, Yunis EJ, Alper CA (1987) There are two C4 genetic loci and a null allele in the chimpanzee. Immunogenetics 26(6): 344–350

Gyllensten UB, Lashkari D, Erlich HA (1990) Allelic diversification at the class II DQB locus of the mammalian major histocompatibility complex. Proc Natl Acad Sci USA 87(5): 1835–1839

Gyllensten UB, Sundvall M, Erlich HA (1991) Allelic diversity is generated by intraexon sequence exchange at the DRB1 locus of primates. Proc Natl Acad Sci USA 88(9): 3686–3690

Hammer RE, Maika SD, Richardson JA, Tang JP, Taurog JD (1990) Spontaneous inflammatory disease in transgenic rats expressing HLA-B27 and human β-2 microglobulin: an animal model of HLA-B27-associated human disorders. Cell 65: 1099–1112

Heise ER, Cook DJ, Schepart BS, Manning CH, McMahan MR, Chedid M, Keever CA (1987) The major histocompatibility complex of primates. Genetica 73(1-2): 53–68

Hou S, Doherty PC, Zijlstra M, Jaenisch R, Katz JM (1992) Delayed clearance of sendai virus in mice lacking class I MHC-restricted CD8$^+$T cells. J Immunol 149: 1319–1325
Hughes AL (1991) Independent gene duplications, not concerted evolution, explain relationships among class I MHC genes of murine rodents. Immunogenetics 33: 367–373
Hughes AL, Nei M (1988) Pattern of nucleotide substitution at major histocompatibility complex class I loci reveals overdominant selection. Nature 335: 167–170
Hughes AL, Nei M (1989) Ancient interlocus exon exchange in the history of the HLA-A locus. Genetics 122: 681
Ito K, Kaer LV, Bonneville M, Hsu S, Murphy DB, Tonegawa S (1990) Recognition of the product of novel MHC TL region gene (27^b) by a mouse γ/δ T cell receptor. Cell 62: 549–561
Jones WC, Barnard AL, Slade A, Mills KHG (1992) Heterogeneity in the recognition of the simian immunodeficiency virus envelope glycoprotein by CD4$^+$T cell clones from immunized macaques. J Immunol 149(9): 3120–3126
Kasahara M, Klein D, Fan W, Gutknecht J (1990) Evolution of the class II major histocompatibility complex alleles in higher primates. immunol Rev (113): 65–82
Kaslow RA, Dugesnoy R, VanRaden M, Kingsley L, Marrari M, Friedman H, Su S, Saah AJ, Detels R, Phair J, Rinaldo C (1990) A1, C277, B8, DR3 HLA antigen combination associated with rapid decline of T-helper lymphocytes in HIV-1 infection. Lancet 335: 927–930
Kato K, Trapani JA, Allopenna J, Dupont B, Yang SY (1989) Molecular analysis of the serologically defined HLA-Aw19 antigens. J Immunol 143: 3371–3378
Kawaguchi H, Zaleska-Rutczynska Z, Figueroa F, O'hUigin C, Klein J (1992) C4 genes of the chimpanzee, gorilla, and orang-utan: evidence for extensive homogenization. Immunogenetics 35(1): 16–23
Keat A (1983) Reiter's syndrome and reactive arthritis in perspective. N Engl J Med 26: 1606–1615
Keever CA, Heise ER (1985a) The major histocompatibility complex of the cynomolgus monkey: absorption analysis of 24 CyLA antisera. Hum Immunol 12(1): 75–90 (RF 005064)
Keever CA, Heise ER (1985b) The major histocompatibility complex of the cynomolgus monkey. II. Polymorphism at three serologically defined loci and correlation of haplotypes with stimulation in MLC and skin graft survival. Hum Immunol 12(3): 143–164
Kenter M, Otting N, Anholts J, Jonker M, Schipper R, Bontrop RE (1992a) Mhc-DRB diversity of the chimpanzee (Pan troglodytes). Immunogenetics 37(1): 1–11
Kenter M, Otting N, Anholts J, Leunissen J, Jonker M, Bontrop RE (1992b) Evolutionary relationships among the primate Mhc-DQA1 and DQA2 alleles. Immunogenetics 36(2): 71–78
Klein J (1987) Origin of major histocompatibility complex polymorphisms: the transsspecies hypothesis. Hum Immunol 19: 155
Kupfermann H, Mayer WE, O'hUigin C, Klein D, Klein J (1992) Shared polymorphism between gorilla and human major histocompatibility complex DRB loci. Hum Immunol 34(4): 267–278
Lawlor DA, Ward FE, Ennis PD, Jackson AP, Parham P (1988) HLA-A and B polymorphism predate the divergence of humans and chimpanzees. Nature 335: 268–271
Lawlor DA, Warren E, Ward FE, Parham P (1990) Comparison of calss I MHC alleles in humans and apes. Immunol Rev 113: 147–185
Lawlor DA, Warren E, Taylor P, Parham P (1991) Gorilla class I major histocompatibillity complex alleles: comparison to human and chimpanzee class I. J Exp Med 174: 1491–1509
Lu SJ, Day NE, Degros L, Lepage V, Wang PC, Chan SH, Simons M, McKnight B, Easton D, Zeng Y, De-the G (1990) Llinkage of a nasopharyngeal carcinoma susceptibility locus to the HLA region. Nature 346: 470–471
Mallal S, Cameron PU, French MA, Dawkins RL (1990) MHC genes and HIV infection (letter: comment). Lancet 335: 1591–1592
Manning CH, Heise ER (1991) Biochemical analysis of class I and class II MHC antigens in cynomolgus macaques by one-dimensional isoelectric focusing. Tissue Antigens 37(2): 56–65
Mayer WE, Jonker M, Klein D, Ivanyi P, van Seventer G, Klein J (1988) Nucleotide sequences of chimpanzee MHC class I alleles: evidence for a trans-species mode of evolution. EMBO J 7: 2765–2774
Miller MD, Lord CI, Stallard V, Mazzara GP, Letvin NL (1990) Gag-specific cytotoxic T

lymphocytes in rhesus monkeys infected with simian immunodeficiency virus of macaques. J Immunol 144: 122–128
Miller MD, Yamamoto H, Hughes AL, Watkins DI, Letvin NL (1991) Definition of an epitope and MHC class I molecule recognized by gag-specific cytotoxic T lymphocytes in SIVmac-infected rhesus monkeys. J Immunol 147(1): 320–329
Mills KH, Kitchin PA, Mahon BP, Barnard AL, Adams SE, Kingsman SM, Kingsman AJ (1990) HIV p24-specific helper T cell clones from immunized primates recognize highly conserved regions of HIV-1. J Immunol 144(5): 1677–1683
Mills KH, Barnard AL, Williams M, Page M, Ling C, Stott EJ, Silvera P, Taffs F, Kingsman AS, Adams SE (1991) Vaccine-induced $CD4^+$ T cells against the simian immunodeficiency virus gag protein. J Immunol 147(10): 3560–3567
Muller D, Koller BH, Whitton JL, LaPan KE, Brigman KK, Frelinger JA (1992) LCMV-specific, class II-restricted cytotoxic T cells in β_2-microglobulin-deficient mice. Science 255: 1576–1578
Murphy-Corb M, Martin LN, Davison-Fairburn B (1989) A formalin- inactivated whole SIV vaccine confers protection in macaques. Science 246: 1293
Neefjes JJ, Ploegh HL (1992) Intracellular transport of MHC class II molecules. Immunol Today 13(5): 179–184
Otting N, Jonker M, Bontrop RE (1988) 2D-gel analysis of the MHC-class II molecules expressed by chimpanzee and rhesus monkey EBV-transformed B cell lines. Rep Annu Rep [Division For Health Research Tno]: 185–186
Otting N, Kenter M, van Weeren P, Jonker M, Bontrop RE (1992) Mhc-DQB repertoire variation in hominoid and Old World primate species. J Immunol 149(2): 461–470
Porcelli S, Brenner MB, Greenstein JL, Balk SP, Terhorst C, Bleicher PA (1989) Recognition of cluster of differentiation 1 antigens by human CD4-CD8 cytolytic T lymphocytes. Nature 341: 447–450
Rada C, Lorenzi R, Powis SJ, Van Den Bogaerde J, Parham P, Howard JC (1990) Concerted evolution of class I genes in the major histocompatibility complex of murine rodents. Proc Natl Acad Sci USA 37: 2167–2171
Schepart BS, Buck DA, Manning CH, Keever CA, McMahan MR, Heise ER (1990) Restriction fragment polymorphism of the cynomolgus monkey major histocompatibility complex. Exp Clin Immunogenet 7(2): 109–122
Schlosstein L, Terasaki PI, Bluestone R, Pearson CM (1973) High association of an HL-A antigen, W27, with ankylosing spondylitis. N Engl J Med 288: 704–706
Slierendregt BL, Otting N, Jonker M, Bontrop RE (1991) RFLP analysis of the rhesus monkey MHC class II DR subregion. Hum Immunol 30(1): 11–17
Slierendregt BL, van Noort JT, Bakas RM, Otting N, Jonker M, Bontrop RE (1992) Evolutionary stability of transspecies major histocompatibility complex class II DRB lineages in humans and rhesus monkeys. Hum Immunol 35(1): 29–39
Steere AC, Dwyder E, Winchester R (1989) Association of chronic lyme arthritis with HLA-DR4 and HLA DR2 alleles. N Engl J Med 323: 219–223
Strominger JL (1989) The $\gamma\delta$T cell receptor and class Ib MHC-related proteins: enigmatic molecules of immune recognition. Cell 57: 895–898
Tarelton RL, Koller BH, Latour A, Postan M (1992) Susceptibility of β_2-microglobulin-deficient mice to Trypanosoma cruzi infection. Nature 356: 338–340
Todd JA, Bell JI, McDevitt HO (1987) *HLA-DQ* β gene contibutes to susceptibility and resistance to insulin-dependent diabetes mellitus. Nature 329: 599–604
Watkins DI, Kannagi M, Stone ME, Letvin NL (1988) Major histocompatibility complex class I molecules of nonhuman primates. Eur J Immunol 18: 1425–1432
Watkins DI, Chen ZW, Hughes AL, Evans MG, Tedder TF, Letvin NL (1990) Evolution of the MHC class I gene of a New World primate from ancestral homologues of human non-classical genes. Nature 346: 60–63
Watkins DI, Chen ZW, Garber TL, Hughes AL, Letvin NL (1991) Segmental exchange between MHC class I genes in a higher primate: recombination in the gorilla between the ancestor of a human non-functional gene and an A locus gene. Immunogenetics 34: 185–191
Watkins DI, McAdam SN, Liu X, Strang CR, Milford EL, Levine CG, Garber TL, Dogon AL, Lord CI, Ghim SH, Troup GM, Hughes AL, Letvin NL (1992a) New recombinant *HLA-B* alleles in a tribe of South American Amerindians suggest that the MHC class I loci can evolve rapidly. Nature 357: 329–333

Watkins DI, Zemmour J, Parham P (1992b) Non-human primate MHC class I sequences, 1992. Immunogenetics 37: 317–330
Yamamoto H, Miller MD, Tsubota H, Watkins DI, Mazzara GP, Stallard V, Panicali DL, Aldovini A, Young RA, Letvin NL (1990a) Studies of cloned simian immunodeficiency virus-specific T lymphocytes. Gag-specific cytotoxic T lymphocytes exhibit a restricted epitope specificity. J Immunol 144(9): 3385–3391
Yamamoto H, Miller MD, Watkins DI, Snyder GB, Chase NE, Mazzara GP, Stallard V, Panicali D, Letvin NL (1990b) Two distinct lymphocyte populations mediate simian immunodeficiency virus envelope-specific target cell lysis. J Immunol 145: 3740–3746
Zemmour J, Parham P (1993) HLA class I nucleotide sequences. Eur J Immunogen 20(1): 29–45
Zhu Z-F, Vincek V, Figueroa F, Schoenbach C, Klein J (1991) Mhc-DRB genes of the pigtail macaque (Macaca nemestrina): implications for the evolution of human DRB genes. Mol Biol Evol 8(5): 563–578
Zijlstra M, Bix M, Simister NE, Loring JM, Raulet DH, Jaenisch R (1990) β_2-Microglobulin deficient mice lack CD4-8$^+$ cytolytic T cells. Nature 344: 742

CD4$^+$ T Cell Responses to Simian Immunodeficiency Virus

K.H.G. MILLS[1,2] and W.C. JONES[1]

1 Role of MHC Restricted T Cells in Immunity to Viral Infections

T cells have classically been divided into two populations on the basis of their surface phenotype, MHC restriction and function. CD8$^+$ T cells recognize endogenous antigen in association with MHC class I molecules and display cytotoxic activity against infected or peptide-pulsed target cells. Although these cytotoxic T lymphocytes (CTL) have usually been described following infection with viruses or intracellular bacteria, they can also be induced with soluble peptides or proteins, especially when the antigen is associated with lipids, which appears to facilitate their transport from intracellular vesicles into the cytoplasm. In contrast, a second population of T cells that express the CD4$^+$ molecule are readily induced with soluble proteins and recognise exogenous processsed antigen in association with MHC class II molecules. Although these cells ahve traditionally been called T helper (Th) cells, in that they provide helper function for B cell production of antibody, recent

[1] Immunobiology Department, National Institute for Biological Standards and Control, South Mimms, Potters Bar, Herts EN6 3QG, UK
[2] Biology Department, St Patrick's College, Maynooth, County Kildare, Ireland

Current Topics in Microbiology and Immunology, Vol. 188

evidence suggests that they play a more diverse role including the regulation and effector function of cellular immune responses against foreign pathogens. Studies with murine T cell clones and, more recently, human T cell clones have shown that $CD4^+$ T cells can be further divided into two subpopulations on the basis of their cytokine production and function (MOSSMANN et al. 1986; ROMAGNANI 1991). One subpopulation, termed Th1 cells, secrete interleukin (IL) 2, interferon (IFN) γ and lymphotoxin and, as $CD8^+$ T cells, display cytotoxic activity against antigen-pulsed target cells. Cytokines secreted by antigen-stimulated Th1 cells mediate delayed-type hypersensitivity and inflammatory activity, stimulate the anti-microbial activity of phagocytic cells and in the mouse have also been shown to provide help for IgG2a antibody responses. In addition, IFN-γ, produced by Th1 cells, $CD8^+$ CTL and natural killer cells, can mediate a direct antiviral effect through the inhibition of viral replication in infected cells (RAGER-ZISMAN and BLOOM 1985). The second population, termed Th2 cells, secrete IL-4, IL-5, IL-6 and IL-10 and are predominantly responsible for helper activity for specific antibody production. The Th2-derived cytokines, IL-4, IL-5 and IL-6 play different roles in B cell proliferation, differentiation and Ig class switching, especially to IgA, IgE, IgG1, whereas IL-10 downregulates Th1 responses (MOSMANN and MOORE 1991).

There is limited evidence from investigations in mice that $CD4^+$ T cells play a role in immunity to retroviruses. Studies with the murine retrovirus Rauscher murine leukaemia virus (RLV) have shown that protective immunity can be adoptively transferred by T cells alone, requiring both $CD4^+$ and $CD8^+$ T cell populations. Immune T cells derived from mice vaccinated with live RLV, could confer protection against viraemia and disease following RLV challenge. In contrast non-immune T cells or immune $CD4^+$ or $CD8^+$ T cells alone had little effect on the course of infection (HOM et al. 1991). Indirect evidence of the role of $CD4^+$ Th cells in the murine antibody response to HIV was demonstrated by SHRIER et al. (1988). HIV env-specific antibodies were generated in normal but not athymic mice following immunisation with a synthetic peptide corresponding to amino acids 598–609 of gp41, which included a T cell epitope and a immunodominant B cell epitope.

2 $CD4^+$ T Cell Responses in SIV-Infected Macaques

In HIV-infected humans abnormalities in the T cell responses to recall antigens, HIV proteins and mitogens have been reported both before and after the onset of $CD4^+$ T cell depletion. Indeed in a study of HIV-infected individuals, progression to disease was found to correlate more closely with a loss in T cell proliferative responses than to $CD4^+$ T cell decline (SHEARER and CLERICI 1992). The loss of responsiveness to foreign antigens was also shown to correlate with changes in the cytokine profile of stimulated peripheral blood mononuclear cells (PBMC) isolated from HIV-infected individuals. Early in infection a strong proliferative response correlate with high IL-2 secretion, reflecting a predominant Th1 profile. As disease progresses however, an increase in cytokines such as IL-4 and IL-10, characteristic of the Th2 subpopulation of CD4$^+$ T cells, is observed. It has consequently been postulated that in order to maintain a healthy state following infection with HIV, a stable predominant Th1 population is necessary.

A limited number of studies on the effects of SIV infection of macaques also suggests that impaired cellular and humoral responses to foreign antigens following SIV infection may reflect functional defects in CD4$^+$ T cells rather than a reduction in CD4$^+$ T cell numbers. Although defective antibody responses have been reported following primary immunisation with a hepatitis B vaccine in HIV-infected humans (CARNE et al. 1987), due to previous immunisations or infections the ability to induce primary immune responses has been poorly addressed. However, using the SIV macaque model, it has been possible to study both humoral and T cell proliferative response in infected animals following primary immunisation with poliovirus and tetanus (MILLS et al. 1993a). In this study, the course of immunodeficiency disease in a group of cynomolgus macaques was followed for up to 3 years after infection with the 32H isolate of SIV_{mac251}. In the second year five SIV-infected animals and three uninfected controls were immunised with poliovirus and tetanus. Whereas strong antibody and proliferative responses to poliovirus and tetanus were observed in the immunised controls, a wide variation in the responses, which was unrelated to circulating CD4$^+$ T cell numbers, was observed in the infected macaques. Two healthy SIV-infected macaques with a normal CD4$^+$ count could mount a proliferative response to poliovirus and tetanus. PBMC from these two macaques also responded by in vitro proliferation to inactivated SIV and a recombinant SIV gag protein. In contrast, PBMC from two macaques with AIDS-like disease symptoms showed poor proliferative responses to tetanus, poliovirus and inactivated SIV and p27. The numbers of circulating CD4$^+$ T cells in one of these macaques was high, whereas in the other it was very low. Finally, one SIV-infected animal with low circulating CD4$^+$ T cell numbers responded to poliovirus and tetanus antigens but not to SIV and p27.

The demonstration of impaired immune function following primary vaccination may explain the susceptibililty of HIV-positive individuals to opportunistic infection. These findings, alongside the demonstration of a transient decline in CD4$^+$ T cell numbers following immunisation in three out of five of the SIV-infected macaques, has important implications for vaccine strategies with HIV-positive patients. Studies in rhesus macaques have also shown that SIV preferentially infects the CD45RO$^+$ memory subpopulation of CD4$^+$ T cells, which may account for the loss of responsiveness to recall antigens (FIRPO et al. 1992). Assessment of the defective function of Th1/Th2 cell subsets resulting from SIV infection by analysis of cytokine profiles should provide important information of the pathogenesis of AIDS. Therefore the SIV macaque model may prove invaluable in the design of immunotherapeutic strategies for the control of lentivirus infections.

3 Induction of SIV-Specific CD4$^+$ T Cells by Immunisation

The development of immunodeficiency and AIDS-like disease symptoms in macaques following inoculation with certain strains of SIV provides an ideal model for the study of the mechanisms of immunity to a pathogenic lentivirus. Although investigations of immune responses in SIV-infected macaques and HIV-infected humans provide circumstantial evidence for the role of T cells in the control of infection, the tropism of HIV and SIV for the CD4$^+$ T cell complicates the study of class II restricted T cells in infection. A study of cellular immunity induced by immunisation prior to live virus challenge should in theory provide more definitive data on the mechanisms of protection against a lentivirus infection. Consistent proliferative responses against HIV proteins have been detected in HIV-infected non-human primates (EICHBERG et al. 1987), and the specificity of CD4$^+$ T cells for HIV proteins has been studied using PBMC or T cell lines and clones derived from infected humans (CLERICI et al. 1989) and immunised non-human primates (MILLS et al. 1990). However, the protective efficacy of these proteins in relation to the identified epitopes can be studied only in the SIV macaque model.

3.1 Whole-Virus Vaccines

Early studies aimed at identifying the viral antigens involved in the protective immune response against SIV focused on the use of inactivated virus vaccines. Macaques were immunised with either formaldehyde, psoralen or β-propiolactone inactivated SIV, or Tween-ether disrupted virus with a variety of adjuvants (CARLSON et al. 1990; DESROSIERS et al. 1989; MURPHEY-CORB et al. 1989; STAHL-HENNIG et al. 1992; STOTT et al. 1990; MILLS et al. 1992). Not surprisingly, PBMC from the majority of immunised animals tested showed positive proliferative responses to semi-purified heat-inactivated SIV (MILLS et al. 1991, 1992; JONES et al. 1992). PBMC from these animals also responded to the recombinant SIV core (p27) protein. However, proliferative responses to the recombinant envelope (env) protein were more variable. In particular, animals immunised with β-propiolactone inactivated

SIV showed weak or inconsistent responses to gp160 (MILLS et al. 1992), whereas immunisation with formaldehyde-inactivated SIV induced detectable proliferation to env in all animals tested (MILLS et al. 1991; Jones and Mills, unpublished). It is suggested that this may reflect a greater loss of the env protein following inactivation with β-propiolactone compared to formaldehyde inactivation, but it may also reflect sequence heterogeneity between the immunogen and the env protein used in the assays. The majority of animals immunised with inactivated SIV also developed moderate levels of anti-SIV neutralising antibodies and were protected against live virus challenge. However, there did not appear to be a correlation between neutralising antibody levels and protection. STAHL-HENNIG et al. (1992) suggested that in animals immunised with Tween-ether treated SIV, positive T cell proliferative responses to SIV, tested on the day of challenge, correlated with protection. However, further studies demonstrated that these vaccines protected against a virus grown in human T cell lines but not against monkey peripheral blood lymphocyte grown virus (CRANAGE et al. 1992; MILLS et al. 1992). The presence of strong antibody responses to cellular components of the semi-purified virus preparations suggested that protection was not due solely to anti-SIV responses (STOTT et al. 1991). Consequently it has not been possible to definitively demonstrate a role for T cell responses in macaques immunised with inactivated SIV.

The mechanism of protection with semi-purified virus, fixed-infected cells or C8166 cells alone appears to involve antibody responses to a protein expressed on C8166 cells (STOTT et al. 1991). However, we have also demonstrated a zenoreactive T cell response to fixed C8166 cells in macaques immunised with SIV-infected C8166 cells or "purified" inactivated SIV (STOTT et al. 1990; Mills et al., unpublished observations). Although the in vitro proliferative responses were strongest against the SIV-infected C8166 cells, a substantial response to uninfected cells was observed with PBMC from many of the animals including those immunised with formaldehyde-inactivated SIV but not in animals immunised with recombinant SIV proteins.

3.2 Recombinant gag Proteins

Because the core proteins of SIV and HIV are more conserved than the surface glycoprotein, it has been suggested that the induction of T cell responses to these proteins may form part of a broadly protective immune response against HIV/SIV (MILLS et al. 1989). The gag protein has been shown to be an important target for class I restricted CTL from HIV-infected humans (MILLS et al. 1989). Furthermore, Th cells specific for the gag protein may provide help for anti-env antibody responses. While much of the available experimental evidence suggests that Th cells and antibody should be directed against the same protein (CALADA and SERCARZ 1988), studies of T and B cell responses with hepatitis B and influenza viruses have shown that

T cells specific for core proteins can help in the production of neutralising antibodies against surface antigens (RUSSEL and LIEW 1980; SCHERLE and GERHARD 1986). In a study in which mice were immunised with synthetic peptides corresponding to regions of the HIV-1 p24 protein, it was reported that these mice were capable of priming antibody responses to the env protein (COATES et al. 1989). Furthermore, recent studies with rhesus macaques have shown that priming with peptides that include Th cell epitopes from conserved regions of the HIV gp120 can enhance the antibody response to gp41 after a boost with a suboptimal dose of gp160 (HOSMALIN et al. 1991).

We have demonstrated consistent T cell proliferation and IL-2 secretion in PBMC from macaques immunised with p27 as Ty virus-like particles (VLPs), or p57 with gp160 in the Syntex adjuvant (MILLS et al. 1991, 1992). Restimulated PBMC from the p57-immunised macaques also displayed class II-restricted cytotoxic activity for autologous B cell lines infected with vaccinia virus recombinants expressing the gag protein or pulsed with gag peptides (MILLS et al. 1992). However neither immunisation protocol was capable of inducing a protective immune response against SIV infection. Furthermore, we were unable to demonstrate an enhanced antibody response to the SIV env protein in the p27-primed macaques. It is possible that intermolecular-intrastructural T cell help reported in other virus systems does not operate between the gag and env proteins of SIV.

3.3 Recombinant env Proteins

In the chimpanzee model immunisation with recombinant env proteins has in certain instances been shown to be capable of protecting against HIV infection (BERMAN et al. 1990). Although neutralising antibodies appear to play an important role in protection (EMINI et al. 1992), there has been little attempt to study the role of cell-mediated immunity in this model. In the SIV macaque model immunisation with purified recombinant SIV mac gp130 or gp160 expressed in either baculovirus, vaccinia virus or CHO cells and injected with the Syntex adjuvant has been shown to induce very strong env-specific PBMC proliferative responses (MILLS et al. 1992,1993b; JONES et al. 1992). Proliferative T cell responses were strongest against a recombinant env protein and weaker against the inactivated virus with a heterogenous env composition. Furthermore, the T cell responses were sustained for a minimum of 20 weeks following immunisation. The failure of these immunogens to protect against challenge with SIV may at first appearances indicate that env-specific T cells do not play a role in protection against SIV infection. However, the animals also had high levels of SIV-neutralising antibodies on the day of challenge. The lack of protection may, however, be due to the presence of sequence variation between the immunogen and the challenge virus. Alternatively, it may reflect the inability of recombinant proteins to induce $CD8^+$ CTL or to generate a protective antibody response. This may have resulted from the failure of the purified recombinant immunogen to enter the endogenous pathway of antigen processing, required for class I restricted CTL induction, or to mimic the conformation of the tertiary or quaternary structure of env in the native virus for the induction of a functionally important antibody response.

Protection against infection with a biological clone of SIV_{mne} has recently been demonstrated in macaques primed with a live vaccinia virus recombinant expressing SIV env, followed by boosting with purified baculovirus expressed gp140 (HU et al. 1992). In this study, moderate T cell proliferative responses were demonstrated against inactivated SIV, as well as high titre anti-SIV neutralising antibodies. In a similar experiment carried out in our laboratory using SIV mac recombinants, $CD8^+$ env-specific cytotoxic T cell responses were demonstrated followed infection with a live vaccinia virus expressing the SIV env protein (JONES and MILLS, unpublished). However macaques primed with vaccinia virus SIV recombinants with or without a boost with purified gp160 were not protected against infection with cloned or uncloned SIV_{mac} challenge viruses (Stott et al., personal communication).

In studies with HIV in chimpanzees, it has been suggested that protection is associated with antibodies to the V3 domain of the env protein (BERMAN et al. 1990; EMINI et al. 1992) Although the HIV V3 equivalent region in SIV is not as variable as that observed in HIV and does not include a linear neutralising antibody epitope, it is possible that this region of the env protein may also be involved in the protective immune response to SIV. It has been suggested that the SIV V3 region forms part of a conformational neutralising antibody domain (JAVAHERIAN et al. 1992). Furthermore, the V3 region in SIV has been shown to be antigenic for T cells (JONES et al. 1992, 1993), and antibodies to overlapping peptides covering the SIV V3 domain have been identified in sera from both SIV-infected and immunised macaques (Silvera et al., personal communication). In a study of the immunogenicity of the V3 region of SIV expressed in Ty-VLPs we found that immunogenicity was greatly affected by the length of the env insert (JONES et al. 1993). Ty-VLPs containing residues 309–346 were highly immunogenic for the Ty component of the antigen but failed to generate SIV env-specific T cells or antibodies. However, when the same animals were boosted with Ty-VLPs containing a longer SIV env insert covering residues 280–384, anti-env antibody and T cell responses were demonstrated. Furthermore, specificity analysis with MHC class II restricted $CD4^+$ T cell lines and clones generated from these animals indicated that T cells were generated against epitopes within the sequence of the original immunogen (JONES et al. 1993). This finding suggests that residues flanking the Ty sequence may affect processing of the env epitopes and illustrates the potential problems of using short polypeptides for the induction of T cell responses. This study also confirmed that the V3 region of SIV does not include a linear neutralising antibody epitope and highlights the risks associated with direct extrapolation from SIV to HIV in the design of a vaccine against AIDS.

4 Specificity of $CD4^+$ T Cells for SIV Proteins

The identification of T cell and antibody epitopes on SIV proteins is an important step towards our understanding of immune recognition of this virus and is essential for the use of the SIV macaque model for the development of a synthetic vaccine against HIV. It has been proposed that the inclusion of Th cell epitopes together with antibody or $CD8^+$ CTL epitopes, with the exclusion of regions of the env protein that have been reported to induce enhancing antibodies, immunosuppression or autoimmunity, may

Table 1. Synthetic peptides of SIV gag and env recognized by macaque $CD4^+$ T cells

Protein	Peptide	Sequence
gag	166–185	FGAEVVPGFQALSEGCTPYD
	186–205	INQMLNCVGDHOAAMQIIRD
	196–215	HOAAMQIIRDIINEEAADWD
	246–265	QIQWMYRQQNPIPVGNIYRR
	256–275	PIPVGNIYRRWIQLGLQKCV
	296–315	SYVDRFYKSLRAEQTDAAVK
	306–325	RAEQTDAAVKNWMTQTLLIQ
env	21–40	YCTQYVTVFYGVPAWRNATI
	41–60	PLFCATKNRDTWGTTQCLPD
	161–180	QEQMIGCKFNMTCLKRDKTL
	171–190	MTGLKRDKTKEYNETWYSTD
	181–200	EYNETWYSTDLVCEQGNSTD
	191–210	LVCEQGNSTDNESRCYMNHC
	201–220	NESRCYMNHCNTSIIQESCD
	211–230	NTSIIQESCDKHYWDTIRFR
	221–240	KHYWDTIRFRYCAPPGYALL
	231–250	YCAPPGYALLRCNDTNYSGF
	261–280	SCTRMMETQTSTWFGFNGTR
	281–300	AENRTYIYWHGRDNRTIISL
	291–310	GRDNRTIISLNKYYNLTMKC
	301–320	NKYYNLTMKCRRPGNKTVLP
	311–330	RRPGNKTVLPVTIMSGLVFH
	321–340	VTIMSGLVFHSQPINDRPKQ
	331–350	SQPINDRPKQAWCWFGGKWK
	341–360	AWCWFGGKWKDAIKEVLQTI
	351–370	DAIKEVKQTIVKHPRYTGTN
	451–470	PREGDLTCNSTVTSLIANID

Overlapping peptides correspond to the SIV_{mac} 32H sequence, recognised by $CD4^+$ T cell lines and clones from macaques immunized with inactivated SIV or recombinant gag or env proteins. Epitopes identified with $CD4^+$ T cell clones are underlined (from MILLS et al. 1991; JONES et al. 1992, 1993).

form the basis of a synthetic vaccine against HIV (ADA 1989). Although certain information has been generated through an examination of the PBMC responses to peptides synthesised from regions containing predicted motifs, we consider that the generation of long-term antigen-specific T cell lines and clones and analysis against overlapping synthetic peptides is a more definitive approach.

4.1 gag-Specific CD4$^+$ T Cells

MHC class II restricted CD4$^+$ T cell lines and clones have been established from macaques immunised with either inactivated SIV or a recombinant p27 protein expressed as yeast-derived VLPs (MILLS et al. 1991). The fine specificity of T cells directed against the gag protein has revealed a number of interesting features of immune recognition of foreign proteins. Considerable overlap in the recognition of epitopes was observed from T cell lines derived from macaques which had been immunised with the p27:Ty-VLPs and inactivated SIV. T cell lines from each macaque recognised up to three distinct regions of the protein, and distinct peptides were recognised by more than one macaque (MILLS et al. 1991; Table 1). Furthermore, a CD4$^+$ p27-specific T cell clone was shown to respond to a single p27 peptide (residues 166–185) which overlaps with a peptide (residues 171–195) recognised by a CD8$^+$ class I-restricted CTL from SIV-infected macaques (YAMAMOTO et al. 1990). Interestingly, it has been shown that immunisation with peptide 171–195 in incomplete Freund's adjuvant induced class I-restricted T cells and anti-peptide antibodies in rhesus macaques (Letvin, personal communication). It is possible that this region also includes a promiscuous Th cell epitope for rhesus macaques.

4.2 env-Specific CD4$^+$ T Cells

In a study of the epitope specificity of a panel of SIV env- specific MHC class II restricted CD4$^+$ T cell lines and clones extensive heterogeneity of recognition of the envelope protein was demonstrated (JONES et al. 1992). CD4$^+$ T cells derived from macaques which had been immunised with either inactivated SIV or recombinant env proteins recognised between one and seven distinct regions of the evn protein (Table 1). Epitopes were recognised in both variable and conserved domains except for the most highly variable regions of the SIV env protein (V1 and V4; ALMOND et al. 1992; OVERBAUGH et al. 1991). Although it was possible that these V1 and V4 regions did not contain T cell epitopes, it is more conceivable that sequence heterogeneity of these regions precluded detection of epitopes where a single sequence recombinant protein was used to establish and maintain the T cell lines and clones.

Although individual peptides were recognised by a number of macaques, which appeared from restriction fragment length polymorphism analysis to have distinct MHC haplotypes, a single peptide did not stimulate the polyclonal T cell lines from all animals (JONES et al. 1992). However the V3 and flanking sequences did appear to be an immunodominant region. T cell lines from eight out of ten macaques recognised one or a pair of overlapping peptides within the region covering amino acids 280–370. Furthermore, in the study in which macaques were immunised with Ty-VLPs containing the SIV env residues 280–384, T cell lines established from three macaques each recognised one to four peptides within the sequence 301–360. Class II-restricted T cell recognition of epitopes corresponding to residues 301–320 and 321–340 was confirmed at the clonal level (JONES et al. 1993). Thus it appears that the V3 region of SIV, as HIV, is a major target for MHC-restricted T cell recognition.

5 Conclusions and Future Prospects

The outcome of the immunisation experiments with candidate SIV vaccines, whilst failing to define a correlation between $CD4^+$ T cell responses or any SIV-specific immune response and protection, have highlighted a number of potential problems in the design of vaccines against HIV. Furthermore, studies on the antigenicity and immunogenicity of SIV proteins, while demonstrating many homologies with HIV, have also revealed a number of important differences. The possible contribution of cellular proteins, derived from the human T cell lines used for virus propagation, in protection against SIV infection has raised important questions as to the nature of these proteins and their relevance to protection against fresh virus isolates. Furthermore, the inability of recombinant SIV env immunogens, as purified proteins or when expressed in live vaccinia recombinants (alone or in combination) to protect against infection with cloned or uncloned SIV mac challenge viruses, despite consistent SIV-specific humoral and cellular responses, is a serious concern. These experiments also illustrate the potential problems of escape from immunity through antigenic variation. The use of single-sequence recombinant env proteins as vaccines may generate immune responses which are not boosted on exposure to a heterologous virus challenge and may be ineffective against highly pathogenic strains. Although it is possible that $CD4^+$ T cells specific for conserved regulatory and structural proteins may prime for anti-env antibody responses, the available data on gag-specific Th cells suggest that this is unlikely. Furthermore, the use of short polypeptides that include highly antigenic regions as immunogens also presents potential problems associated with the effect of flanking regions on antigen processing and, more importantly, universal recognition by a population with diverse MHC haplotypes. One of the most promising vaccine approaches in protection against SIV, which has serious problems in relation to safety for its application to HIV in man, has arisen through the recent demonstration that a transient infection of macaques with a SIV mac clone containing a nef deletion conferred resistance against infection with the parental clone and the pathogenic uncloned virus (DANIEL et al. 1992). Although it is tempting to speculate that cell-mediated immunity may be involved, the mechanism of protection remains to be defined.

The limited studies on cellular responses in SIV-infected macaques suggest that SIV infection results in defective CD4$^+$ T cell function similar to that reported in HIV-infected individuals. A decline in Th activity and cytokine secretion may contribute to a reduction in immune effector functions responsible for controlling the virus in the early stages of infection. The effects of distinct CD4$^+$ T cell subpopulation induction on protection has only begun to be addressed. We have preliminary data to suggest that it is possible to define macaque T cell subpopulations with cytokine profiles typical of Th1 and Th2 cells. We have defined MHC class II-restricted SIV env-specific T cells that secrete IL-2, IFN-γ and tumour necrosis factor but not IL-4 or IL-6 (Th1) and T cells that secrete IL-4 and IL-6 (Th2). T cell clones that produce IFN-γ, IL-4 and IL-6 have also been identified; these may form part of a Th0 population (JONES et al. 1993). Whereas Th2 cells may be important in the induction of mucosal IgA responses, Th1 cells may be more important in the control of disease in the SIV-infected macaques. The SIV macaque model should prove invaluable in the development of methods for selectively generating or enhancing Th1 or Th2 responses and thereby help in defining their role in prevention or control of lentivirus infections.

References

Ada G (1989) Prospects for a vaccine against AIDS. Nature 339: 331–332

Almond N, Jenkins A, Slade A, Heath A, Cranage M, Kitchin P (1991) Population sequence analysis of a simian immunodeficiency virus (32H isolate of SIV mac 251) a virus stock used for international vaccine studies. AIDS Res Hum Retroviruses 8: 77–88

Berman PW, Groopman JE, Gregory T, Riddle L, Nakamura GR, Champe MA, Porter JP, Wurm FM, Hershberg RD, Cobb EK, Eichberg JW (1990) Protection of chimpanzees from infection by HIV-1 after vaccination with recombinant glycoprotein gp120 but not gp160. Nature 345: 622–625

Calada F, Sercarz EE (1988) Preferential pairing of T-B specificities in the same antigen: the concept of directional help. Vaccine 6: 94–98

Carlson J, McGraw TP, Kedddie E, Yee OJL, Rosenthal A, Langlois AJ, Dickover R, Donovan R, Luciw PA, Jennings MB, Gardner B (1990) Vaccine protection against simian immunodeficiency virus infection. AIDS Res Hum Retroviruses 6: 1239–1246

Carne CA, Weller IVD, Waite J, Briggs M, Pearce F, Adler MW, Tedder RS (1987) Impaired responsiveness of homosexual men with HIV antibodies to plasma derived hepatitis B vaccine. Br Med J 294: 866–868

Clerici M, Stock NJ, Zaja RA, Boswell RN, Bernstien DC, Mann DL, Shearer GM, Berzofsky JA (1989) Interleukin-2 production used to detect antigenic peptide recognition by T helper lymphocytes from asymptomatic HIV-seropositive individuals. Nature 339: 383–385
Coates ARM, Sia DY, Mascagni P, Toth GK, Haaheim LR, Gibbon WA (1989) Induction of antibody to HIV-1 envelope protein by p24 core T cell epitopes (Abstr MCP 40) 5th International Conference on AIDS, Montreal
Cranage MP, Ashworth LAE, Greenaway PJ (1992) AIDS vaccine developments. Nature 355: 685
Daniel MD, Kirchhoff F, Czazal SC, Sehgal, Desrosiers RC (1992) Protective effects of a live attenuated SIV vaccine with a deletion in the nef gene. Science 258: 1938–1941
Desrosiers RC, Wyand MS, Kodoma T, Ringler DJ, Arthur LO, Sehgal PK, Letvin NL, King NW, Daniel MD (1989) Vaccine protection against simian immunodeficiency virus infection. Proc Natl Acad Sci USA 86: 6353–6357
Eichberg JW, Zarling JM, Alter HJ, Levy JA, Berman PW, Gregory T, Lasky LA, McClure J, Cobb KE, Moran PA, Hu S-L, Kennedy RC, Chanh TC, Dreesman GR (1987) T cell responses to human immunodeficiency virus (HIV) and its recombinant antigens in HIV infected chimpanzees. J Virol 61: 3804–3808
Emini EA, Scheilf WA, Nunberg J, Conley A, Eda Y, Tokiyoshi S, Putney SD, Matsushita S, Cobb KE, Jett RM, Eichberg JW, Murthy KK (1992) Prevention of HIV-1 infection in chimpanzees by gp120 V3 domain-specific monoclonal antibody. Nature 335: 738–740
Firpo PP, Scheibel AM, Clark EA (1992) Macaque $CD4^+$ T-cell subsets: influence of activation on infection by simian immunodeficiency viruses (SIV). AIDS Res Hum Retroviruses 8: 357–366
Hom RC, Finberg RW, Mullaney S, Ruprecht RM (1991) Protective cellular retroviral immunity requires both $CD4^+$ and $CD8^+$ T cells. J Virol 65: 220–224
Hosmalin A, Nara PL, Zweig M, Lerche NW, Cease KB, Gard EA, Markham PD, Putney SD, Daniel MD, Desrosiers RC, Berzofsky JA (1991) Priming with T helper cell epitope enhances the antibody response to envelope glycoprotein of HIV-1 in primates. J Immunol 146: 1667–1673
Hu S-L, Abrams K, Barber GN, Moran P, Zarling JM, Langlois AJ, Kuller L, Morton WR, Benveniste RE (1992) Protection of macaques against SIV infection by subunit vaccines of SIV envelope glycoprotein gp160. Science 255: 456–459
Javaherian K, Langlois AJ, Schmidt S, Kauffman M, Cates N, Langedijk JPM, Moleon RH, Desrosiers RC, Burns DPW, Bolognesi DP, LaRosa GJ, Putney SD (1992) The principle neutralization determinant of simian immunodeficiency virus differs from that of human immunodeficiency virus type 1. Proc Natl Acad Sci USA 89: 1418–1422
Jones WC, Barnard AL, Slade A, Mills KHG (1992) Heterogeneity in the recognition of the simian immunodeficiency virus envelope glycoprotein by $CD4^+$ T cell clones from immunized macaques. J Immunol 149: 3120–3216
Jones WC, Page M, Silvera P, Ling C, Layton GT, Adams SE, Stenner NF, Stott J, Almond N, Rud E, Mills KHG (1993) Induction of T cells and antibodies specific for the V3 region of SIV in macaques: effect of V3 insert size on immunogenicity when presented in a recombinant expression system. (submitted)
Mills KHG, Nixon DF, McMicheal AJ (1989) T cell strategies in AIDS vaccines. MHC restricted T cell responses to HIV proteins. AIDS 3 Suppl 1: S101–S110
Mills KHG, Kitchin PA, Mahon BP, Barnard AL, Adams SE, Kingsman SM, Kingsman AL (1990) HIV p24-specific helper T cell clones from immunized primates recognize highly conserved regions of HIV-1. J Immunol 144: 1677–1683
Mills KHG, Barnard AL, Williams M, Pag M, Ling C, Stott EJ, Silvera P, Taffs F, Kingsman A, Adams S, Almond N, Kitchin P, Jones WC (1991) Vaccine-induced $CD4^+$ T cells against the simian immunodeficiency virus gag protein. Epitope specificity and relevance to protective immunity. J Immunol 147: 3560–3567
Mills KHG, Page M, Chan WL, Kitchin P, Stott EJ, Taffs F, Jones W, Rose J, Ling C, Silvera P, Corcoran T, Flanagan B, Burny A, Bex F, Delchambre M, Van Opstal O, Fabry L, Thiriart C, Delers A, DeWilde M, Bruck C (1992) Protection against SIV infection in macaques by immunization with inactivated virus from the BK28 molecular clone, but not with BK28-derived recombinant env and gag proteins. J Med Primatol 21: 50–58
Mills KHG, Jones WC, Mahon B, Katrak K, Taffs F, Stott EJ, Page M (1993a) Cellular and humoral immune responses to poliovirus and tetanus following primary vaccination of

SIV-infected macaques. In: Racz P (ed) Animals models for HIV infection. Karger Basel, pp 75–85
Mills KHG, Page M, Kitchin P, Chan L, Jones WC, Silvera P, Corcoran T, Flanagan B, Ling C, Thiriart C, DeWilde M, Bruck C, Rud E, Clark B, Stott EJ (1993b) Immunization of macaques with SIV env recombinants: specificity of T cell and antibody responses and evaluation of protective efficacy. J Med Primatol 22: 104–109
Mosmann TR, Moore KW (1991) The role of IL-10 in crossregulation of Th1 and Th2 responses. Immunol Today 13: 49–53
Mosmann TR, Cherwinski H, Bond MW, Giedlin MA, Coffman RL (1986) Two types of murine helper T cell clones: definition according to profiles of lymphokine activities and secreted proteins. J Immunol 136: 2348–2357
Murphey-Corb M, Matin LN, Davison-Fairburn B, Montelaro RC, Miller M, West M, Ohkawa S, Baskin GB, Zhang J-Y, Putney SD, Allison AC, Eppstein DA (1989) A formalin-inactivated SIV vaccine confers protection in macaques. Science 246: 1293–1297
Overbaugh J, Rudensey LM, Papenhausen MD, Beneveniste RE, Morton WR (1991) Variation in the simian immunodeficiency virus is confined to V1 and V4 during progression to simian AIDS. J Virol 65: 7025–7031
Rager-Zisman B, Bloom BR (1985) Interferons and natural killer cells. Br Med Bull 41: 22–27
Romagnani S (1991) Human Th1 and Th2 subsets: doubt no more. Immunol Today 13: 256–257
Russel SM, Liew FY (1980) Cell cooperation in antibody responses to influenza virus-priming of helper T cells by internal components of the virion. Eur J Immunol 10: 791–796
Scherle PA, Gerhard W (1986) Functional analysis of influenza-specific helper T cell cloned in vivo. T cells specific for internal viral proteins provide cognate help for B cell responses to haemagglutinin. J Exp Med 164: 1114–1128
Shearer GM, Clerici M (1992) T helper cell immune dysfunction in asymptomatic HIV-1 seropositive individuals: the role of Th1/Th2 cross regulation. Prog Chem Immunol 54: 21–43
Shrier RD, Gnann JW, Langlois A, Schriver K, Nelson JA, Oldstone MBA (1988) B- and T-lymphocyte responses to an immunodominant epitope of human immunodeficiency virus. J Virol 62: 2531–2536
Stahl-Hennig C, Voss G, Nick S, Petry H, Fuchs D, Wachter H, Coulibaly C, Luke W, Hunsmann G (1992) Immunization with Tween-ether treated SIV adsorbed onto aluminum hydroxide protects monkeys against experimental SIV infection. Virology 186: 588–596
Stott EJ, Chan WL, Mills KHG, Page M, Taffs F, Cranage M, Greenaway P, Kitchin P (1990) Preliminary report: protection of cynomolgus macaques against simian immunodeficiency virus by fixed infected cells. Lancet 1: 1538–1541
Stott EJ, Kitchin PA, Page M, Flanagan B, Taffs F, Chan WL, Mills KHG, Silvera P, Rodgers A (1991) Anti-cell antibody in macaques. Nature 353: 393
Yamamoto H, Miller MD, Tasuboto H, Watkins DI, Mazzara GP, Stallard V, Panicali DL, Aldovini A, Young RA, Letvin NL (1990) Studies of cloned immunodeficiency virus-specific T lymphocytes: gag-specific cytotoxic T lymphocytes exhibit a restricted epitope specificity. J Immunol 145: 3385–3391

The SIV_{mac} Specific Cytotoxic T Lymphocyte Response in the Acutely Infected Rhesus Monkey

N.L. LETVIN, K.A. REIMANN, Y. YASUTOMI, D.J. RINGLER, and H. YAMAMOTO

1 Introduction

Early virologic events following infection may have important consequences for the eventual clinical course of HIV-induced disease. Moreover, the immune responses that block initial virus spread are likely to be central mechanisms for containing HIV infections. It is therefore important to elucidate the virologic and immunologic events that occur immediately following HIV infection. However, since the acutely infected person comes to clinical attention weeks or even months following viral exposure, if at all, it is difficult to study the immunologic mechanisms responsible for the early clearance of HIV.

Cytotoxic T lymphocytes (CTL) have been shown in a number of viral infections to play a crucial role in blocking the early spread of virus (OLDSTONE et al. 1986; QUINNAN et al. 1982; YAP et al. 1978). The contribution of virus-specific CTL to the containment of an HIV infection is currently being explored (CULMANN et al. 1989; NIXON et al. 1988; PLATA et al. 1987; WALKER et al. 1987, 1988). The presumption that an effective, early, virus-specific CTL response may limit the spread of HIV provides much of the rationale for developing vaccines which induce HIV-specific CTL

Harvard Medical School, New England Regional Primate Research Center, Southborough, MA 01772-9102, USA

Current Topics in Microbiology and Immunology, Vol. 188

(KOFF and HOTH 1988). It is therefore important to gain an understanding of the virus-specific CD8 + CTL response in acute HIV infections.

The SIV_{mac} infected rhesus monkey provides a unique system for studying early events in the immunopathogenesis of AIDS (LETVIN 1990). SIV_{mac} has significant nucleotide sequence homology to HIV, infects CD4-bearing lymphocytes and macrophages, and induces a disease in macaque monkeys with significant similarities to HIV-induced disease in man (CHAKRABARTI et al. 1987; KANNAGI et al. 1988; WATANABE et al. 1990; LETVIN et al. 1985). Here we review recent studies from our laboratory employing the SIV_{mac} infected rhesus monkey to assess immunopathogenic events in acute AIDS virus infections. Particular attention has been given in these studies to the role of $CD8^+$, MHC class I restricted virus-specific CTL in containing early virus spread and in inducing clinical disease.

2 Immune Cellular Changes Coincident with Early Clearance of SIV_{mac} Antigenemia

A period of plasma antigenemia occurs in humans shortly after an acute HIV infection (DAAR et al. 1991; CLARK et al. 1991). We sought to determine the virologic and immunologic events that occur in both the peripheral blood and lymph nodes during this early antigenemia and the immune mechanisms responsible for its clearance. Eight naive rhesus monkeys were infected by intravenous inoculation with one to ten animal infectious doses of uncloned SIV_{mac}251. SIV_{mac} p27 could be detected in the plasma of these animals after day 7 following virus inoculation. By day 21, six of the eight monkeys had cleared this antigenemia; by day 30, seven of the eight monkeys had no detectable plasma p27.

These acutely infected animals were assessed for evidence of an emerging SIV_{mac} specific antibody. Indirect immunofluorescence, employing chronically SIV_{mac} infected H9 cells as an indicator cell population, demonstrated seroconversion by day 14 in this cohort of animals. High titer antibody responses were clearly demonstrable by day 28 in all animals. Thus the emergence of an anti-SIV_{mac} antibody response clearly correlated with the clearance of plasma antigenemia in these animals.

Cellular immune responses were then characterized in these acutely infected monkeys. Total peripheral blood lymphocyte (PBL) counts initially decreased coincident with the onset of plasma antigenemia; however, they rose in the ensuing 2 weeks to a level higher than pre inoculation values. PBL subset analyses were performed to characterize further this lymphocytosis. By approximately day 10 following infection, the percentage of $CD4^+$ PBL declined and remained persistently low. Both the percentage and the absolute number of $CD8^+$ cells rose in monkeys coincident with the

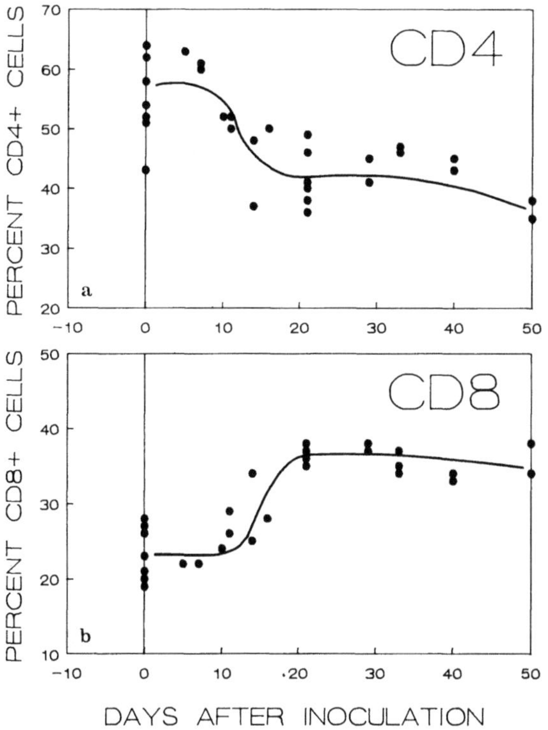

Fig. 1a, b. Changes in CD4:CD8 lymphocyte ratio associated with acute infection with SIV also occur in lymph node. Lymph node biopsies were obtained from eight rhesus monkeys before or immediately following acute infection with SIV and immunphenotyped for CD4 (**a**) and CD8 (**b**)

clearing of antigenemia. This increase in circulating $CD8^+$ lymphocytes then persisted following infection.

Lymph nodes were also sampled during the first days to weeks following infection with SIV_{mac}, and phenotypic changes in the T cell compartment of these nodes were assessed (Fig. 1). Coincident with the brief period of antigenemia, a drop in the percentage of $CD4^+$ lymph node T cells was demonstrated. This decline in the percentage of $CD4^+$ cells persisted. At the same time as this decline in the percentage of $CD4^+$ lymph node T cells, a rise in the percentage of $CD8^+$ lymph node T cells was documented. Since lymphadenopathy was clinically apparent in these animals by approximately 1 week following SIV_{mac} infection, this rise in the percentage of $CD8^+$ cells probably also represented an absolute increase in lymph node $CD8^+$ cells. These observations indicate that an increase in circulating and lymph node $CD8^+$ T cells correlated with the clearance of acute antigenemia in this cohort of monkeys.

We sought to characterize further this $CD8^+$ lymphocyte population by phenotypic analysis. PBL were examined before, during, and following antigenemia, assessing their state of activation (MHC class II expression), memory status (CD45RA expression), and adhesion potential (LFA-1 expression). Interestingly, no significant change in $CD8^+$ PBL expression of

MHC class II, CD45RA, or LFA-1 correlated with the clearance of plasma antigenemia. However, in studies of T cells from lymph nodes of these animals, both the expression of MHC class II and that of LFA-1 molecules were increased on $CD8^+$ lymphocytes following antigenemia. The expression of CD45RA was decreased in lymph node $CD8^+$ T cells following antigenemia. Thus, the $CD8^+$ lymphocytes had the phenotype of activated memory cells.

3 SIV_{mac} Specific CTL Generation Following Acute Infection

These studies indicated the emergence of a $CD8^+$ lymphocyte population in the lymph nodes and peripheral blood of the acutely infected animals coincident with the clearance of plasma antigenemia and replicating virus in lymph nodes. It was therefore of interest to determine whether these cells play a role in suppressing viremia in the monkeys. $CD8^+$ lymphocytes of SIV_{mac} and HIV-infected individuals can suppress AIDS virus replication in autologous PBL (WALKER et al. 1986; KANNAGI et al. 1988). We determined the time course for the generation of this virus inhibitory response in PBL from three of these prospectively studied monkeys. Peak SIV_{mac} replication was observed in both unfractionated and $CD8^+$ lymphocyte-depleted PBL obtained from these monkeys between days 6 and 12 following virus inoculation. Interestingly, the ability of the $CD8^+$ cells to inhibit SIV_{mac} replication was detectable as early as day 4 following experimental infection. This effect became much more pronounced 12–16 days following infection, coincident with both the inversion of the normal PBL CD4:CD8 ratio and the increase in absolute numbers of circulating $CD8^+$ lymphocytes. Since the $CD8^+$ lymphocytes which suppress AIDS virus replication have phenotypic characteristics of CTL (TSUBOTA et al. 1989), these observations suggest that CTL may be demonstrable in PBL of rhesus monkeys quite early following SIV_{mac} infection.

We sought to determine at what time following SIV_{mac} infection virus-specific CTL precursors are present in PBL of the rhesus monkey. To address this question we made use of our previous observation that SIV_{mac} gag recognition is limited to a single peptide fragment of the gag protein (residues 182–190, contained in the 25 amino acid gag peptide 11 and 12 amino acid peptide 11C) in those SIV_{mac} infected rhesus monkeys that express the MHC class I gene product *Mamu-A*01* (YAMAMOTO et al. 1990; MILLER et al. 1991). Two *Mamu-A*01*$^+$ rhesus monkeys were assessed prior to and every 2 days following inoculation with ten animal infectious doses of SIV_{mac} for SIV_{mac} gag peptide 11 specific CTL precursors in their PBL (Table 1). PBL were cultured in interleukin 2 (IL-2) supplemented medium with mitomycin C treated autologous PBL which had been pulsed with peptide 11 or an

Table 1. Precursors of SIV_{mac} gag peptide-specific cells were detectable in PBL of SIV_{mac} infected rhesus monkeys 4–6 days after virus inoculation

Day[a]	Stimulation[b]	Target[c]	Percentage specific lysis[d]							
			Mm 238–85				Mm 467–91			
			20:1	10:1	5:1	3:1	20:1	10:1	5:1	3:1
0	gag peptide 6	gag paptide 6	0	0	0	0	0	0	0	0
		gag paptide 11	1	0	1	0	0	1	0	0
	gag peptide 11	gag paptide 6	0	0	0	0	1	2	0	0
		gag peptide 11	3	0	0	0	2	0	1	0
2	gag peptide 6	gag peptide 6	0	0	0	0	0	1	0	0
		gag peptide 11	0	1	0	0	1	1	0	1
	gag peptide 11	gag peptide 6	5	1	1	0	1	0	0	0
		gag peptide 11	2	1	1	1	0	0	0	0
4	gag peptide 6	gag peptide 6	1	0	0	0	1	0	1	0
		gag peptide 11	4	2	0	0	3	0	0	0
	gag peptide 11	gag peptide 6	6	2	1	0	3	0	1	0
		gag peptide 11	29	20	15	8	3	1	0	0
6	gag peptide 6	gag peptide 6	1	1	0	0	6	5	4	1
		gag peptide 11	0	0	0	0	4	3	1	1
	gag peptide 11	gag peptide 6	4	1	1	0	11	8	6	3
		gag peptide 11	18	5	6	1	69	66	58	55

[a] PBL were obtained from a *Mamu-A*01*$^+$ rhesus monkey before 2, 4, and 6 days following SIV_{mac} infection.

[b] Effector cells were generated from density gradient isolated PBL of each infected monkey on the indicated day. Cells were cultured for 3 days at 2.5×10^6/ml with SIV_{mac} gas peptide 11 (VPGFQALSEGCTPYDINQMLNCVGD) or peptide 6 (WCHIHAEEKVKHTEEAKQIVQRHLVV) (5 µg/ml). IL-2 (400 U/ml) was added to the culture 3 days later. Peptide-pulsed autologous PBL were added as feeder and stimulator cells on day 6 of culture. Cytolytic assays were performed on day 11 of culture.

[c] Target cells were autologous B lymphoblastoid cell lines incubated with peptide 6 or 11 at a final concentration of 50 µg/ml for 16 h at 37 °C.

[d] Target cells were labeled with ^{51}Cr and incubated for 5 h with effector cells at the noted effector: target ratios.

irrelevant SIV_{mac} gag peptide (peptide 6). After 11 days in culture PBL were assessed for peptide-specific killing activity (Table 1). Peptide stimulation did not expand peptide 11 specific killing activity by PBL from these monkeys before or 2 days following SIV_{mac} inoculation. However, by 4 days after virus inoculation in one and 6 days after virus inoculation in the other monkey, peptide 11 stimulated PBL lysis of peptide 11 pulsed autologous target cells was readily demonstrable. Therefore, virus-specific CTL precursors were present in the PBL of these rhesus monkeys as early as 4–6 days following virus infection.

4 SIV_{mac} Specific CTL in the AIDS-Associated Skin Rash in Rhesus Monkeys

Although CTL play an important role in containing the spread of viral infections, recent studies suggest that they can also contribute to the pathologic manifestations of such infections. CTL have been shown to mediate the encephalitis induced by lymphocytic choriomeningitis virus infections in mice (BUCHMEIER et al. 1980). They have also been implicated in the liver damage which occurs in hepatitis B virus infections (BERTOLETTI et al. 1991). A skin rash occurs as a common clinical manifestation of infections with HIV or SIV_{mac} (JAMES et al. 1985; RINGLER et al. 1986). This rash is a common clinical sign of disease in many acutely HIV-infected humans. It is also seen in both acutely and chronically SIV_{mac} infected rhesus monkeys. Immunohistologic characterization of this exanthem has indicated that the underlying pathologic process is a dermal and epidermal infiltration of $CD8^+$ lymphocytes with variable epidermal injury (RINGLER et al. 1987).

To explore the function of the $CD8^+$ lymphocytes in skin rashes of SIV_{mac} infected rhesus monkeys, short-term cell lines were generated in parallel from skin-derived and peripheral blood derived lymphocytes of two animals with erythematous maculopapular eruptions comprised predominantly of $CD8^+$ lymphocytes. After concanavalin A stimulation and expansion for approximately 3 weeks in IL-2 supplemented medium, the skin-derived and peripheral blood derived lymphocytes had similar phenotypic profiles. The lymphocytes from both compartments were more than 70% $CD3^+$ and more than 65% $CD8^+$. However, the cells derived from the skin rashes exhibited significantly greater killing of target cells expressing structural proteins of SIV_{mac}. For example, in monkey 244, at an effector:target ratio of 12:1, SIV_{mac} gag-specific lysis by skin-associated lymphocytes was 58% while lysis by peripheral blood derived lymphocytes was 14%.

The cloning efficiency of SIV_{mac} specific effector T cells in skin rash-associated and peripheral blood derived lymphocytes was also compared in three virus-infected animals. After 2 weeks in cultures initiated with one or three cells per well under standard cloning conditions, the percentage of wells containing cells that lysed SIV_{mac} env- or gag-expressing autologous targets was determined. The cloning frequency of SIV_{mac} specific effector cells was higher in the skin-derived lymphocytes than in the PBL in each of three monkeys examined. These studies suggest that SIV_{mac} specific effector T cells are concentrated in the skin rash-associated lymphocytes of virus-infected monkeys.

The effector function of these skin-associated lymphocytes was then further characterized. Ten lymphocyte clones from monkey 152 that lysed autologous env-expressing target cells were assessed for their antigen specificity and the MHC class I restriction of their env recognition (Table 2). All of these clones lysed env- but not gag- or pol-expressing autologous

Table 2. MHC class I restricted, env-specific CTL are readily cloned from skin rash-associated lymphocytes of SIV_{mac} infected monkey 152 (percentage specific release[a])

Clone	Autologous			MHC class I mismatched[b]						K562
	152 env	152 gag	152 pol	165 env	244 env	108 env	174 env	179 env	458 env	
03061–4	55	0	0	3	16	21	6	0	12	2
–119	50	0	0	0	0	0	7	0	0	0
–131	62	0	0	0	0	0	0	0	0	0
03201–43	33	5	3	0	0	2	0	0	1	0
–62	46	1	2	0	0	0	7	0	4	3
–88	55	0	1	3	0	10	11	0	3	1
–115	31	0	2	2	0	14	12	0	6	1
–130	45	0	0	2	0	0	1	0	0	1
–155	55	0	5	1	0	4	6	0	3	0
–162	25	1	0	0	0	2	5	2	2	0

[a] Target cells were *Herpesvirus papio* transformed B lymphoblastoid cell lines infected with a recombinant vaccinia virus expressing either the entire envelope (env), gag, or polymerase (pol) gene of SIV_{mac}. Target cells (10^4) were ^{51}Cr labeled and incubated for 5 h with approximately 10^4 cloned effector cells. Spontaneous release, determined individually for each target cell population, varied from 15% to 28%. Percentage of specific release was calculated as (experimental release – spontaneous release)/(100% release – spontaneous release).

[b] Effector and target cells were assessed by one-dimensional isoelectric focusing to characterize their expressed MHC class I molecules.

target cells. Moreover, these cloned effector lymphocytes did not lyse any of six env-expressing allogeneic target cells. These cloned lymphocytes also did not lyse the natural killer cell sensitive target cells K562. Therefore, these cloned effector cells were SIV_{mac} env-specific , MHC class I restricted CTL. The phenotype of these cloned, env-specific effector lymphocytes was also determined. They were all $CD3^+$, $CD8^+$, and $CD4^+$; they also expressed the adhesion molecule CD29.

Eleven lymphocyte clones from monkey 244 that lysed autologous gag- but not env- or pol-expressing target cells were assessed for their MHC class I restriction. We have previously shown that monkey 244, as all other monkeys that express the MHC class I gene product *Mamu-A*01*, generate gag-specific PBL CTL activity restricted by *Mamu-A*01* (Yamamoto et al. 1990; Miller et al. 1991). The skin-derived, gag-specific CTL clones, as the PBL-derived gag-specific CTL, lysed gag-expressing *Mamu-A*01*$^+$ but not *Mamu-A*01*$^-$ target cells. We also sought to define the epitope-specificity of the gag recognition of these cloned CTL. We have previously demonstrated that gag-specific recognition by PBL CTL is restricted in *Mamu-A*01*$^+$ rhesus monkeys to a single epitope of the gag protein spanning amino acid residues 182–190 of the gag protein, contained in peptide 11 (Yamamoto et al. 1990; Miller et al. 1991). All of the skin rash derived gag-specific CTL from this *Mamu-A*01*$^+$ monkey demonstrated precisely the same gag epitope recognition, lysing peptide 11 pulsed target cells but not target cells pulsed with flanking peptide sequences. We have, thus, shown that SIV_{mac} specific CTL are present within skin rash associated $CD8^+$ lymphocyte populations.

5 Discussion

Events following acute intravenous infection of rhesus monkeys with SIV_{mac} parallel to a remarkable degree those seen in the few acutely infected HIV-infected humans who have been studied. Within weeks of infection lymphadenopathy and skin rash are seen in both. Moreover, antigenemia occurs within days to weeks following SIV_{mac} or HIV infection. In both, a virus-specific antibody and MHC class I restricted $CD8^+$ CTL response are generated. Lymphocyte subset changes, both a decline in $CD4^+$ and an increase in $CD8^+$ circulating lymphocyte populations, are seen. Finally, the early, high virus load that occurs in both settings is cleared.

The clearance of plasma viral antigen correlates with both the emergence of an anti-SIV_{mac} antibody response and a SIV_{mac} specific CTL response. Therefore, these studies do not provide evidence that CTL rather than antibodies are responsible for the clearance of this early burst of viral replication. However, the rapid emergence of a virus-specific, $CD8^+$ effector cell in both PBL and lymph nodes and the dramatic phenotypic changes in the $CD8^+$ lymphocyte populations in these animals provide further evidence for the likely importance of the cell-mediated immune response in clearing the early AIDS virus infection.

Although CTL in HIV-infected individuals probably represents an important mechanism for controlling HIV infections, data suggest that the persistence of virus-specific CTL during chronic infections may actually contribute to disease pathogenesis. CTL specific for HIV can be detected in lungs of patients with lymphocytic alveolitis, an HIV-related disorder frequently observed clinically before the onset of opportunistic infections. These $CD8^+$ CTL mediate cytotoxic activity against HIV-infected, MHC-compatible alveolar macrophages in vitro (PLATA et al. 1987). Clinical and pulmonary function abnormalities are correlated in affected individuals with the presence of HIV-specific CTL in lungs (AUTRAN et al. 1988). Similarly, an infiltration of lymphocytes, predominantly $CD8^+$, has been described in parotid and salivary glands of HIV-infected patients (SMITH et al. 1988). This inflammatory lesion, indistinguishable histologically from that seen in Sjögren's syndrome, may result from an infiltration of HIV-specific effector lymphocytes. The present study indicates that AIDS virus-specific CTL also are found in the skin and therefore play a role in the pathogenesis of the skin lesions which occur in AIDS.

References

Autran B, Mayaud CM, Raphael M, Plata F, Denis M, Bourguin A, Guillon JM, Debre P, Akoun G (1988) Evidence for a cytotoxic T-lymphocyte alveolitis in human immunodeficiency virus-infected patients. AIDS 2: 179–183

Bertoletti A, Ferrari C, Fiaccadori F, Penna A, Margolskee R, Schlicht HJ, Fowler P, Guilhot S, Chisari FV (1991) HLA class I restricted human cytotoxic T cells recognize endogenously synthesized hepatitis B virus nucleocapsid antigen. Proc Natl Acad Sci USA 88(23): 10445–10449

Buchmeier MJ, Welsh RM, Dutko FJ, Oldstone MBA (1980) The virology and immunology of lymphocyte choriomeningitis virus infection. Adv Immunol 30: 275–291

Chakrabarti L, Guyader M, Alizon M, Daniel MD, Desrosiers RC, Tiollais P, Sonigo P (1987) Sequence of simian immunodeficiency virus from macaque and its relationship to other human and simian retroviruses. Nature 328: 543–547

Clark SJ, Saag MS, Decker WD, Campbell-Hill S, Roberson JL, Veldkamp PJ, Kappes JC, Hahn BH, Shaw GM (1991) High titers of cytopathic virus in plasma of patients with symptomatic primary HIV-1 infection N Engl J Med 324: 954–960

Culmann B, Gomard E, Kieny MP, Guy B, Dreyfus F, Saimot AG, Sereni D, Levy JP (1989) An antigenic peptide of the HIV-1 Nef protein recognized by cytotoxic T-lymphocytes of seropositive individuals in association with different HLA-B molecules. Eur J Immunol 19: 2383–2386

Daar ES, Moudgil T, Meyer RD, Ho DD (1991) Transient high levels of viremia in patients with primary human immunodeficiency virus type 1 infection. N Engl J Med 324: 961–964

James WD, Redfield RR, Lupton GP, Meltzer MS, Berger TG, Rodman OG, Markham PD, Sarngadharan MG, Salahuddin SZ, Gallo RC (1985) A popular eruption associated with human T cell lymphotropic virus type III disease. J Am Acad Dermatol 13: 563–566

Kannagi M, Yetz JM, Letvin NL (1985) In vitro growth characteristics of simian T lymphotropic virus type III. Proc Natl Acad Sci USA 82: 7053–7057
Kannagi M, Chalifoux LV, Lord CI, Letvin NL (1988) Suppression of simian immunodeficiency virus replication in vitro by $CD8^+$ lymphocytes. J Immunol 140: 2237–2242
Koff WC, Hoth DF (1988) Development and testing of AIDS vaccines. Science 241: 426–432
Letvin NL (1990) Animal models for AIDS. Immunol Today 11: 322–326
Letvin NL, Daniel MD, Seghal PK, Desrosiers RC, Hunt RD, Waldron LM, MacKey JJ, Schmidt DK, Chalifoux LV, King NW (1985) Induction of AIDS-like disease in macaques with T-cell tropic retrovirus STLV-III. Science 230: 71–73
Miller MD, Yamamoto H, Hughes AL, Watkins DI, Letvin NL (1991) Definition of an epitope and an MHC class I molecule recognized by gag-specific CTL in SIV_{mac} infected rhesus monkeys. J Immunol 147: 320–329
Nixon DF, Townsend ARM, Elvin JG, Rizza CR, Gallwey J, McMichael AJ (1988) Gag-specific cytotoxic T-lymphocytes defined with recombinant vaccinia virus and synthetic peptides. Nature 336: 484–487
Oldstone MBA, Blount P, Southern PJ, Lampert PW (1986) Cytoimmunotherapy for persistent virus infection reveals a unique clearance pattern from the central nervous system. Nature 321: 239–243
Plata F, Autran B, Pedroza-Martins L, Wain-Hobson S, Raphael M, Mayaud C, Denis M, Guillion JM, Debre P (1987) AIDS virus-specific cytotoxic T-lymphocytes in lung disorders. Nature 328: 348–351
Quinnan GV, Kirmani N, Rook AH, Manischewitz JF, Jackson L, Moreschi G, Santos GW, Saral R, Burns WH (1982) Cytotoxic T cells in cytomegalovirus infection N Engl J Med 307: 7–13
Ringler DJ, Murphy GF, King NW (1986) An erythematous maculopapular eruption in macaques infected with an HTLV-III-like virus (STLV-III). J Invest Dermatol 87: 674–677
Ringler DJ, Hancock WW, King NW, Letvin NL, Daniel MD, Desrosiers RG, Murphy GF (1987) Immunophenotypic characterization of the cutaneous exanthem of SIV-infected rhesus monkeys: apposition of degenerative Langerhans cells and cytotoxic lymphocytes during the development of an acquired immunodeficiency syndrome. Am J Pathol 126: 199–207
Smith FG, Rajdeo H, Panesar N, Bhuta K, Stahl R (1988) Benign lymphoepithelial lesion of the parotid gland in intravenous drug users. Arch Pathol Lab Med 112: 742–745
Tsubota H, Lord CI, Watkins DI, Morimoto C, Letvin NL (1989) A cytotoxic T lymphocyte inhibits AIDS virus replication in peripheral blood lymphocytes. J Exp Med 169: 1421–1434
Walker CM, Moody DJ, Stites DP, Levy JA (1986) $CD8^+$ lymphocytes can control HIV infection in vitro by suppressing virus replication Science 234: 1563–1566
Walker BD, Chakrabarti S, Moss B, Paridis TJ, Flynn T, Durno AG, Blumberg RS, Kaplan JC, Hirsh MS, Schooley RT (1987) HIV-specific cytotoxic T lymphocytes in seropositive individuals. Nature 328: 345–348
Walker BD, Flexner C, Paradis TJ, Fuller TC, Hirsch MS, Schooley RT, Moss B (1988) HIV-1 reverse transcriptase is a target for cytotoxic T-lymphocytes in infected individuals. Science 240: 64–66
Watanabe M, Ringler DJ, Nakamura M, DeLong PA, Letvin NL (1990) Simian immunodeficiency virus inhibits bone marrow hematopoietic progenitor cell growth. J Virol 64: 656–663
Yamamoto H, Miller MD, Tsubota H, Watkins DI, Mazzara GP, Stallard V, Panicali DL, Aldovini A, Young RA, Letvin NL (1990) Studies of cloned simian immunodeficiency virus-specific T lymphocytes: gag-specific cytotoxic T lymphocytes exhibit a restricted epitope specificity. J Immunol 144: 3385–3391
Yap KI, Ada GL, McKenzie IPC (1978) Transfer of specific cytotoxic T-lymphocytes protects mice inoculated with influenza virus. Nature 273: 238–240

Envelope Sequence Variation, Neutralizing Antibodies, and Primate Lentivirus Persistence

D.P.W. BURNS and R.C. DESROSIERS

1 Introduction

Many of the critical features of the neutralizing antibody response to HIV and SIV infection have yet to be elucidated. It is not known to what extent the primary neutralizing antibody response of the host is directed at linear versus complex epitopes or to V3 versus non-V3 determinants. Neutralization escape mutants appear to arise during the course of infection of single individuals, but the variable regions primarily responsible for this escape have not been defined. Most importantly, we do not know the full impact of the emergence of neutralization-resistant variants on the maintenance of persistent infection and the evolution of the chronic disease course. In this review we summarize the evidence for immune selection of escape mutants among the more primitive ungulate lentiviruses and discuss corresponding information in SIV and HIV primate systems. In doing this, we describe the

New England Regional Primate Research Center, Harvard Medical School, P.O. Box 9102, One Pine Hill Drive, Southborough, MA 01772-9102, USA

Current Topics in Microbiology and Immunology, Vol. 188

nature of sequence variation in SIV and HIV envelope and discuss the selective forces which may be driving sequence change in each of the variable domains. Despite gaps in existing knowledge, we conclude that neutralizing antibodies are important in limiting HIV and SIV replication in vivo. Specific selection of sequence variants that are resistant to neutralization argues strongly that neutralizing antibodies do serve to limit HIV and SIV replication during the lengthy asymptomatic stage of infection.

2 Immune Selection Among the Ungulate Lentiviruses

During the past two decades antigenic variation during persistent infection and the concept of humoral immune selection have been studied in three lentivirus systems: visna virus, caprine arthritis-encephalitis virus (CAEV), and equine infectious anemia virus (EIAV). Results from these lentivirus systems suggest that the host neutralizing antibody response selects for neutralization-resistant mutants during the course of infection of a single individual. These findings suggest that antigenic variation is likely to contribute to persistent viral replication in chronically infected animals.

Early studies with visna virus demonstrated that antigenic variants emerge in vivo during persistent infection of sheep (NARAYAN et al. 1977b, 1978; PÉTURSSON et al. 1976). Antigenic variants emerged in infected cell cultures treated with neutralizing antibodies (DUBOIS-DALCQ et al. 1979; NARAYAN et al. 1977a, 1981) and in long-term persistently infected sheep (LUTLEY et al. 1983; THORMAR et al. 1983). Antigenic variants in these studies were typed by a standard virus neutralization assay using polyclonal antisera (THORMAR 1963). More detailed study of these variants revealed that differences in antigenicity were due to single nucleotide changes in the envelope gene (CLEMENTS et al. 1980; SCOTT et al. 1979). Common genomic changes, as detected by oligonucleotide fingerprinting, were found between antigenic variants isolated independently from two persistently infected sheep and from antibody-treated cell cultures. Cumulative genomic changes were also found among antigenic variants from a single infected sheep (CLEMENTS et al. 1982).

In general, there seems to be agreement among visna researchers that antigenic variants emerge over time in persistently infected sheep; however, there is disagreement about the role of these variants in the pathogenesis of the disease. Some researchers believe that virulent mutant viruses arise at unpredictable times during the course of persistent infection, and that these mutants cause new pathological lesions which cumulatively lead to clinical disease (GRIFFIN et al. 1978; NARAYAN et al. 1978). Others believe that antigenic drift (defined as consistent isolation of variant viruses) is not essential for virus persistence or for the development of clinically evident central nervous system lesions, and that antigenic variants do not have a role in the pathogenesis of the disease. These latter conclusions are based on studies of long-term persistently infected sheep in which variants emerged at low frequencies and did not replace the parental strain (LUTLEY et al. 1983; THORMAR et al. 1983).

Disparate results in these early visna studies may be understood at least in part through awareness of the methods that were used. Since the polymerase chain reaction technique was not yet available for the earlier visna studies, variant viruses had to be isolated in cell culture. Thus, major forms of pathogenic and antigenically altered variants may have been missed during virus isolation due to cell culture selection (MEYERHANS et al. 1989; OVERBAUGH et al. 1988). Also, the antigenic typing method that was used in the early studies of persistently infected sheep was capable of detecting only major antigenic shifts. Therefore, subtle antigenic changes were not detected. Monoclonal antibodies have been used recently to demonstrate more subtle antigenic differences among visna variants. Results with these monoclonals suggest that visna virus variants may emerge more fequently than predicted by the earlier serological tests, and that the extent of phenotypic changes in the envelope glycoprotein may be greater than previously thought (STANLEY et al. 1987). The use of monoclonal antibodies should be useful in the future for clarifying the effects of specific sequence variations on immune recognition.

In addition to visna virus, antigenic variation during persistent infection has also been examined in the CAEV system. In general, CAEV is not as well studied as visna because most animals do not produce detectable virus-neutralizing antibodies to the virus (KLEVJER–ANDERSON and MCGUIRE 1982). By immunizing goats with virus and large amounts of inactivated *Mycobacterium tuberculosis*, neutralizing antibodies were first produced against CAEV, and antigenic variants were first identified (NARAYAN et al. 1984). In a study of six CAEV-infected goats an antigenic variant was isolated from one goat at 78 weeks postinfection, demonstrating for the first time that antigenic variation occurs during persistent infection of an individual goat (ELLIS et al. 1987). As with visna virus, there is a disparity of opinion regarding the role of such CAEV variants in pathogenesis. One camp believes that antigenic variants do not have a role in the severity of disease (ELLIS et al. 1987) while the other believes that the variants are associated with severity of disease (CHEEVERS et al. 1991; MCGUIRE et al. 1988).

Unlike the slow progressive diseases produced by visna and CAEV, the disease produced by EIAV is unique. Equine infectious anemia is characterized by recurrent episodes of fever, hemolytic anemia, bone marrow depression, lymphoproliferation, and immune complex glomerulonephritis. Bursts of viremia which occur during these febrile episodes are separated by time periods of several weeks or months allowing sequential isolations of virus from individual animals (MONTELARO et al. 1984). The EIAV system is therefore well suited to study immune selection and to investigate the role of antigenic variants in the pathogenesis of lentivirus disease.

Like visna virus and CAEV, early studies in the EIAV system demonstrated the existence of antigenic variants (KONO et al. 1971). Due to the cyclical nature of the clinical disease, it was somewhat easier in this system to study antigenic variation over time in chronically infected animals. Taking advantage of the periodic relapses, KONO et al. (1973b) were able to isolate virus at different time points postinfection representative of different clinical episodes. Serum samples collected at the same time points as the virus isolations were used in neutralization assays to detect antigenic differences. The neutralization tests demonstrated that successive virus isolates were not susceptible to previously formed neutralizing antibody but were susceptible to antibody produced some weeks later (KONO et al. 1973b). Thus, there were delayed neutralizing antibody responses against variant viruses in individual animals. Late virus isolates were also antigenically different from each other. Each of five variant virus isolates was inoculated into five different horses, and immune sera against the variants were obtained at various time points postinfection. Neutralization tests with these variant sera demonstrated that each virus reacted strongly with its homologous antiserum but only weakly or not at all with heterologous sera (KONO et al. 1973b). These cross-neutralization experiments provided strong evidence for immune selection in persistently infected horses.

In order to study the nature of these antigenic variants, oligonucleotide mapping experiments were later used by other researchers to show genomic differences betwen EIAV variants (PAYNE et al. 1984). Techniques such as sodium dodecyl sulfate–polyacrylamide gel electrophoresis and peptide mapping were used to demonstrate electrophoretic mobility differences and structural variations among antigenic variants; these changes were localized to the virion surface glycoproteins gp90 and gp45 (MONTELARO et al. 1984). Glycopeptide mapping analyses revealed alterations in glycosylation patterns of these virion surface proteins, and a monoclonal antibody against gp90 was further able to distinguish between variants of EIAV by recognizing more subtle antigenic changes (SALINOVICH et al. 1986). DNA sequencing experiments also revealed regions of clustered base substitutions in the envelope gene of EIAV isolates recovered during sequential clinical episodes from a single infected pony (PAYNE et al. 1987a).

Studies designed to identify neutralization epitopes on EIAV envelope glycoproteins subsequently led to the discovery that these variable regions serve as targets for neutralizing antibodies. From a panel of 18 monoclonal antibodies directed against EIAV surface glycoproteins, four had neutralizing activity and were found to be directed against the gp90 glycoprotein. The neutralizing monoclonal antibodies were believed to be directed against variable regions of gp90 since they displayed unique patterns of reactivity with different isolates of EIAV (HUSSAIN et al. 1987). Subfragments of the

EIAV envelope protein were used in western blot analyses, and it was found that these NMAbs did in fact map to a variable region of the EIAV envelope gene. The variable region was a relevant immunogenic domain in vivo since sera from EIAV-infected horses displayed variable patterns of reactivity against this region (PAYNE et al. 1989).

Based on the experiments described above, there seems to be a substantial amount of evidence in the EIAV system supporting the emergence of neutralization-resistant variants during the course of persistent infection. Studies of antigenic variants obtained from successive virus isolations suggest that the EIAV variants may have a role in the cyclical nature of clinical episodes and persistence of the virus. Equine infectious anemia, however, remains a complex disease. For example, in the early studies by KONO et al. (1973b) one of five horses was found to be viremic, even during afebrile stages of infection. In another study by PAYNE et al. (1987b), two Shetland ponies were infected in parallel with identical inocula. Successive virus isolations from these two animals revealed that each isolate was structurally unique. Peptide and oligonucleotide mapping experiments indicated that the variation observed in these two animals was random and noncumulative (PAYNE et al. 1987b), contrasting with findings in the visna virus system. In a study of 16 isolates from four EIAV-infected horses, KONO (1988) also concluded that antigenic variation is random and has no fixed pattern of variation. Interpretation of these results is complicated by the use of uncloned virus stocks that had not been plaque purified and the lack of sophisticated cloning and sequencing strategies that are currently available.

Other conflicting reports in the EIAV system reflect the complexity of this lentivirus disease. One study has suggested that variation may emerge as a result of pressure from the cellular, rather than the humoral, branch of the immune system, and that variants may also emerge from nonimmunological selection processes (CARPENTER et al. 1987). Cellular immunity was also implicated in studies of horses which recovered from EIA. In these studies it was found that recovered horses often resisted challenge with an immunologically heterologous virulent virus even though the horses did not have detectable levels of neutralizing antibodies to the challenge virus (KONO 1973, 1988; KONO et al. 1973a).

Overall, evidence for the emergence of antigenic variants in these three lentiviral systems is quite convincing. The role of antigenic variants in the pathogenesis of disease is not as clear. The strongest evidence for immune selection lies in the EIAV system where delayed neutralizing antibody responses against variant viruses have been demonstrated in individual animals, and where reciprocal neutralization tests have demonstrated type-specific responses against variant strains of the virus. The studies by CARPENTER et al. (1987) raise the interesting possibility of cellular (rather than or in addition to humoral) immune selection and selection from nonimmunological processes (such as cell tropism, replication rate, etc.). In light of the complexity of lentiviral disease, there may indeed be multiple selective forces operating at any one time in infected animals, and some of these selective forces may be stronger than others. It is entirely possible that the humoral branch of the immune system may not be the major selective force for change in infected animals in some circumstances or for some variable domains. As molecular clones become available for use in these systems, more detailed studies should provide further insight into the selective processes operating in vivo.

3 SIV Neutralization Epitopes

The SIV_{mac251} strain of virus was first isolated in tissue culture in 1984 (DANIEL et al. 1985). The SIV_{mac251} isolate has been shown to induce AIDS reproducibly in rhesus monkeys. In one study, four of six rhesus monkeys (*Macaca mulatta*) inoculated with this virus died with AIDS by 160 days post-inoculation with signs including wasting, opportunistic infections, primary retroviral encephalitis, and decreased $CD4^+$ cell counts (LETVIN et al. 1985). In another study of 16 animals inoculated with the SIV_{mac251} isolate, 50% died rapidly with AIDS in less than 1 year. The other 50% became persistently infected and lived for several years before developing a more protracted form of disease which also closely resembled AIDS in humans. Some of the infected animals developed very high levels of neutralizing antibodies with titers as high as 1:20480–1:81 920 (DANIEL et al. 1987). In general, animals with high antibody titers (as measured by both neutralization and enzyme-linked immunosorbent assay) tend to be long-term survivors (DANIEL et al. 1987; KANNAGI et al. 1986).

The SIV envelope protein is produced from a spliced, subgenomic-length mRNA species (PARK et al. 1991) as a glycosylated precursor polyprotein gp160 (KANKI et al. 1985). The SIV gp160 precursor is posttranslationally cleaved into an external gp120 protein (SU) and a transmembrane protein (TM) (VERONESE et al. 1989). Natural SIV transmembrane protein has a size of 41 kDa, but many isolates have truncated forms of 28–34 kDa. Truncated forms of the transmembrane protein appear to be selected by growth of virus in human cell lines (KODAMA et al. 1989).

Several groups have begun to identify and characterize neutralization epitopes in the envelope of SIV. One weak type-specific neutralizing determinant was identified in a variable region of the transmembrane protein of SIV_{mac}; this determinant was identified and mapped by analyzing the sensitivity of a neutralizing monoclonal antibody to changes at amino acid residues 106–110 of the TM (631–635 of entire envelope; KODAMA et al. 1991). KENT et al. (1992) and COLLIGNON et al. (1993a) have identified a number of anti-gp120 monoclonal antibodies that neutralize SIV_{mac} infectivity.

Most of those that were found to neutralize infectivity recognized conformational or discontinuous epitopes; these did not react with denatured protein on western blots and/or did not react with *Escherichia coli* produced subfragments or synthetic peptides. Several, however, reacted with peptides around amino acids 170–190 of gp120 (TMTGLKRDKTKEYNETWYSTC). BENICHOU et al. (1992) also identified two neutalizing mAbs that react with SIV_{mac} env amino acids 170–190. Using peptides to elicit antibodies, TORRES et al. (1993) found that amino acids 410–430 in a variable region of gp120 contain a neutralization epitope. Using synthetic peptides that span most of the envelope region of HIV-2, BJORLING et al. (1991) have mapped a number of neutralizing epitopes of HIV-2. HIV-2, SIV_{mac}, and SIV_{sm} are relatively closely related and form a single subgroup of primate lentiviruses distinct from other SIVs and HIV-1. Neutralizing domains mapped with this approach include amino acids 119–137, 310–340, 480–510, 595–614, and 714–729. The 595–614 epitope in the HIV-2 study of BJORLING et al. (1991) corresponds to the neutralizing determinant in SIV_{mac} identified by KODAMA et al. (1989). However, the 170–190 and 410–430 neutralization determinants identified in SIV_{mac} did not score in the HIV-2 study of BJORLING et al. The 310–330 epitope of BJORLING et al. corresponds in location to V3 of HIV-1, a major target of neutralizing antibodies of HIV-1 (see below). These "V3" results of BJORLING et al. are difficult to reconcile, at least at the present time, with extensive studies of other groups who have found that the "V3" corresponding region of SIV_{mac} does not serve as a target of neutralizing antibodies (JAVAHERIAN et al. 1992). A summary of the reported targets of SIV_{mac}/HIV-2 neutralization is presented in Fig. 1.

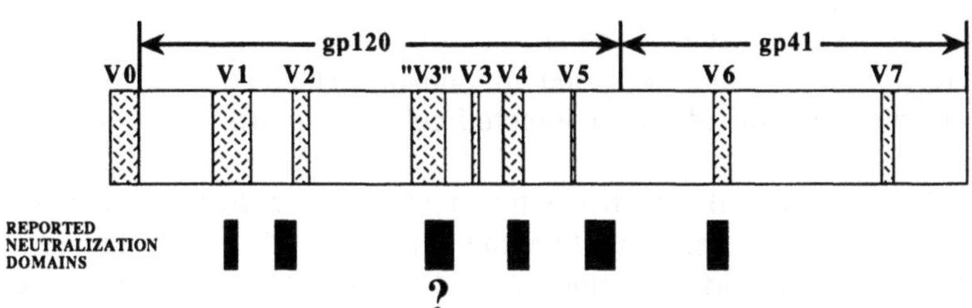

Fig. 1. Neutralization domains reported for SIV_{mac}/HIV-2/SIV_{sm}. *Stippled areas* variable domains. *V0*, 1–22 (signal peptide); *V1*, 105–145; *V2*, 188–203; *"V3"*, 311–344; *V3*, 371–377; *V4*, 404–426; *V5*, 474–478; *V6*, 622–637; *V7*, 790–810. Cleavage site to form gp120 and gp41, 525/526. Reported neutralization domains for SIV_{mac} are 170–190, 410–430, and 631–636. Reported neutralization domains for HIV-2 are 119–131, 325–355, 490– 520, and 615–635. Landmark locations are based on the amino acid sequence of SIV_{mac239}. The question mark for 325–355 indicates discrepant results (see text). *"V3"* is the sequence in SIV_{mac}/HIV-2/SIV_{sm} which corresponds to V3 in HIV-1, but which has not varied in several studies with SIV_{mac} (see text)

4 HIV-1 Neutralization Epitopes

Neutralizing antibodies against HIV-1 in sera from AIDS and pre-AIDS patients were originally reported in 1985 by several groups (CLAVEL et al. 1985; HO et al. 1985; MCDOUGAL et al. 1985; ROBERT-GUROFF et al. 1985; WEISS et al. 1985). Although detectable, the neutralizing titers reported by these groups were low, even though several different methods were used to measure neutralization. WEISS et al. (1985) conducted experiments which included the use of homologous serum and virus isolates to determine whether antigenic variation was at least partially responsible for the low neutralizing titers. The neutralization titers were still low despite these attempts. Based on analogies with other enveloped viruses including other retroviruses, the external envelope glycoprotein was thought to be the target of neutralizing antibodies in these original studies (ROBERT-GUROFF et al. 1985; WEISS et al. 1985). Later, it was shown that neutralizing antibodies from HIV-1-infected persons (MATTHEWS et al. 1986) or immunized animals (LASKY et al. 1986; ROBEY et al. 1986) were in fact largely directed against the gp120 protein.

By immunizing goats with a recombinant segment of the HIV-1 envelope protein (PB1) or with deglycosylated gp120 protein, PUTNEY et al. (1986) were able to raise neutralizing antibodies against HIV-1. The neutralizing titers of these sera were equivalent to those of sera raised against native gp120. As with the neutralization titers of human sera, the titers of these experimental animal sera were low (ranging from 1:20 to 1:60 in two different types of neutralization assays). By immunizing goats with HIV-1 envelope protein (r160-III$_B$) produced in insect cells from a *Baculovirus* expression vector, higher neutralizing antibody titers were generated against HIV-1 (RUSCHE et al. 1987). Neutralizing titers of this new goat sera ranged from 1:190 to 1:1600 in three different types of neutralization assays (RUSCHE et al. 1987). It was thought that either the presentation of antigen as an entire 160-kDa protein or the specific glycosylation pattern of insect cells was responsible for the increased antigenicity. The new goat sera was also able to block fusion of HIV-1 infected cells in culture (RUSCHE et al. 1987).

The PB1 subfragment which elicited neutralizing activity in animals encompassed most of the carboxy-terminal half of gp120 and contained a stretch of 180 amino acids (residues 295–474; RATNER et al. 1985). By immunizing goats with a tetanus toxoid coupled synthetic peptide, PALKER et al. (1988) were able to map the neutralization epitope to a region spanning amino acids residues 303–321. The synthetic peptide generated both neutralizing and fusion-inhibiting antibodies against HIV-1, and the sera were type specific; antibodies raised against the HTLV-III$_B$ peptide neutralized and inhibited fusion against HTLV-III$_B$ but not against HTLV-III$_{RF}$ of HTLV-III$_{MN}$ strains (PALKER et al. 1988).

Using a slightly different approach, RUSCHE et al. (1988) also mapped the PB1 region of the HIV-1 envelope. Using the goat sera described above against PB1 (PUTNEY et al. 1986) and r160-IIIB (RUSCHE et al. 1987) in addition to goat sera raised against three subfragments of PB1, it was found that the amino-terminal one-third contained the relevant epitope. Smaller synthetic peptides covering this region were then tested for their ability to block fusion-inhibition activity of anti-PB1 and anti-r160 sera. In this way the epitope was narrowed to a 24-amino acid peptide (termed RP135, amino acids 307–330; RUSCHE et al. 1988). Goat serum raised against this peptide was capable of inhibiting fusion of HIV-1 infected cells. The peptide was also able to block fusion-inhibiting activity of serum from a chimpanzee infected with HTLV-III_B. By changing two amino acids within this region, the ability of this region to block fusion-inhibition was lost (RUSCHE et al. 1988), indicating that this region is type specific.

Working independently, GOUDSMIT et al. (1988b) also mapped a neutralization epitope within the PB1 region. This group had originally located the epitope by using the pepscan technique with rabbit sera raised against the C-terminal portion of the HTLV-III_B gp120 protein. The PEPSCAN technique is a method of scanning antigens for immunogenic regions in which a series of short, overlapping peptides are synthesized on solid supports and reacted with sera in a microtiter plate format (GEYSEN et al. 1984, 1987). By pepscanning with the rabbit sera, GOUDSMIT et al. (1988b) identified four domains on the envelope, one of which bound to all chimpanzee sera that neutralized HTLV-III_B. The peptide which bound the chimpanzee sera represented amino acids 307–321 of the HTLV-III_B envelope. As the others, GOUDSMIT et al. (1988b) found this epitope to be type specific; chimpanzee sera which did not neutralize HTLV-III_B also did not bind this peptide. By comparing sequences of clones of HTLV-III_B and HTLV-III_{RF}, these researchers deduced that residues RIQR and AFV, flanking a highly conserved β-turn (GPGR), were responsible for the type-specific response (GOUDSMIT et al. 1988b). A subsequent study of the binding region indicated that at least five consecutive amino acids including the β-turn GPGR were essential for antibody binding (MELOEN et al. 1989).

The epitope mapped in the PB1 region lies between cysteine residues 303 and 338 (RATNER et al. 1985) which are joined by a disulfide bridge (LEONARD et al. 1990). This region is also the third variable region of the HIV-1 envelope (STARCICH et al. 1986). For these reasons, this region is commonly referred to as the "cysteine loop" or "V3 loop." Based on the three studies which mapped a neutralization epitope to this region, the loop is also commonly referred to as the "principal neutralizing determinant" (PND) of HIV-1. While it was essential to use experimental animal sera from goats and chimpanzees to map this region, human sera from HIV-1 infected persons has been shown to bind to peptides in the PND region (BROLIDEN et al. 1992; DEVASH et al. 1990; GOUDSMIT et al. 1988a; KENEALY et al. 1989; PALKER et al. 1988; PROFY et al. 1990).

To determine the extent of sequence divergence in the variable loop region, a large study was undertaken in which clones from 245 random field isolates of HIV-1 were analyzed (LAROSA et al. 1990). One of the most significant findings from this study was the amount of conservation in the center of the loop region. The central proline and the two flanking glycines were highly conserved; the GPG tripeptide sequence was found in 237 of the 245 sequences analyzed. Conservation at six other sites was also noted where particular amino acids were found in 80% or more of the sequences. There was also a tendency to conserve amino acid charge at certain positions, indicating that one constraining influence may be a structural motif. The PND sequences of HTLV-III_B and LAV-1 (BRU) were among some of the rarer sequences in comparison to the consensus PND (LAROSA et al. 1990). This is important since the HTLV-III_B and LAV-1 (BRU) strains have been frequently used in HIV-1 serological studies.

By constructing peptides of the loop region with varying sequences it was later found that broadly neutralizing antibodies could be raised against peptides containing the central consensus sequence GPGRAF (JAVAHERIAN et al. 1990). In a more recent study, both broad and type-specific antibodies were also generated against the V3 loop (LAMAN et al. 1992). Thus, type-specific neutralizing antibodies may be raised against the more variable regions of the PND which surround the center of the loop, while group-specific neutralizing antibodies may be raised against more conserved regions, such as the central GPGRAF sequence (JAVAHERIAN et al. 1990). These studies of the PND indicate that the V3 loop is variable, but that there are constraints at paticular amino acid positions. Such conserved regions, which are most likely necessary for preserving protein structure and function of the virus, may be exploited in order to raise more broadly reactive neutralizing antibodies.

In addition to the V3 epitopes described above, there have been other linear neutralization epitopes found on the HIV-1 envelope (Table 1) (ÅKERBLOM et al. 1990; BROLIDEN et al. 1990, 1992; CHANH et al. 1986; DALGLEISH et al. 1988; DURDA et al. 1990; FUNG et al. 1992; GNANN et al. 1987a,b; HO et al. 1987, 1988; KENEALY et al. 1989; KENNEDY et al. 1986; MATSUSHITA et al. 1988; SCHRIER et al. 1988; SKINNER et al. 1988; SUN et al. 1989; THOMAS et al. 1988). Some of these additional reports confirm the PND determinants described above, while others identify additional or different regions of the envelope which are important for neutralization. The epitopes are listed in Table 1 by their amino acid residues according to numbering systems of published sequences for HIV-1 (MYERS et al. 1988; RATNER et al. 1985; WAIN-HOBSON et al. 1985). The epitopes are classified as variable or conserved based on the alignment by STARCICH et al. (1986). Although it remains in Table 1, one of the linear neutralization epitopes (group 2, region C2) has recently been called into question by another research goup. RONCO et al. (1991) have recently attempted to repeat the experiments of HO et al. (1988) using peptides corresponding to the C2 region and were unsuccessful in their attempts to generate neutralizing antibodies against HIV-1.

All of these additional linear epitopes were identified by the same strategies used for mapping the PND. Basically, animals were immunized

Table 1. HIV-1 Linear Neutralization Epitopes

Group[a]	Amino acids	Sequence[b]	Region[c]	Protein	Reactivity with human sera	Reference
1	169–183	HTLV-IIIB	V2	gp120	NR	Fung et al. (1992)
2	254–274	HTLV-IIIB	C2	gp120	–	Ho et al. (1988)
3	279–472	LAV-1	V3	gp120	NR	Thomas et al. (1988)
	288–472	HTLV-IIIB	V3	gp120	+	Kenealy et al. (1989)
	298–314	HTLV-IIIB	V3	gp120	NR	Ho et al. (1987)
	303–321	HTLV-IIIB	V3	gp120	+	Palker et al. (1988)
	304–323	HTLV-IIIB	V3	gp120	NR	Akerblom et al. (1990); Broliden et al. (1990)
	304–318	HTLV-IIIB	V3	gp120	+	Broliden et al. (1992)
	307–330	HTLV-IIIB	V3	gp120	NR	Matsushita et al. (1988); Rusche et al. (1988); Skinner et al. (1988)
	307–321	HTLV-IIIB	V3	gp120	+	Goudsmit et al. (1988b)
	308–322	HTLV-IIIB	V3	gp120	NR	Durda et al. (1990)
4	423–437	HXB2	CD4 Binding	gp120	NR	Sun et al. (1989)
5	458–484	HTLV-IIIB	V5	gp120	NR	Ho et al. (1987)
6	489–508	HTLV-IIIB	C4	gp120	+	Broliden et al. (1992)
7	503–532	HTLV-IIIB	Slightly variable	gp120/gp41	NR	Chanh et al. (1986)
	508–537	LAV-1	Slightly variable	gp120/gp41	NR	Chanh et al. (1986)
8	598–609	LAV-1	Conserved	gp41	+	Gnann et al. (1987b); Schrier et al. (1988)
	603–609	LAV-1	Conserved	gp41	+	Gnann et al. (1987a)
9	616–632	HTLV-IIIB	Conserved	gp41	NR	Ho et al. (1987)
10	647–671	HTLV-IIIB	Slightly variable	gp41	+	Broliden et al. (1992)
11	728–751	HTLV-IIIB	Conserved	gp41	NR	Ho et al. (1987)
	732–746	HTLV-IIIB	Conserved	gp41	+	Broliden et al. (1992)
	735–752	HTLV-IIIB	Conserved	gp41	+	Chanh et al. (1986); Dalgleish et al. (1988); Kennedy et al. (1986)
	740–757	LAV-1	Conserved	gp41	NR	Chanh et al. (1986)

NR, Not reported.
[a] Overlapping epitopes mapped by different research groups.
[b] Sequence used for numbering amino acids.
[c] Variable and conserved regions are designated by a "V" or "C" and are numbered throughout gp120.

with either envelope proteins, fragments, or peptides, and the sera were pepscanned to identify antigenic regions. Monoclonal antibodies were isolated from some of these immunized animals and were mapped through the use of synthetic peptides. Sera raised against peptides were tested for neutralizing activity, and peptides were measured for their ability to block fusion inhibition activity of HIV-1 positive sera. Some of the antigenic peptides were also tested for their ability to be recognized by human sera from HIV-1 infected patients.

From Table 1 one can see that there have been many linear neutralization epitopes identified in the HIV-1 envelope. These epitopes are located in both gp120 and gp41 regions of the envelope and may be found in both variable and conserved regions. Approximately half of the regions identified have been reported to be reactive with human HIV-1 positive sera, indicating that these epitopes are relevant immunogenic domains.

Studies on HIV-1 neutralization indicate that sera from infected patients can neutralize a broad range of isolates (BERKOWER et al. 1989; HO et al. 1987; ROBERT-GUROFF et al. 1985; WEISS et al. 1986). One explanation for the broad specificity is that HIV-1 infected individuals mount type-specific responses to a large number of distinct antigenic variants (CHENG-MAYER et al. 1988; LOONEY et al. 1988; PROFY et al. 1990). Alternatively, there may be conserved or group-specific determinants on the envelope protein (BERKOWER et al. 1989; EVANS et al. 1989; PROFY et al. 1990). Studies of V3 loop diversity have shown that both type- and group-specific neutalizing antibodies can be raised against variable and conserved regions of the PND (JAVAHERIAN et al. 1990; LAMAN et al. 1992). In a recent report, HARIHARAN et al. (1993) describe a unique approach to study the cross-reactive nature of the neutralizing antibody response during HIV-1 infection. Murine anti-idiotypic monoclonal antibodies were generated against cross-reactive anti-gp120 polyclonal antibodies purified from HIV-1 positive sera. The anti-idiotypic antibodies were used to affinity-purify antibodies from HIV-1 positive sera. The idiotype-positive antibodies purified from the HIV-1 positive sera were found to be heterogeneous; they had unique epitope specificities and distinct neutralizing activities. Most of the idiotype-specific antibodies were directed against CD4 binding regions of gp120 and showed a high degree of cross-reactivity, while a small proportion showed specificity for the V3 region of gp120 and exhibited limited cross-neutralizing activities (HARIHARAN et al. 1993).

5 Conformational Neutralization Determinants

As discussed in the above section, one of the most commonly used approaches to identify neutralization epitopes in the HIV-1 system has been the use of synthetic peptides. While such approaches are useful for identifying simple linear epitopes, these methods do not detect complex conformational or discontinuous neutralization determinants which may be more important for immunogenicity during natural infection. Such approaches also do not take into consideration the influence of variable regions and distant regions of the envelope glycoprotein. Recently, researchers have begun to recognize these important issues and are now using approaches designed to detect the more complex determinants. In reviewing the literature, it is important to discriminate between neutralizing antibodies raised to purified env protein, fragments or peptides in adjuvant and those induced by natural infection. Peptides, protein fragments, and even env protein purified from mammalian cells may not retain native structure or may retain it only partially. The nature of neutralizing antibody induced by partially or fully denatured protein or fragments may be quite different from that resulting from natural infection. It is important to remember that the ultimate goal is to understand the nature of the neutralizing response resulting from natural infection.

Several conformation-dependent neutralizing monoclonal antibodies have been described recently in the literature. Ho et al. isolated a human (Ho et al. 1991b) and a murine (Ho et al. 1991a) monoclonal antibody (15e and G3-4, respectively) which neutralized HIV-1 virus but did not react with synthetic or recombinant peptides of the HIV-1 envelope. Interestingly, these monoclonals were capable of neutralizing a number of diverse HIV-1 isolates and blocking the binding of gp120 to soluble CD4 (Ho et al. 1991a, b). Tilley et al. (1991) also isolated a human monoclonal antibody (1125H) which neutralized divergent strains of HIV-1 and blocked gp120-CD4 interactions. This monoclonal antibody was classified as conformation dependent since reduction of the disulfide bonds destroyed its ability to recognize envelope glycoproteins (Tilley et al. 1991).

By utilizing a large panel of HIV-1 envelope mutants from a previous study, Thali et al. (1991) were able to map determinants for recognition by a conformational human monoclonal antibody (F105). Like other conformational neutralizing monoclonal antibodies, this monoclonal antibody was able to neutralize diverse isolates of HIV-1 and block gp120-CD4 binding (Posner et al. 1991). For the mapping experiments, recombinant viruses expressing "conformationally correct" mutant envelope proteins were selected and scored for their ability to be recognized and neutralized by the F105 monoclonal antibody. From these experiments, it was determined that several discontinuous regions of the envelope were important for antibody recognition, including amino acids 256, 257, 368–370, 421, and 470–484 (Thali et al. 1991). These positions are located in conserved regions C2, C3, and C4 (Starcich et al. 1986; Lasky et al. 1987). All of these positions were previously found to be important for CD4 binding (Olshevsky et al. 1990) and are believed to be located near each other on the native envelope glycoprotein (Thali et al. 1991).

In addition to monoclonal antibodies, polyclonal antisera have also been used to demonstrate the existence of conformation-dependent conserved neutralization epitopes. In a large study of 51 baboons, broadly neutralizing antibody responses were generated by immunizing animals with native, glycosylated gp120 protein of HIV-1 but *not* with denatured, nonglycosylated gp120 protein. Repeated boostings with the native protein were found to further increase the broadness of the immune response (HAIGWOOD et al. 1992). Native protein conformation was also important in a study of SIV_{mac} neutralization. By immunizing animals with native, glycosylated gp110 protein of SIV, high-titer neutralizing antisera were obtained. Immunization with denatured envelope proteins, fragments, and synthetic peptides generated low or undetectable levels of neutralizing antibodies (JAVAHERIAN et al. 1992).

Using pooled human sera positive for HIV antibody, STEIMER et al. (1991) also found that conformation-dependent antibodies were capable of neutralizing a range of isolates and were responsible for the CD4 blocking activity observed in HIV-1 positive sera. In this study, the conformation-dependent antibodies were separated from the antibodies specific for linear determinants by affinity chromatography. Two different forms of envelope protein specific for the SF2 strain of HIV-1 were presented on successive affinity columns. The first column contained a denatured, nonglycosylated form of the envelope (env $2\text{–}3_{SF2}$) to bind linear epitopes, while the second column contained a native, glycosylated form (rgp 120_{SF2}) to bind conformational epitopes. Antibodies recovered from the bound and unbound fractions of these columns were then tested for their ability to neutralize HIV-SF2 and to block CD4 binding activity. Although bound fractions contained only a small proportion of the total neutralizing activity, the antibody fraction which bound to the native protein (the conformational fraction) neutralized HIV-SF2. Control experiments indicated that this conformation-dependent neutralizing activity was not due to incomplete removal of V3-directed neutralizing antibodies.

In the study described above by STEIMER et al. (1991), the neutralizing activity of the linear and conformational fractions combined was not equivalent to that of the unfractionated material, and there was a significant amount of neutralizing activity left in the unbound fraction after adsorption with the denatured and native proteins. It is not surprising that some neutralizing activity is lost during purification; however, one explanation for the large discrepancy between the amount of neutralizing activity between the fractionated and unfractionated material may be synergistic neutralization; binding of antibodies to the CD4 binding domain of gp120 may facilitate exposure of the V3 epitope, resulting in an overall enhancement of neutralization (BUCHBINDER et al. 1992; MCKEATING et al. 1992; TILLEY et al. 1992). However, this still does not explain the large amoung of neutralizing activity left in the unbound fraction after adsorption with the denatured and native proteins (STEIMER et al. 1991). This discrepancy may be explained by the fact that a significant fraction of neutralizing antibodies may have recognized epitopes other than those which were presented on the cloned envelope proteins on the affinity columns, such as epitopes of gp120 oligomers as present in the virion.

By using another antibody adsorption technique, BERKOWER et al. (1991) were not only able to demonstrate the presence of conformation-dependent neutralizing antibodies in human sera, but they were able to map these antibodies to a region of the envelope spanning amino acids 342–511 (numbering system of MYERS et al. 1988). This region is distinct from the V3 loop and encompasses constant region 4 (LASKY et al. 1987) and flanking variable regions V4 and V5 (STARCICH et al. 1986). BERKOWER et al. (1991) first localized this epitope to amino acid residues 1–511 of gp120 based on experiments which showed that gp160 but not gp41 adsorbs neutralizing antibodies. Fragment 1–471 of gp120 failed to adsorb neutralizing antibodies, indicating that the critical residues are 472–511. However, another fragment which included these residues (fragment 472–857) also failed to adsorb neutralizing antibodies. By constructing an SIV/HIV hybrid protein, Berkower et al. (1991) were able to determine that the failure of these sequences to adsorb was due to protein conformation. The hybrid protein, which consisted of amino acids 1–399 of SIV and 342–511 of HIV-1, was found to remove neutralizing antibodies with nearly the same efficiency as the full-length gp120 of HIV-1; however, the gp120 protein of SIV could not remove neutralizing antibodies. Thus, it was concluded that the 342–511 region contains a conformation-dependent neutralizing epitope of HIV-1 (BERKOWER et al. 1991).

To determine whether hypervariable regions are important in the generation of neutralizing antibodies, a study was undertaken in which hypervariable regions were deleted from the envelope protein (HAIGWOOD et al. 1990). The deleted envelope proteins were used to immunize experimental animals, and the sera were tested for neutralizing activity. The antigens used in this study were more complex than simple peptides and recombinant fragments, but they were not as complex as native proteins; none of the deletion variants, for example, were able to bind CD4. By measuring the neutralization titers of sera raised against a series of these deletion variants, it was found that at least two other hypervariable regions (other than V3) were important for eliciting neutralizing antibodies; one of these epitopes is in the amino-terminal half of gp120 (VI or V2) while the other is in the carboxy-terminal half (V4 or V5). When all five variable regions were deleted, there were no neutralizing antibodies raised, even though the protein was still highly immunogenic by enzyme-linked immunosorbent assay. By removing one or more of the variable regions, there were no broadly specific immune responses generated in any of the animals. Thus, there were no conserved neutralization epitopes unmasked in this study, at least in the conformational context in which they were presented (HAIGWOOD et al. 1990).

In another study, variable region 2 of HIV-1 has also been reported to contain a conformation-dependent neutralization epitope (Fung et al. 1992).

A murine monoclonal antibody (G3-136) was mapped through the use of synthetic peptides to amino acids 169–183 of gp120 (numbering system of RATNER et al. 1985). Although this monoclonal antibody was mapped using peptides, it is referred to as conformation-dependent because it only weakly recognizes reduced forms of gp120 in radioimmunoprecipitation assays. Monoclonal antibody G3-136 neutralized certain laboratory and primary isolates of HIV-1 and was partially able to inhibit gp120-CD4 binding (FUNG et al. 1992).

Although the V3 loop is often referred to as the PND of HIV-1, there is no clear evidence to date which suggests that linear neutralization epitopes are more important than conformational epitopes during the natural course of HIV-1 infection in humans. V3 may indeed be the PND for denatured HIV-1/IIIB/BRU envelope protein delivered in adjuvant, but there is no evidence to indicate that it is the principal target of neutralizing antibodies in sera of infected persons. It is not yet known to what extent neutralizing antibodies in patient sera are directed at linear versus conformational epitopes, variable versus conserved, or V3 versus other domains. Binding studies have shown that antibodies against both the V3 region (HOLMBÄCK et al. 1993) and discontinuous sites that can block CD4 binding (MOORE and HO 1993) are present in HIV-1 positive sera. Some of the studies described above and the recent report of MOORE and HO (1993) suggest that conformational and discontinuous epitopes may indeed predominate over linear epitopes. Since neutralizing antibodies represent only a very small fraction of the total antibody response against gp120, and since neutralization titers of human sera are low, it has been difficult to quantify these different types of immune responses. The situation is further clouded by the fact that neutralization assays are run with one or a few strains of virus which are not representative of the complex mix of virus genotypes present in an infected individual.

6 Envelope Sequence Variation

Comparison of the nucleotide sequences of the AIDS retroviruses has revealed extensive genomic variability, particularly in the envelope (*env*) regions. Such variability can be seen in published alignments of HIV-1 molecular clones (ALIZON et al. 1986; STARCICH et al. 1986; WILLEY et al. 1986) and among the many other HIV and SIV *env* sequences in GenBank. Studies on the fidelity of HIV-1 reverse transcriptase have revealed that this enzyme is highly error prone (PRESTON et al. 1988; ROBERTS et al. 1988; WEBER and GROSSE 1989). Thus, the accumulation of point mutations resulting from mistakes by reverse transcriptase is a major source of genetic variability. Deletions, insertions, inversions, and recombinant forms have also been observed (DOUGHERTY and TEMIN 1986, 1988; HU and TEMIN 1990a, b; PATHAK and TEMIN 1990a, b; PULSINELLI and TEMIN 1991; ZHANG and TEMIN 1993).

Studies with independent and sequential isolates of HIV-1 have demonstrated genetic heterogeneity between isolates from different persons, within single virus isolates, and with sequential isolates over time in infected individuals (BALFE et al. 1990; GOODENOW et al. 1989; HAHN et al. 1986; MEYERHANS et al. 1989; SAAG et al. 1988; SIMMONDS et al. 1990a, b, 1991; WONG-STAAL et al. 1985). One question which arises from these studies of sequential isolates is whether such heterogeneity is generated by rapid evolution of HIV-1 over time in infected individuals, or whether such diversity may have been originally present in the infecting virus inoculum. Studies on sequential isolates of HIV-1 are usually complicated further by lack of information on the time of infection in these individuals, making precise measurement of nucleotide substitution rates difficult or impossible in this system. Characterization of the precise rate and nature of nucleotide substitutions is best made using molecularly cloned virus of known sequence in an experimental animal model system in which the exact time of infection is known.

In our laboratory we have studied envelope gene variation in rhesus monkeys infected with molecularly cloned SIV_{mac239} (BURNS and DESROSIERS 1991). We have documented the extent to which genetic changes accumulate in *env* with time of in vivo infection and have characterized the nature of the nucleotide substitutions. Our results show that molecularly cloned SIV rapidly evolves into a complex mix of genotypes during infection of a single individual. All 27 clones recovered from two rhesus monkeys were different from each other and differed from the starting clone. The average fixation rate of nucleotide substitutions of 8.5 per 1000 per year in SU is approximately 10^6 times greater than that of the average mammalian gene (HAYASHIDA and MIYATA 1983; LI and GOJOBORI 1983) and roughly equivalent to rates which have been reported for influenza virus (BUONAGURIO et al. 1986; HAYASHIDA et al. 1985) and for type C retroviruses in cell culture (DOUGHERTY and TEMIN 1988). Amino acid substitutions were fixed at 2.1 per 100 per year. This corresponds to 11 amino acid substitutions per gp120 per year, not counting deletions and insertions.

Five variable regions have been identified in late time-point clones from two rhesus monkeys studied in our laboratory that had strong stable antibody responses and that remained persistently infected for longer than 2 years. (BURNS and DESROSIERS 1991). Four of these five variable regions of SIV_{mac} (V1, V2, V4, and V5) correspond exactly to variable regions V1, V2, V4, and V5 in gp120 of HIV-1 (STARCICH et al. 1986). Although there is extensive divergence between the envelope sequences of HIV-1 and SIV_{mac}, these and other results (MORRISON et al. 1993) suggest that there is at least rough correspondence in the location of functional domains within the envelope proteins of these viruses.

Analysis of the data from our sequencing studies strongly indicates that variation in five discrete segments results from selective forces operating in vivo (BURNS and DESROSIERS 1991). The fixation rate of nonsynonymous (amino acid changing) nucleotide substitutions (6.8×10^{-3} per site per year) was found to be approximately 4.3 times higher than the fixation rate of synonymous mutations (1.6×10^{-3} per site per year). Within the five variable regions, a remarkable 98% (182/186) of the nucleotide substitutions changed the amino acid, and statistically significant differences between observed frequencies and frequencies expected from random genetic drift were found. The difference in nonsynonymous versus synonymous fixation rates in the SIV_{mac} env gene contrasts markedly with rates for eukaryotic genes. LI et al. (1985), for example, have shown that the nonsynonymous mutation rate for eukaryotic genes is only one-fifth the synonymous mutation rate. Quite clearly this usual predominance of synonymous mutations results from severe limitations on the numbers of amino acid changes that will still allow optimal protein function. As with cellular genes, the HIV-1 *gag* gene shows a bias toward synonymous mutations (SIMMONDS et al. 1990b). These observations demonstrate that variant SIV envelopes with amino acid changes in discrete variable domains have a significant selective advantage in vivo. The host immune response is likely to be at least one major source of the selective pressure for change.

One surprising result from our study was that the region corresponding to the variable V3 cysteine loop of HIV-1 was found not to be variable in SIV_{mac239}; this region was found to be quite conserved among the 27 clones analyzed (BURNS and DESROSIERS 1991). Furthermore, using the same peptide-based strategy that was used for showing that the V3 cysteine loop of HIV-1 is a neutralizing epitope, researchers have recently found that linear peptides corresponding to the SIV_{mac251} and SIV_{mac239} cysteine loop do not elicit neutralizing antibody against SIV (JAVAHERIAN et al. 1992). This contrasts markedly with the V3 region of HIV-1 envelope, which is variable and can serve as a target of neutralizing antibodies (see above). HIV-1 V3 is also a major determinant of cell tropism (HWANG et al. 1991; O'BRIEN et al. 1990: SHIODA et al. 1991; WESTERVELT et al. 1991). It needs to be pointed out, however, that when independent SIV_{mac}, SIV_{sm}, and HIV-2 isolates in the sequence database are compared, the "V3" corresponding region does indeed appear to be variable. Furthermore, by studying different classes of experimentally infected rhesus monkeys (rhesus monkeys which die rapidly without ever mounting a strong antibody response and rhesus monkeys in which virus strains with altered cell and tissue tropism evolve), KODAMA et al. (1993) were able to demonstrate significant variation in the region of SIV_{mac} corresponding to V3 of HIV-1. Important gaps in our understanding of HIV-1 V3 and SIV "V3" clearly exist, and it is premature to conclude that there are fundamental differences in the functional role of these corresponding regions. Might this region serve as a neutralization epitope for some strains (perhaps those with a particular tropism) but not others? What selective forces are primarily responsible for driving change in these corresponding regions?

In addition to our study on SIV_{mac} (BURNS and DESROSIERS 1991), several other laboratories have published variation studies using molecular clones of SIV_{smm} (JOHNSON et al. 1991), SIV_{mne} (OVERBAUGH et al. 1991), and SIV_{agm} (BAIER et al. 1991). In general, there are many similarities among these four systems. Overall, SIV undergoes rapid mutation during in vivo infection; among all four systems, the average mutation rate was on the order of 10^{-2} per site per year. The SIV rates are remarkably similar to estimated rates for HIV-1 (BALFE et al. 1990; HAHN et al. 1986). There was a clustering of mutations into distinct variable regions in the SIV envelope, and the region corresponding to the V3 loop of HIV-1 was conserved in all of the four systems. In all cases, the preferred nucleotide substitutions were G → A and A → G, and there was strong selection for nonsynonymous mutations. Within the variable regions of SIV_{mac}, SIV_{mne}, and SIV_{agm}, the percentage of nucleotide subsitutions which changed the amino acid was extremely high (98, 98 and 92, respectively).

As with SIV_{mac} (BURNS and DESROSIERS 1991), the SIV_{mne} clone (OVERBAUGH et al. 1991) often undergoes amino acid changes in and around potential sites for N-linked glycosylation, and changes in the V4 region of SIV_{mne} often result in substitution of charged amino acids. As with SIV_{mac}, V1 of SIV_{mne} is also the most variable, and some of the clones have short, in-frame deletions of one or two amino acids in this region. Relative to the parental clone, some SIV_{mne} variants have a nine amino acid insertion in V1. The SIV_{mac} system is distinct among the SIV systems in having additional in-frame deletions in V4 (BURNS and DESROSIERS 1991).

Unlike the other SIV systems, variants in the SIV_{smm} system have a high number of defective mutations (JOHNSON et al. 1991). The defective mutations are not due to frameshift mutations but are due almost exclusively to the generation of stop codons at tryptophan codons. While the stop codons in SIV_{smm} variants are found in both SU and TM, there is a slightly higher number in the TM region. In the SIV_{mac} system, truncated forms of the transmembrane protein have been associated with growth on human cell lines (KODAMA et al. 1989) and with growth on simian macrophages in culture (MORI et al. 1992).

Although there have been no variability studies published on accidental infections with cloned HIV-1 in humans, there has been a series of reports published on a group of 18 hemophiliacs infected with a single batch of factor VIII (LUDLAM et al. 1985). Although the amount of diversity present in the original batch of factor VIII is unknown, it has been found that interpatient sequence variability among these individuals infected from a common source is markedly less than that among individuals infected from different sources (SIMMONDS et al. 1990a). Researchers have performed double-nested polymerase chain reaction on blood samples from this group of hemophiliacs and have analyzed the sequence data obtained from the amplified material.

Overall, there are many similarities with the SIV data. In sequencing the V3 and V4/V5 regions, the estimated mutation rate was found to be 5.0×10^{-3} and 4.0×10^{-3} per site per year, respectively (BALFE et al. 1990). Very strong selection for amino acid change was found among these sequences, and alterations in the number and position of potential N-linked glycosylation sites was a common feature. Duplications and deletions of 3–15 bp were also observed (SIMMONDS et al. 1990a). Among 42 kb of nucleotide sequence obtained from *gag* and *env*, there were only two defective mutations (BALFE et al. 1990).

7 Emergence of HIV-1 Neutralization-Resistant Variants

As in the ungulate lentivirus systems, the concept of humoral immune selection has been explored in the HIV-1 system to determine whether neutralizing antibodies of the host immune response select for neutralization-resistant mutants, and whether antigenic variation has a role in the maintenance of persistent infection during chronic infection. Immune selection studies with the ungulate lentiviruses have been complicated by the lack of infectious molecular clones. While molecular clones are available in the HIV-1 system, they are infectious only for chimpanzees, which do not develop AIDS disease. Human experiments are complicated by the fact that a complex mix of genotypes is already present in infected individuals. One exception may be the infected laboratory worker, but any results from this case have not been published. Despite the many obstacles, attempts are being made to study HIV-1 immune selection in infected cell cultures in vitro, in infected chimpanzees, and in infected humans.

There have been several reports on the emergence of neutralization-resistant HIV-1 variants in vitro. One such variant was described in a 1986 paper in which an escape mutant was selected by culturing the virus in the constant presence of human neutralizing antiserum (ROBERT-GUROFF et al. 1986). A clone of the resistant strain revealed that resistance was conferred by a single mutation in the transmembrane protein (REITZ et al. 1988). By using monoclonal instead of polyclonal antibodies to select for mutants, neutralization-resistant mutations have also been selected in the V3 loop of gp120 (MASUDA et al. 1990; MCKEATING et al. 1989). One of these studies identified a neutralization-resistant mutation at amino acid residue 308 of the V3 loop (MCKEATING et al. 1989). Naturally occurring mutations which alter antigenicity have been found at this location in virus isolated from infected patients (WOLFS et al. 1991).

One of the few in vivo studies on the emergence of HIV-1 neutralization-resistant variants has been presented by NARA et al. (1990). In this study, variants resistant to neutralization by preexisting antibody were found in two chimpanzees experimentally infected with HIV-1. The chimpanzees were inoculated with a parental stock of virus prepared from HTLV-IIIB infected H9 cells (ARTHUR et al. 1989). One chimpanzee was naive for HIV-1 infection, while the other animal had been previously vaccinated with the gp120 protein of HTLV-III_B. Both chimpanzees had been used previously in human hepatitis studies but were negative for HIV-1. The neutralization-resistant variants were isolated from the two chimpanzees at 6, 16, and 32 weeks postinfection. Sera taken before, during, and after these virus isolations were used in homologous neutralization tests. Although the neutralizing titers were low (1:4–1:64), delayed neutralizing antibody responses were observed for the 6-, 16-, and 32-week variants in each of the two animals. Interestingly, the response against the variants was delayed by 18 weeks in the animal which had been immunized with gp120 prior to virus inoculation. Five of the six virus variants from the two animals were eventually neutralized by sera from the homologous chimpanzee, and all six variants scored positive for neutralization against other HIV-1 sera, indicating that all of the variant virus isolates were capable of being neutralized (NARA et al. 1990).

In one study on immune selection in an infected human, TREMBLAY and WAINBERG (1990) obtained sequential virus and serum samples from one asymptomatic HIV-1 infected individual. Ten different serum samples were collected from this patient at time points ranging from 6 to 108 weeks after the initiation of the study, and five different virus isolates were obtained at 6–72 weeks. All of the ten serum samples were tested for their ability to neutralize each of the five virus isolates from this patient. Each of the virus isolates was able to be neutralized only by serum samples taken either around or after the time of virus isolation. Thus, there were delayed neutralizing antibody responses against the variant viruses in this individual. Each of the five virus isolates was eventually able to be neutralized by late time point sera, indicating that each of the viruses was capable of being neutralized (TREMBLAY and WAINBERG 1990).

In four other studies on immune selection in humans, HIV-1 isolates were obtained which could not be neutralized by the patients' own sera, indicating that neutralization-resistant mutants had emerged in these individuals (ALBERT et al. 1990; HOMSY et al. 1990; MONTEFIORI et al. 1991; VON GEGERFELT et al. 1991). In two of these studies (ALBERT et al. 1990; VON GEGERFELT et al. 1991), the neutralization assay which was used has recently been called into question. The assay, which is based on the detection of viral p24 gag protein, did not control for anti-gag antibodies which may have been present in the patients' sera. Such anti-gag antibodies can interfere with the detection of viral p24 gag protein and can result in unreliable neutralization titers (BURNS and DESROSIERS 1992). In two other studies (HOMSY et al. 1990; MONTEFIORI et al. 1991), variant viruses were not tested against control sera to determine to what extent they were capable of being neutralized. Also, sera from individuals infected with variant virus were not available for reciprocal testing. Therefore, it is difficult to conclude from any of the four studies whether mutants specifically resistant to neutralization by the patient sera had actually emerged in these HIV-1 infected individuals.

At the present time, there is little evidence to document the concept of immune selection of neutralization resistant variants in HIV-1 infected people. There are a few studies which show that neutralizing antibodies can select for resistant mutants in tissue culture systems, and there are only two substantiated reports on the emergence of resistant variants in vivo. One of these reports presented evidence from two chimpanzees, one of which had been vaccinated with gp120 protein prior to infection. The other report presented evidence on only one asymptomatic HIV-1 infected human patient. The neutralization assays in these reports were performed in one direction only since additional sera from individuals naturally or experimentally infected with variant virus were not available for reciprocal testing of neutralizing activity. This is problematic since different virus stocks show considerable variation in the efficiency with which they score in neutralization tests (LAYNE et al. 1991; MCKEATING et al. 1991). Such inherent differences in the efficiencies with which virus stocks score in neutralization tests can be due to a variety of factors, including differences in infectivity titers, methods of stock preparation, replication rate, particle/infectivity ratios, gp120 packing density, gp120 structure, gp120-gp41 affinity, gp120-CD4 affinity, design of the assay, and other ill-defined parameters. It has been noted previously that some strains of HIV-1 (e.g., RF) appear considerably resistant to neutralization, while others (e.g., SF2) are considerably more sensitive to neutralization (BERMAN et al. 1992; MCKEATING et al. 1991; WEISS et al. 1986). Thus, the significance of poor neutralization of variant virus in these previous experiments with HIV-1 (NARA et al. 1990; TREMBLAY and WAINBERG 1990) is not entirely clear since it was not possible to demonstrate reciprocal effects with reciprocal sera.

8 Emergence of SIV Neutralization-Resistant Variants

Rhesus monkeys infected with molecularly cloned SIV_{mac239} provide a useful model system for studying immune selection. More than 50% of rhesus monkeys infected with cloned virus develop strong antibody responses to the virus and become persistently infected for 1 or more years prior to the development of AIDS (KESTLER et al. 1990). By cloning viral envelope genes from rhesus monkeys over time and obtaining sequential serum samples from them, we have been able to study not only the evolution of envelope sequences but also the emergence of neutralization-resistant variants and to verify the selectivity of the neutralizing antibody responses.

In a recent study we investigated whether neutralizing-antibody responses are one selective force responsible at least in part for the observed pattern of sequence variation in the SIV envelope (BURNS et al. 1993). Variant *env* sequences termed 1–12 and 8–22 obtained 69 and 93 weeks after infection of a rhesus monkey with cloned SIV_{mac239} were recombined

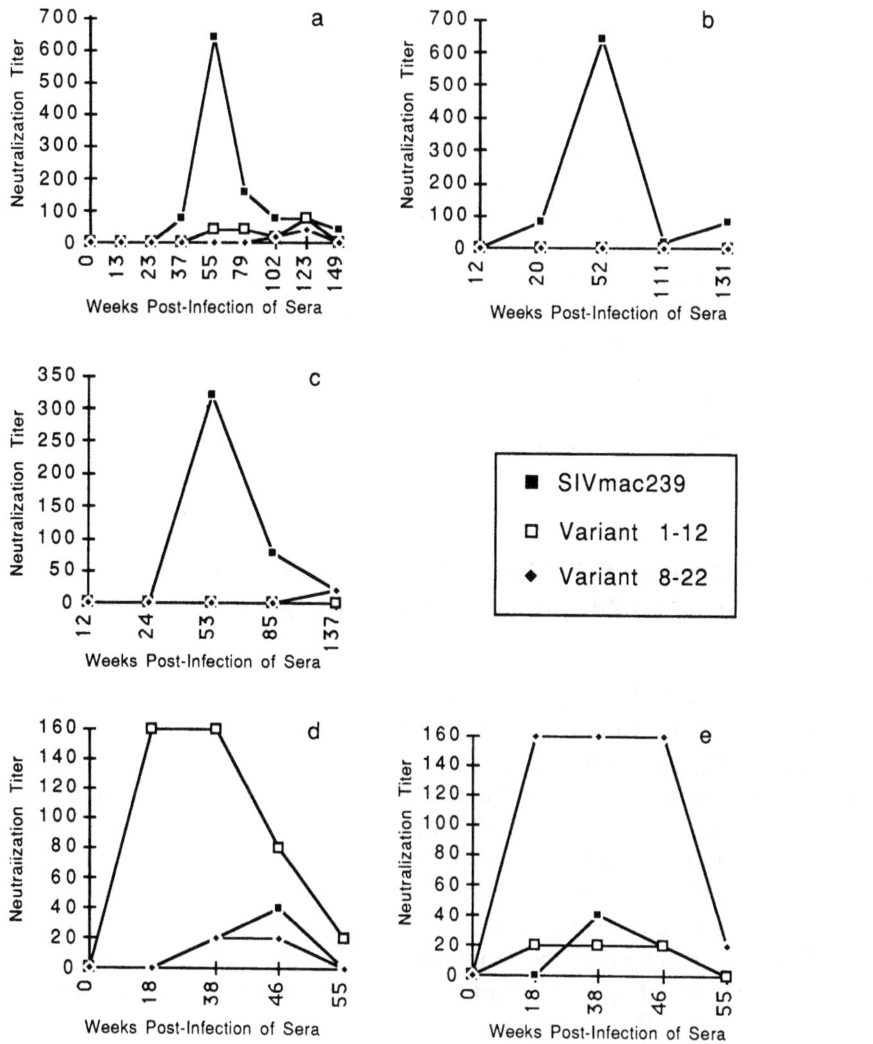

Fig. 2a–e. Neutralization titers of sequential serum samples for five SIV-infected rhesus monkeys. Mm243-86 (**a**), Mm 135-88 (**b**), and Mm206-86 (**c**) were all infected with cloned SIV_{mac239}. Mm206-89 (**d**) was infected with cloned variant 1–12. Mm262-89 (**e**) was infected with cloned variant 8–22

into the parental SIV_{mac239} genome, and variant viruses were generated by transfection of cultured cells with cloned DNA (BURNS et al. 1993; BURNS and DESROSIERS 1991). The 1–12 and 8–22 variants each differ from the parental SIV_{mac239} at 20 amino acid positions in gp120, and the mutations cluster into five discrete variable regions (BURNS and DESROSIERS 1991). Sequential sera from the animal infected with cloned SIV_{mac239} from which the 1–12 and 8–22 variants were derived (Mm243-86) showed much higher neutralizing antibody titers to cloned SIV_{mac239} than to cloned 1–12 and 8–22 variants (Fig. 2a). Sequential sera from two other animals infected with parental virus (Mm135-88 and Mm206-86) were also tested. Sera from these two animals also neutralized the parental virus SIV_{mac239} much better than the two variant viruses (Fig. 2b, c).

As an internal control for immune selection, reciprocal neutralization studies were performed with variant-specific sera, i.e., sera from rhesus monkeys infected with variant virus. Results from these cross-neutralization tests revealed that each variant was neutralized best by its homologous antisera (Fig. 2d, e). Sera from Mm206-89, for example, which was inoculated with the 1–12 virus neutralized 1–12 better than any other virus. Similarly, sera from Mm262-89 which was inoculated with the 8–22 virus neutralized 8–22 better than any other virus. These reciprocal neutralization experiments demonstrate unambiguously that at least some of the genetic variants that arise during the course of persistent infection in vivo are resistant to neutralization by antisera capable of neutralizing the input virus.

9 Use of Cloned SIV Variants and Monoclonal Antibodies to Map Neutralization Determinants

Cloned variants 1–12 and 8–22 (discussed above) along with cloned variant 3–18 (BURNS and DESROSIERS 1991) and molecularly cloned viruses SIV_{mac239}, SIV_{mac251}, and SIV_{mac142} (NAIDU et al. 1988) have been used in radioimmunoprecipitation assays with a panel of six neutralizing monoclonal antibodies (COLLIGNON et al. 1993, in preparation). These six neutralizing monoclonal antibodies were generated to native vaccinia recombinant antigen (SIV_{mac251} gp150, molecular clones BK28 and J5). All of these neutralizing monoclonal antibodies recognized conformational or discontinuous epitopes; five of the six did not recognize denatured protein on western blots, and none of the six recognized synthetic peptides. Radioimmunoprecipitation analyses revealed distinct patterns of reactivity against the cloned viruses. Relative to the parental clone, variants 1–12 and 8–22 reacted poorly or not at all with this panel of neutralizing monoclonal antibodies. Thus, sequence changes that accumulated in gp120 during persistent infection dramatically influenced recognition by these neutralizing antibodies generated to native gp120.

Recombinant viruses which were constructed by exchanging variable regions between SIV_{mac251} and variant 1–12 were used to define domain(s) important for recognition. Radioimmunoprecipitation analysis with these recombinants demonstrated that sequence changes in V4/V5 were primarily responsible for the loss of recognition of the 1–12 variant (COLLIGNON et al. 1993, in preparation). Site-specific mutants were then used to define precise changes that eliminate recognition by these neutralizing antibodies (Choi et al., in preparation). Changing amino acid 409 N to D, deletion of KPKE, and deletion of KEQH in V4 each resulted in loss of recognition by all five monoclonal antibodies. SIVs with these natural sequence changes are still replication competent and viable. Changing 417 A to T or 417/418 AN to TK in V4 or 477 Q to K in V5 did not alter recognition.

10 Summary

Studies in ungulate lentivirus systems clearly indicate that neutralization escape variants emerge over time in chronically infected animals. Studies in the EIAV system, in particular, have provided strong evidence that the humoral branch of the immune system is at least one selective force acting on an array of viral variants. In previous studies with the ungulate lentiviruses, molecularly cloned virus was never used, and plaque-purified virus was only sometimes used; the genetic determinants responsible for antigenic variation and immune selection were not determined. While molecular clones are available for HIV-1, immune selection studies have been hampered in this system by the fact that HIV-1 is infectious only for chimpanzees, which do not develop disease and are available in only limited numbers. Experiments on immune selection in humans are generally complicated by lack of knowledge on the time of infection and the genetic make-up of the infecting virus.

Our studies on SIV immune selection summarized in this review provide definitive evidence that neutralization-resistant variants emerge in an individual during persistent infection by primate lentiviruses. By cloning viral envelope genes from rhesus monkeys over time and obtaining sequential serum samples from them, we have been able to study not only the evolution of envelope sequences but also the emergence of neutralization-resistant variants. Reciprocal neutralization studies were performed using parental and variant specific sera, and immune selection was demonstrated using molecularly cloned virus of defined sequence.

During the course of persistent infection with SIV and HIV, there is clear selective pressure for change in discrete variable regions of envelope. The host neutralizing antibody response appears to be at least one of the selective forces driving sequence change in envelope since one result of the sequence variation is the emergence of neutralization escape mutants. This indicates that neutralizing antibodies do serve to limit HIV and SIV replication during the lengthy asymptomatic stage of infection. The coincidence of neutralization domains of HIV and/or SIV with variable regions V1, V2, V3, V4, V5, and V6 suggests a direct relationship between neutralization domains and the emergence of sequence variants. However, different selective forces may be responsible all or in part for driving sequence changes in some variable domains (summarized in Table 2). For example, alterations in cell and/or tissue tropism may be responsible at least in part for driving change in V3 and the cytotoxic T-lymphocyte response may be responsible for driving change in the signal peptide (V0; HENDERSON et al. 1992; WEI and CRESSWELL 1992). Continued experimentation, particularly in the SIV system, will be needed for a more detailed understanding of the selective forces driving change in variable regions and for identification of variable domains most responsible for escape from neutralization.

Although the V3 loop is often referred to as the PND of HIV-1, there is no clear evidence to suggest that V3 epitopes predominate over conformational epitopes as targets of neutralizing antibody responses during the natural course of HIV-1 infection in humans. It is not yet known to what extent neutralizing antibodies in patient sera are directed at linear versus conformational epitopes, variable versus conserved, or V3 versus other domains. Attempts to identify neutralization epitopes in the SIV system indicate that conformational determinants may be more important than linear determinants. Studies with cloned SIV variants and neutralizing monoclonal antibodies also indicate that conformational or discontinuous epitopes are important targets of neutralizing antibodies to SIV_{mac}. Sequence changes that accumulate in gp120 during persistent infection in vivo dramatically influence recognition by neutralizing antibodies. Our experiments have defined specific, naturally occurring sequence changes in the fourth variable region of SIV that result in loss of recognition by at least one class of neutralizing antibody.

Table 2. Selective forces associated with sequence change in variable regions of envelope of SIV and HIV

Variable region	Possible selective force
V0 (signal peptide)	CTL
V1	Neut Ab
V2	Neut Ab
V3	Neut Ab, cell tropism, replicative potential
V4	Neut Ab
V5	Neut Ab
V6	Neut Ab
V7	?

CTL, Cytotoxic T-lymphocyte; Neut Ab, neutralizing antibody.

Continued study of this defined system will hopefully allow a clearer understanding of the role of antigenic variation and immune selection in viral persistence and the pathogenesis of AIDS.

References

Åkerblom L, Hinkula J, Broliden P-A, Mäkitalo B, Fridberger T, Rosen J, Villacres-Eriksson M, Morein B, Wahren B (1990) Neutralizing cross-reactive and non-neutralizing monoclonal antibodies to HIV-1 gp120. AIDS 4: 953–960

Albert J, Abrahamsson B, Nagy K, Aurelius E, Gaines H, Nyström G, Fenyö EM (1990) Rapid development of isolate-specific neutralizing antibodies after primary HIV-1 infection and consequent emergence of virus variants which resist neutralization by autologous sera. AIDS 4: 107–112

Alizon M, Wain-Hobson S, Montagnier L, Sonigo P (1986) Genetic variability of the AIDS virus: nucleotide sequence analysis of two isolates from African patients. Cell 46: 63–74

Arthur LO, Bess JW Jr, Waters DJ, Pyle SW, Kelliher JC, Nara PL, Krohn K, Robey WG, Langlois AJ, Gallo RC, Fischinger PJ (1989) Challenge of chimpanzees (Pan troglodytes) immunized with human immunodeficiency virus envelope glycoprotein gp120. J Virol 63: 5046–5053

Baier M, Dittmar MT, Cichutek K, Kurth R (1991) Development in vivo of genetic variability of simian immunodeficiency virus. Proc Natl Acad Sci USA 88: 8126–8130

Balfe P, Simmonds P, Ludlam CA, Bishop JO, Brown AJL (1990) Concurrent evolution of human immunodeficiency virus type 1 in patients infected from the same source: rate of sequence change and low frequency of inactivating mutations. J Virol 64: 6221–6233

Benichou S, Legrand R, Nakagawa N, Faure T, Traincard F, Vogt G, Dormont D, Tiollais P, Kieny M-P, Madaule P (1992) Identification of a neutralizing domain in the external envelope glycoprotein of simian immunodeficiency virus. AIDS Res Hum Retroviruses 8:1165–1170

Berkower I, Smith GE, Giri C, Murphy D (1989) Human immunodeficiency virus 1: predominance of a group-specific neutralizing epitope that persists despite genetic variation. J Exp Med 170: 1681–1695

Berkower I, Murphy D, Smith CC, Smith GE (1991) A predominant group-specific neutralizing epitope of human immunodeficiency virus type 1 maps to residues 342 to 511 of the envelope glycoprotein gp120. J Virol 65: 5983–5990

Berman PW, Matthews TJ, Riddle L, Champe M, Hobbs MR, Nakamura GR, Mercer J, Eastman DJ, Lucas C, Langlois AJ, Wurm FM, Gregory TJ (1992) Neutralization of multiple laboratory and clinical isolates of human immunodeficiency virus type 1 (HIV-1) by antisera raised against gp120 from the MN isolate of HIV-1. J Virol 66: 4464–4469

Bjorling E, Broliden K, Bernardi D, Utter G, Thorstensson R, Chiodi F, Norrby E (1991) Hyperimmune antisera against synthetic peptides representing the glycoprotein of human immunodeficiency virus type 2 can mediate neutralization and antibody-dependent cytotoxic activity. Proc Natl Acad Sci USA 88: 6082–6086

Broliden P-A, Ljunggren K, Hinkula J, Norrby E, Åkerblom L, Wahren B (1990) A monoclonal antibody to human immunodeficiency virus type 1 which mediates cellular cytotoxicity and neutralization. J Virol 64: 936–940

Broliden P-A, von Gegerfelt A, Clapham P, Rosen J, Fenyö E-M, Wahren B, Broliden K (1992) Identification of human neutralization-inducing regions of the human immunodeficiency virus type 1 envelope glycoproteins. Proc Natl Acad Sci USA 89: 461–465

Buchbinder A, Karwowska S, Gorny MK, Burda ST, Zolla-Pazner S (1992) Synergy between human monoclonal antibodies to HIV extends their effective biologic activity against homologous and divergent strains. AIDS Res Hum Retroviruses 8: 425–427

Buonagurio DA, Nakada S, Parvin JD, Krystal M, Palese P, Fitch WM (1986) Evolution of human influenza A viruses over 50 years: rapid, uniform rate of change in NS gene. Science 232: 980–982

Burns DPW, Desrosiers RC (1991) Selection of genetic variants of simian immunodeficiency virus in persistently infected rhesus monkeys. J Virol 65: 1843–1854

Burns DPW, Desrosiers RC (1992) A caution of the use of SIV/HIV gag antigen detection systems in neutralization assays. AIDS Res Hum Retroviruses 8: 1189–1192

Burns DPW, Collignon C, Desrosiers RC (1993) Simian immunodeficiency virus mutants resistant to serum neutralization arise during persistent infection of rhesus monkeys. J Virol 67: 4104–4113

Carpenter S, Evans LH, Sevoian M, Chesebro B (1987) Role of the host immune response in selection of equine infectious anemia virus variants. J Virol 61: 3783–3789

Chanh TC, Dreesman GR, Kanda P, Linette GP, Sparrow JT, Ho DD, Kennedy RC (1986) Induction of anti-HIV neutralizing antibodies by synthetic peptides. EMBO J 5: 3065–3071

Cheevers WP, Knowles DP Jr, Norton LK (1991) Neutralization-resistant antigenic variants of caprine arthritis-encephalitis lentivirus associated with progressive arthritis. J Infect Dis 164: 679–685

Cheng-Mayer C, Homsy J, Evans LA, Levy JA (1988) Identification of human immunodeficiency virus subtypes with distinct patterns of sensitivity to serum neutralization. Proc Natl Acad Sci USA 85: 2815–2819

Clavel F, Klatzmann D, Montagnier L (1985) Deficient LAV_1 neutralizing capacity of sera from patients with AIDS or related syndromes. Lancet I: 879–880

Clements JE, Pedersen FS, Narayan O, Haseltine WA (1980) Genomic changes associated with antigenic variation of visna virus during persistent infection. Proc Natl Acad Sci USA 77: 4454–4458

Clements JE, D'Antonio N, Narayan O (1982) Genomic changes associated with antigenic variation of visna virus. II. Common nucleotide sequence changes detected in variants from independent isolations. J Mol Biol 158: 415–434

Collignon C, Delers A, Bruck C, De Wilde M, Mullins J, Desrosiers RC, Bex F, Burny A, Thiriart C (1993) Identification of neutralizing epitopes on the SIV gp120 external glycoprotein. (submitted for publication)

Dalgleish AG, Chanh TC, Kennedy RC, Kanda P, Clapham PR, Weiss RA (1988) Neutralization of diverse HIV-1 strains by monoclonal antibodies raised against a gp41 synthetic peptide. Virology 165: 209–215

Daniel MD, Letvin NL, King NW, Kannagi M, Sehgal PK, Hunt RD, Kanki PJ, Essex M, Desrosiers RC (1985) Isolation of T-cell tropic HTLV-III-like retrovirus from macaques. Science 228: 1201–1204

Daniel MD, Letvin NL, Sehgal PK, Hunsmann G, Schmidt DK, King NW, Desrosiers RC (1987) Long-term persistent infection of macaque monkeys with the simian immunodeficiency virus. J Gen Virol 68: 3183–3189

Devash Y, Matthews TJ, Drummond JE, Javaherian K, Waters DJ, Arthur LO, Blattner WA, Rusche JR (1990) C-Terminal fragments of gp120 and synthetic peptides from five HTLV-III strains: prevalence of antibodies to the HTLV-III-MN isolate in infected individuals. AIDS Res Hum Retroviruses 6: 307–316

Dougherty JP, Temin HM (1986) High mutation rate of a spleen necrosis virus-based retrovirus vector. Mol Cell Biol 168: 4387–4395

Dougherty JP, Temin HM (1988) Determination of the rate of base-pair substitution and insertion mutations in retrovirus replication. J Virol 62: 2817–2822

Dubois-Dalcq M, Narayan O, Griffin DE (1979) Cell surface changes associated with mutation of visna virus in antibody-treated cell cultures. Virology 92: 353–366

Durda PJ, Bacheler L, Clapham P, Jenoski AM, Leece B, Matthews TJ, McKnight A, Pomerantz R, Rayner M, Weinhold KJ (1990) HIV-1 neutralizing monoclonal antibodies induced by a synthetic peptide. AIDS Res Hum Retroviruses 6: 1115–1123

Ellis TM, Wilcox GE, Robinson WF (1987) Antigenic variation of caprine arthritis-encephalitis virus during persistent infection of goats. J Gen Virol 68: 3145–3152

Evans DJ, McKeating J, Meredith JM, Burke KL, Katrak K, John A, Ferguson M, Minor PD, Weiss RA, Almond JW (1989) An engineered poliovirus chimaera elicits broadly reactive HIV-1 neutralizing antibodies. Nature 339: 385–388

Fung MSC, Sun CRY, Gordon WL, Liou R-S, Chang TW, Sun WNC, Daar ES, Ho DD (1992) Identification and characterization of a neutralization site within the second variable region of human immunodeficiency virus type 1 gp120. J Virol 66: 848–856

Geysen HM, Meloen RH, Barteling SJ (1984). Use of peptide synthesis to probe viral antigens for epitopes to a resolution of a single amino acid. Proc Natl Acad Sci USA 81: 3998–4002

Geysen HM, Rodda SJ, Mason TJ, Tribbick G, Schoofs PG (1987) Strategies for epitope analysis using peptide synthesis. J Immunol Methods 102: 259–274

Gnann JW Jr, Nelson JA, Oldstone MBA (1987a) Fine mapping of an immunodominant domain in the transmembrane glycoprotein of human immunodeficiency virus. J Virol 61: 2639–2641
Gnann JW Jr, Schwimmbeck PL, Nelson JA, Truax AB, Oldstone BA (1987b) Diagnosis of AIDS by using a 12-amino acid peptide representing an immunodominant epitope of the human immunodeficiency virus. J Infect Dis 156: 261–267
Goodenow M, Huet T, Saurin W, Kwok S, Sninsky J, Wain-Hobson S (1989) HIV-1 isolates are rapidly evolving quasispecies: evidence for viral mixtures and preferred nucleotide substitutions. J Acquir Immune Defic Syndr 2: 344–352
Goudsmit J, Boucher CAB, Meloen RH, Epstein LG, Smit L, van der Hoek L, Bakker M (1988a) Human antibody response to a strain-specific HIV-1 gp120 epitope associated with cell fusion inhibition. AIDS 2: 157–164
Goudsmit J, Debouck C, Meloen RH, Smit L, Bakker M, Asher DM, Wolff AV, Gibbs CJ Jr, Gajdusek DC (1988b) Human immunodeficiency virus type 1 neutralization epitope with conserved achitecture elicits early type-specific antibodies in experimentally infected chimpanzees. Proc Natl Acad Sci USA 85: 4478–4482
Griffin DE, Narayan O, Adams RJ (1978) Early immune responses in visna, a slow viral disease of sheep. J Infect Dis 138: 340–350
Hahn BH, Shaw GM, Taylor ME, Redfield RR, Markham PD, Salahuddin SZ, Wong-Staal F, Gallo RC, Parks ES, Parks WP (1986) Genetic variation in HTLV-III/LAV over time in patients with AIDS or at risk for AIDS. Science 232: 1548–1553
Haigwood NL, Shuster JR, Moore GK, Lee H, Skiles PV, Higgins KW, Barr PJ, George-Nascimento C, Steimer KS (1990) Importance of hypervariable regions of HIV-1 gp120 in the generation of virus neutralizing antibodies. AIDS Res Hum Retroviruses 6: 855–869
Haigwood NL, Nara PL, Brooks E, van Nest GA, Ott G, Higgins KW, Dunlop N, Scandella CJ, Eichberg JW, Steimer KS (1992) Native but not denatured recombinant human immunodeficiency virus type 1 gp120 generates broad-spectrum neutralizing antibodies in baboons. J Virol 66: 172–182
Hariharan K, Nara PL, Caralli VM, Norton FL, Haigwood N, Kang C-Y (1993) Analysis of the cross-reactive anti-gp120 antibody population in human immunodeficiency virus-infected asymptomatic individuals. J Virol 67: 953–960
Hayashida H, Miyata T (1983) Unusual evolutionary conservation and frequent DNA segment exchange in class I genes of the major histocompatibility complex. Poc Natl Acad Sci USA 80: 2671–2675
Hayashida H, Toh H, Kikuno R, Miyata T (1985) Evolution of influenza virus genes. Mol Biol Evol 2: 289–303
Henderson RA, Michel H, Sakaguchi K, Shabanowitz J, Appella E, Hunt DF, Engelhard VH (1992) HLA-A2.1-associated peptides from a mutant cell line: a second pathway of antigen presentation. Science 255: 1264–1266
Ho DD, Rota TR, Hirsch MS (1985) Antibody to lymphadenopathy-associated virus in AIDS. N Engl J Med 312: 649–650
Ho DD, Sarngadharan MG, Hirsch MS, Schooley RT, Rota TR, Kennedy RC, Chanh TC, Sato VL (1987) Human immunodeficiency virus neutralizing antibodies recognize several conserved domains on the envelope glycoproteins. J Virol 61: 2024–2028
Ho DD, Kaplan JC, Rackauskas IE, Gurney ME (1988) Second conserved domain of gp120 is important for HIV infectivity and antibody neutralization. Science 239: 1021–1023
Ho DD, Fung MSC, Cao Y, Li XL, Sun C, Chang TW, Sun N-C (1991a) Another discontinuous epitope on glycoprotein gp120 that is important in human immunodeficiency virus type 1 neutralization is identified by a monoclonal antibody. Proc Natl Acad Sci USA 88: 8949–8952
Ho DD, McKeating JA, Li XL, Moudgil T, Daar ES, Sun N-C, Robinson JE (1991b) Conformational epitope on gp120 important in CD4 binding and human immunodeficiency virus type 1 neutralization identified by a human monoclonal antibody. J Virol 65: 489–493
Holmbäck K, Kusk P, Hulgaard EF, Bugge TH, Scheibel E, Lindhardt BØ (1993) Autologous antibody response against the principal neutralizing domain of human immunodeficiency virus type 1 isolated from infected humans. J Virol 67: 1612–1619
Homsy J, Meyer M, Levy JA (1990) Serum enhancement of human immunodeficiency virus (HIV) infection correlates with disease in HIV-infected individuals. J Virol 64: 1437–1440
Hu W-S, Temin HM (1990a) Genetic consequences of packaging two RNA genomes in one retroviral particle: pseudodiploidy and high rate of genetic recombination. Proc Natl Acad Sci USA 87: 1556–1560

Hu W-S, Temin HM (1990b) Retroviral recombination and reverse transcription. Science 250: 1227–1233
Hussain KA, Issel CJ, Schnorr KL, Rwambo PM, Montelaro RC (1987) Antigenic analysis of equine infectious anemia virus (EIAV) variants by using monoclonal antibodies: epitopes of glycoprotein gp90 of EIAV stimulate neutralizing antibodies. J Virol 61: 2956–2961
Hwang WS, Boyle TJ, Lyerly HK, Cullen BR (1991) Identification of the envelope V3 loop as the primary determinant of cell tropism in HIV-1. Science 253: 71–74
Javaherian K, Langlois AJ, LaRosa GJ, Profy AT, Bolognesi DP, Herlihy WC, Putney SD, Matthews TJ (1990) Broadly neutralizing antibodies elicited by the hypervariable neutralizing determinant of HIV-1. Science 250: 1590–1593
Javaherian K, Langlois AJ, Schmidt S, Kaufmann M, Cates N, Langedijk JPM, Meloen RH, Desrosiers RC, Burns DPW, Bolognesi DP, LaRosa GJ, Putney SC (1992) The principal neutralization determinant of simian immunodeficiency virus differs from that of human immunodeficiency virus type 1. Proc Natl Acad Sci USA 89: 1418–1422
Johnson PR, Hamm TE, Goldstein S, Kitov S,Hirsch VM (1991) The genetic fate of molecularly cloned simian immunodeficiency virus in experimentally infected macaques. Virology 185: 217–228
Kang C-Y, Nara P, Chamat S, Caralli V, Ryskamp T, Haigwood N, Newman R, Köhler H (1991) Evidence for non-V3-specific neutralizing antibodies that interfere with gp120/CD4 binding in human immunodeficiency virus 1-infected humans. Proc Natl Acad Sci USA 88: 6171–6175
Kanki PJ, McLane MF, King NW Jr, Letvin NL, Hunt RD, Sehgal P, Daniel MD, Desrosiers RC, Essex M (1985) Serologic identification and characterization of a macaque T-lymphotropic retrovirus closely related to HTLV-III. Science 228: 1199–1201
Kannagi M, Kiyotaki M, Desrosiers RC, Reimann KA, King NW, Waldron LM, Letvin NL (1986) Humoral immune responses to T cell tropic retrovirus simian T lymphotropic virus type III in monkeys with experimentally induced acquired immune deficiency-like syndrome. J Clin Invest 78: 1229–1236
Kenealy WR, Matthews TJ, Ganfield M-C, Langlois AJ, Waselefsky DM, Petteway SR Jr (1989) Antibodies from human immunodeficiency virus-infected individuals bind to a short amino acid sequence that elicits neutralizing antibodies in animals. AIDS Res Hum Retroviruses 5: 173–182
Kennedy RC, Henkel RD, Pauletti D, Allan JS, Lee TH, Essex M, Dreesman GR (1986) Antiserum to a synthetic peptide recognizes the HTLV-III envelope glycoprotein. Science 231: 1556–1559
Kent KA, Rud E, Corcoran T, Powell C, Thiriart C, Collignon C, Stott EJ (1992) Identification of two neutralizing and 8 non-neutralizing epitopes on simian immunodeficiency virus envelope using monoclonal antibodies. AIDS Res Hum Retroviruses 8: 1147–1151
Kestler H, Kodama T, Ringler D, Marthas M, Pedersen N, Lackner A, Regier D, Sehgal P, Daniel M, King N, Desrosiers R (1990) Induction of AIDS in rhesus monkeys by molecularly cloned simian immunodeficiency virus. Science 248: 1109–1112
Klevjer-Anderson P, McGuire T (1982) Neutralizing antibody response of rabbit and goats to carpine arthritis-encephalitis virus. Infect Immun 38: 455–561
Kodama T, Wooley DP, Naidu YM, Kestler III HW, Daniel MD, Li Y, Desrosiers RC (1989) Significance of premature stop codons in env of simian immunodeficiency virus. J Virol 63: 4709–4714
Kodama T, Burns DPW, Silva DP, Veronese FD, Desrosiers RC (1991) Strain-specific neutralizing determinant in the transmembrane protein of simian immunodeficiency virus. J Virol 65: 2010–2018
Kodama T, Mori K, Kawahara T, Ringler DJ, Desrosiers RC (1993) Analysis of SIV sequence variation in tissues of rhesus macaques with simian AIDS. J Virol 67: 6522–6534
Kono Y (1973) Development of immunity after immunization with avirulent, attenuated and virulent equine infectious anemia viruses. In: Bryans JT, Gerber H (eds) Proceedings of the 3rd International Conference on Equine Infectious Diseases Paris. Karger, Basel
Kono Y (1988) Antigenic variation of equine infectious anemia virus as detected by virus neutralization. Arch Virol 98: 91–97
Kono Y, Kobayashi K, Fukunaga Y (1971). Serological comparison among various strains of equine infectious anemia virus. Arch Gesamte Virusforsch 34: 202–208
Kono Y, Fukunaga Y, Kobayashi K (1973a) Resistance of horses infected chronically with equine infectious anemia virus against reinfection. Natl Inst Anim Health Q (Tokyo) 13: 173–181

Kono Y, Kobayashi K, Fukunaga Y (1973b) Antigenic drift of equine infectious anemia virus in chronically infected horses. Arch Gesamte Virusforsch 41: 1–10
Laman JD, Schellekens MM, Abacioglu YH, Lewis GK, Tersmette M, Fouchier RAM, Langedijk JPM, Claassen E, Boersma WJA (1992) Variant-specific monoclonal and group-specific polyclonal human immunodeficiency virus type 1 neutralizing antibodies raised with synthetic peptides from the gp120 third variable domain. J Virol 66: 1823–1831
LaRosa GJ, Davide JP, Weinhold K, Waterbury JA, Profy AT, Lewis JA, Langlois AJ, Dreesman GR, Boswell RN, Shadduck P, Holley LH, Karplus M, Bolognesi DP, Matthews TJ, Emini EA, Putney SD (1990) Conserved sequence and structural elements in the HIV-1 principal neutralizing determinant. Science 249: 932–935
Lasky LA, Groopman JE, Fennie CW, Benz PM, Capon DJ, Dowbenko DJ, Nakamura GR, Nunes WM, Renz ME, Berman PW (1986) Neutralization of the AIDS retrovirus by antibodies to a recombinant envelope glycoprotein. Science 233: 209–212
Lasky LA, Nakamura G, Smith DH, Fennie C, Shimasaki C, Patzer E, Berman P, Gregory T, Capon DJ (1987) Delineation of a region of the human immunodeficiency virus type 1 gp120 glycoprotein critical for interaction with the CD4 receptor. Cell 50: 975–985
Layne SP, Merges MJ, Spouge JL, Dembo M, Nara PL (1991) Blocking of human immunodeficiency virus infection depends on cell density and viral stock age. J Virol 65: 3293–3330
Leonard CK, Spellman MW, Riddle L, Harris RJ, Thomas JN, Gregory TJ (1990) Assignment of intrachain disulfide bonds and characterization of potential glycosylation sites of the type 1 recombinant human immunodeficiency virus envelope glycoprotein (gp120) expressed in Chinese hamster ovary cells. J Biol Chem 265: 10373–10382
Letvin NL, Daniel MD, Sehgal PK, Desrosiers RC, Hunt RD, Waldron LM, MacKey JJ, Schmidt DK, Chalifoux LV, King NW (1985) Induction of AIDS-like disease in macaque monkeys with T-cell tropic retrovirus STLV-III. Science 230: 71–73
Li W-H, Gojobori T (1983) Rapid evolution of goat and sheep globin genes following gene duplication. Mol Biol Evol 1: 94–108
Li W-H, Wu C-I, Luo C-C (1985) A new method for estimating synonymous and nonsynonymous rates of nucleotide substitution considering the relative likelihood of nucleotide and codon changes. Mol Biol Evol 2: 150–174
Looney DJ, Fisher AG, Putney SD, Rusche JR, Redfield RR, Burke DS, Gallo RC, Wong-Staal F (1988) Type-restricted neutralization of molecular clones of human immunodeficiency virus. Science 241: 357–359
Ludlam CA, Steel CM, Cheingsong-Popov R, McClelland DBL, Tucker J, Tedder RS, Weiss RA, Philip I, Prescott RJ (1985) Human T-lymphotropic virus type III (HTLV-III) infection in seronegative haemophiliacs after transfusion of factor VIII. Lancet II: 233–236
Lutley R, Pétursson G, Pálsson PA, Georgsson G, Klein J, Nathanson N (1983) Antigenic drift in visna: virus variation during long-term infection of icelandic sheep. J Gen Virol 64: 1433–1440
Masuda T, Matsushita S, Kuroda MJ, Kannagi M, Takatsuki K, Harada S (1990) Generation of neutralization-resistant HIV-1 in vitro due to amino acid interchanges of third hypervariable env region. J Immunol 145: 3240–3246
Matsushita S, Robert-Guroff M, Rusche J, Koito A, Hattori T, Hoshino H, Javaherian K, Takatsuki K, Putney S (1988) Characterization of a human immunodeficiency virus neutralizing monoclonal antibody and mapping of the neutralizing epitope. J Virol 62: 2107–2114
Matthews T, Langlois A, Robey W, Chang N, Gallo R, Fischinger P, Bolognesi D (1986) Restricted neutralization of divergent human T-lymphotropic virus type III isolated by antibodies to the major envelope glycoprotein. Proc Natl Acad Sci USA 83: 9709–9713
McDougal JS, Cort SP, Kennedy MS, Cabridilla CD, Feorino PM, Francis DP, Hicks D, Kalyanaraman VS, Martin LS (1985) Immunoassay for the detection and quantitation of infectious human retrovirus, lymphadenopathy-associated virus (LAV). J Immunol Methods 76: 171–183
McGuire TC, Norton LK, O'Rourke KI, Cheevers WP (1988) Antigenic variation of neutralization-sensitive epitopes of caprine arthritis-encephalitis lentivirus during persistent arthritis. J Virol 62: 3488–3492
McKeating JA, Gow J, Goudsmit J, Pearl LH, Mulder C, Weiss RA (1989) Characterization of HIV-1 neutralization escape mutants. AIDS 3: 777–784
McKeating JA, McKnight A, Moore JP (1991) Differential loss of envelope glycoprotein gp120 from virions of human immunodeficiency virus type 1 isolates: effects on infectivity and neutralization. J Virol 65: 852–860

McKeating JA, Cordell J, Dean CJ, Balfe P (1992) Synergistic interaction between ligands binding to the CD4 binding site and V3 domain of human immunodeficiency virus type I gp120. Virology 191: 732–742
Meloen RH, Liskamp RM, Goudsmit J (1989) Specificity and function of the individual amino acids of an important determinant of the human immunodeficiency virus type 1 that induces neutralizing activity. J Gen Virol 70: 1505–1512
Meyerhans A, Cheynier R, Albert J, Seth M, Kwok S, Sninsky J, Morfeldt-Månson L, Asjö B, Wain-Hobson S (1989) Temporal fluctuations in HIV quasispecies in vivo are not reflected by sequential HIV isolations. Cell 58: 901–910
Montefiori DC, Zhou J, Barnes B, Lake D, Hersh EM, Masuho Y, Lefkowitz LB Jr (1991) Homotypic antibody responses to fresh clinical isolates of human immunodeficiency virus. Virology 182: 635–643
Montelaro RC, Parekh B, Orrego A, Issel CJ (1984) Antigenic variation during persistent infection by equine infectious anemia virus, a retrovirus. J Biol Chem 259: 10539–10544
Moore JP, Ho DD (1993) Antibodies to discontinuous or conformationally sensitive epitopes on the gp120 glycoprotein of human immunodeficiency virus type 1 are highly prevalent in sera of infected humans. J Virol 67: 863–875
Mori K, Ringler DJ, Kodama T, Desrosiers RC (1992) Complex determinants of macrophage tropism in *env* of simian immunodeficiency virus. J Virol 66: 2067–2075
Morrison HG, Kirchhoff F, Desrosiers RC (1993) Evidence for the cooperation of gp120 amino acids 322 and 448 in SIVmac entry. Virology 195: 167–174
Myers G, Josephs SF, Rabson AB, Smith TF, and F Wong-Staal (eds) (1988) Human retroviruses and AIDS. Los Alamos National Laboratory, Los Alamos
Naidu YM, Kestler HW III, Li Y, Butler CV, Silva DP, Schmidt DK, Troup CD, Sehgal PK, Sonigo P, Daniel MD, Desrosiers RC (1988) Characterization of infectious molecular clones of simian immunodeficiency virus (SIVmac) and human immunodeficiency virus type 2: persistent infection of rhesus monkeys with molecularly cloned SIVmac. J Virol 62: 4691–4696
Nara PL, Smit L, Dunlop N, Hatch W, Merges M, Waters D, Kelliher J, Gallo RC, Fischinger PJ, Goudsmit J (1990) Emergence of viruses resistant to neutralization by V3-specific antibodies in experimental human immunodeficiency virus type 1 IIIB infection of chimpanzees. J Virol 64: 3779–3791
Narayan O, Griffin DE, Chase J (1977a) Antigenic shift of visna virus in persistently infected sheep. Science 197: 376–378
Narayan O, Griffin DE, Silverstein AM (1977b) Slow virus infection: replication and mechanisms of persistence of visna virus in sheep. J Infect Dis 135: 800–806
Narayan O, Griffin DE, Clements JE (1978) Virus mutation during 'slow infection:' temporal development and characterization of mutants of visna virus recovered from sheep. J Gen Virol 41: 343–352
Narayan O, Clements JE, Griffin DE, Wolinsky JS (1981) Neutralizing antibody spectrum determines the antigenic pofiles of emerging mutants of visna virus. Infect Immun 32: 1045–1050
Narayan O, Sheffer D, Griffin DE, Clements J, Hess J (1984) Lack of neutralizing antibodies to caprine arthritis-encephalitis lentivirus in persistently infected goats can be overcome by immunization with inactivated Mycobacterium tuberculosis. J Virol 49: 349–355
O'Brien WA, Koyangi Y, Namazie A, Zhao J-Q, Diagne A, Idler K, Zack JA Chen ISY (1990) HIV-1 tropism for mononuclear phagocytes can be determined by regions of gp120 outside the CD4-binding domain. Nature 348: 69–73
Olshevsky U, Helseth E, Furman C, Li J, Haseltine W, Sodroski J (1990) Identification of individual human immunodeficiency virus type 1 gp120 amino acids important for CD4 receptor binding. J Virol 64: 5701–5707
Overbaugh J, Donahue PR, Quackenbush SL, Hoover EA, Mullins JI (1988) Molecular cloning of a feline leukemia virus that induces fatal immunodeficiency disease in cats. Science 239: 906–910
Overbaugh J, Rudensey LM, Papenhausen MD, Benveniste RE, Morton WR (1991) Variation in simian immunodeficiency virus env is confined to V1 and V4 during progression to simian AIDS. J Virol 65: 7025–7031
Palker TJ, Clark ME, Langlois AJ, Matthews TJ, Weinhold KJ, Randall RR, Bolognesi DP, Haynes BF (1988) Type-specific neutralization of the human immunodeficiency virus with antibodies to env-encoded synthetic peptides. Proc Natl Acad Sci USA 85: 1932–1936

Park I-W, Steen R, Li Y (1991) Characterization of multiple mRNA species of simian immunodeficiency virus from macaques in a CD4$^+$ lymphoid cell line. J Virol 65: 2987–2992

Pathak VK, Temin HM (1990a) Broad spectrum of in vivo forward mutations, hypermutations, and mutational hotspots in a retroviral shuttle vector after a single replication cycle: deletions and deletions with insertions. Proc Natl Acad Sci USA 87: 6024–6028

Pathak VK, Temin HM (1990b) Broad spectrum of in vivo forward mutations, hypermutations, and mutational hotspots in a retroviral shuttle vector after a single replication cycle: substitutions, frameshifts, and hypermutations. Proc Natl Acad Sci USA 87: 6019–6023

Payne S, Parekh B, Montelaro RC, Issel CJ (1984) Genomic alterations associated with persistent infections by equine infectious anaemia virus, a rertrovirus. J Gen Virol 65: 1395–1399

Payne SL, Fang F-D, Liu C-P, Dhruva BR, Rwambo P, Issel CJ, Montelaro RC (1987a) Antigenic variation and lentivirus persistence: variations in envelope gene sequences during EIAV infection resemble changes reported for sequential isolates of HIV. Virology 161: 321–331

Payne SL, Salinovich O, Nauman SM, Issel CJ, Montelaro RC (1987b) Course and extent of variation of equine infectious anemia virus during parallel persistent infections. J Virol 61: 1266–1270

Payne SL, Rushlow K, Dhruva BR, Issel CJ, Montelaro RC (1989) Localization of conserved and variable antigenic domains of equine infectious anemia virus envelope glycoproteins using recombinant env-encoded protein fragments produced in Escherichia coli. Virology 172: 609–615

Pétursson G, Nathanson N, Georgsson G, Panitch H, Pálsson PA (1976) Pathogenesis of visna. I. Sequential virologic, serologic, and pathologic studies. Lab Invest 35: 402–412

Posner MR, Hideshima T, Cannon T, Mukherjee M, Mayer KH, Byrn RA (1991) IgG human monoclonal antibody that reacts with HIV-1/gp120, inhibits virus binding to cells, and neutralizes infection. J Immunol 146: 4325–4332

Preston BD, Poiesz BJ, Loeb LA (1988) Fidelity of HIV-1 reverse transcriptase. Science 242: 1168–1171

Profy AT, Salinas PA, Eckler LI, Dunlop NM, Nara PL, Putney SD (1990) Epitopes recognized by the neutralizing antibodies of an HIV-1-infected individual. J Immunol 144: 4641–4647

Pulsinelli GA, Temin HM (1991) Characterization of large deletions occurring during a single round of retrovirus vector replication: novel deletion mechanism involving errors in strand transfer. J Virol 65: 4786–4797

Putney SD, Matthews TJ, Robey WG, Lynn DL, Robert-Guroff M, Mueller WT, Langlois AJ, Ghrayeb J, Petteway SR Jr, Weinhold KJ, Fischinger PJ, Wong-Staal F, Gallo RC, Bolognesi DP (1986) HTLV- III/LAV-neutralizing antibodies to an E. Coli-produced fragment of the virus envelope. Science 234: 1392–1395

Ratner L, Haseltine W, Patarca R, Livak KJ, Starcich B, Josephs SF, Doran ER, Rafalski JA, Whitehorn EA, Baumeister K, Ivanoff L, Petteway SR Jr, Pearson ML, Lautenberger JA, Papas TS, Ghrayeb J, Chang NT, Gallo RC, Wong-Staal F (1985) Complete nucleotide sequence of the AIDS virus, HTLV-III. Science 313: 277–284

Reitz MS Jr, Wilson C, Naugle C, Gallo RC, Robert-Guroff M (1988) Generation of a neutralization-resistant variant of HIV-1 is due to selection for a point mutation in the envelope gene. Cell 54: 57–63

Robert-Guroff M, Brown M, Gallo RC (1985) HTLV-III-neutralizing antibodies in patients with AIDS and AIDS-related complex. Nature 316: 72–74

Robert-Guroff M, Reitz MS Jr, Robey WG, Gallo RC (1986) In vitro generation of an HTLV-III variant by neutralizing antibody. J Immunol 137: 3306–3309

Roberts JD, Bebenek K, Kunkel TA (1988) The accuracy of reverse transcriptase from HIV-1. Science 242: 1171–1173

Robey WG, Arthur LO, Matthews TJ, Langlois A, Copeland TD, Lerche NW, Oroszlan S, Bolognesi DP, Gilden RV, Fischinger PJ (1986) Prospect for prevention of human immunodeficiency virus infection: purified 120-kDa envelope glycoprotein induces neutralizing antibody. Proc Natl Acad Sci USA 83: 7023–7027

Ronco J, Charbit A, Dedieu J-F, Mancini M, Michel M-L, Henin Y, O'Callaghan D, Kaczorek M, Girard M, Hofnung M (1991) Is there a neutralization epitope in the second conserved domain of HIV-1 envelope protein? AIDS Res Hum Retroviruses 7: 1–2

Rusche JR, Lynn DL, Robert-Guroff M, Langlois AJ, Lyerly HK, Carson H, Krohn K, Ranki A, Gallo RC, Bolognesi DP, Putney SD, Matthews TJ (1987) Humoral immune response to the entire human immunodeficiency virus envelope glycoprotein made in insect cells. Proc Natl Acad Sci USA 84: 6924–6928

Rusche JR, Javaherian K, McDanal C, Petro J, Lynn DL, Grimaila R, Langlois A, Gallo RC, Arthur LO, Fischinger PJ, Bolognesi DP, Putney SD, Matthews TJ (1988) Antibodies that inhibit fusion of human immunodeficiency virus-infected cells bind a 24-amino acid sequence of the viral envelope, gp120. Proc Natl Acad Sci USA 85: 3198–3202

Saag MS, Hahn BH, Gibbons J, Li Y, Parks ES, Parks WP, Shaw GM (1988) Extensive variation of human immunodeficiency virus type-1 in vivo. Nature 334: 440–444

Salinovich O, Payne SL, Montelaro RC, Hussain KA, Issel CJ, Schnorr KL (1986) Rapid emergence of novel antigenic and genetic variants of equine infectious anemia virus during persistent infection. J Virol 57: 71–80

Schrier RD, Gnann JW Jr, Langlois AJ, Shriver K, Nelson JA, Oldstone MBA (1988) B- and T-lymphocyte responses to an immunodominant epitope of human immunodeficiency virus. J Virol 62: 2531–2536

Scott JV, Stowring L, Haase AT, Narayan O, Vigne R (1979) Antigenic variation in visna virus. Cell 18: 321–327

Shioda T, Levy JA, Cheng-Mayer C (1991) Macrophage and T cell-line tropisms of HIV-1 are determined by specific regions of the envelope gp120 gene. Nature 349: 167–169

Simmonds P, Balfe P, Ludlam CA, Bishop JO, Brown AJL (1990a) Analysis of sequence diversity in hypervariable regions of the external glycoprotein of human immunodeficiency virus type 1. J Virol 64: 5840–5850

Simmonds P, Balfe P, Peutherer JF, Ludlam CA, Bishop JO, Brown AJL (1990b) Human immunodeficiency virus-infected individuals contain provirus in small numbers of peripheral mononuclear cells and at low copy numbers. J Virol 64: 864–872

Simmonds P, Zhang LQ, McOmish F, Balfe P, Ludlam CA, Brown AJL (1991) Discontinuous sequence change of human immunodeficiency virus (HIV) type 1 env sequences in plasma viral and lymphocyte-associated proviral populations in vivo: implications for models of HIV pathogenesis. J Virol 65: 6266–6276

Skinner MA, Ting R, Langlois AJ, Weinhold KJ, Lyerly HK, Javaherian K, Matthews TJ (1988) Characteristics of a neutralizing monoclonal antibody to the HIV envelope glycoprotein. AIDS Res Hum Retroviruses 4: 187–197

Stanley J, Bhaduri LM, Narayan O, Clements JE (1987) Topographical rearrangements of visna virus envelope glycoprotein during antigenic drift. J Virol 61: 1019–1028

Starcich BR, Hahn BH, Shaw GM, McNeely PD, Modrow S, Wold H, Parks ES, Parks WP, Josephs SF, Gallo RC, Wong-Staal F (1986) Identification and characterization of conserved and variable regions in the envelope gene of HTLV-III/LAV, the retrovirus of AIDS. Cell 45: 637–648

Steimer KS, Scandella CJ, Skiles PV, Haigwood NL (1991) Neutralization of divergent HIV-1 isolates by conformation-dependent human antibodies to gp120. Science 254: 105–108

Sun N-C, Ho DD, Sun CRY, Liou R-S, Gordon W, Fung MSC, Li X-L, Ting RC, Lee T-H, Chang NT, Chang T-W (1989) Generation and characterization of monoclonal antibodies to the putative CD4-binding domain of human immunodeficiency virus type 1 gp120. J Virol 63: 3579–3585

Thali M, Olshevsky U, Furman C, Gabuzda D, Posner M, Sodroski J (1991) Characterization of a discontinuous human immunodeficiency virus type 1 gp120 epitope recognized by a broadly reactive neutralizing human monoclonal antibody. J Virol 65: 6188–6193

Thomas EK, Weber JN, McClure J, Clapham PR, Singhal MC, Shriver MK, Weiss RA (1988) Neutralizing monoclonal antibodies to the AIDS virus. AIDS 2: 25–29

Thormar H (1963) Neutralization of visna virus by antisera from sheep. J Immunol 90: 185–192

Thormar H, Barshatzky MR, Arnesen K, Kozlowski PB (1983) The emergence of antigenic variants is a rare event in long-term visna virus infection in vivo. J Gen Virol 64: 1427–1432

Tilley SA, Honnen WJ, Racho ME, Hilgartner M, Pinter A (1991) A human monoclonal antibody against the CD4-binding site of HIV-1 gp120 exhibits potent, broadly neutralizing activity. Res Virol 142: 247–259

Tilley SA, Honnen WJ, Racho ME, Chou T-C, Pinter A (1992). Synergistic neutralization of HIV-1 by human monoclonal antibodies against the V3 loop and the CD4 binding site of gp120. AIDS Res Hum Retroviruses 8: 461–467

Torres JV, Malley A, Banapour B, Anderson DE, Axthelm MK, Gardner MB, Benjamin E (1993) An epitope on the surface envelope glycoprotein (gp130) of simian immunodeficiency virus

(SIVmac) involved in viral neutralization and T cell activation. AIDS Res Hum Retroviruses 9: 423–430
Tremblay M, Wainberg MA (1990) Neutralization of multiple HIV-1 isolates from a single subject by autologous sequential sera. J Infect Dis 162: 735–737
Veronese FD, Joseph B, Copeland TD, Oroszlan S, Gallo RC, Sarngadharan MG (1989) Identification of simian immunodeficiency virus SIVmac env gene products. J Virol 63: 1416–1419
Von Gegerfelt A, Albert J, Morfeldt-Månson L, Broliden K, Fenyö EM (1991) Isolate-specific neutralizing antibodies in patients with progressive HIV-1-related disease. Virology 185: 162–168
Wain-Hobson S, Sonigo P, Danos O, Cole S, Alizon M (1985) Nucleotide sequence of the AIDS virus, LAV. Cell 40: 9–17
Weber J, Grosse F (1989) Fidelity of human immunodeficiency virus type I reverse transcriptase in copying natural DNA. Nucleic Acids Res 17: 1379–1393
Wei ML, Cresswell P (1992) HLA-A2 molecules in an antigen processing mutant cell contain signal sequence-derived peptides. Nature 356: 443–444
Weiss RA, Clapham PR, Cheingsong-Popov R, Dalgleish AG, Carne CA, Weller IVD, Tedder RS (1985) Neutralization of human T-lymphotropic virus type III by sera of AIDS and AIDS-risk patients. Nature 316: 69–72
Weiss RA, Clapham PR, Weber JN, Dalgleish AG, Lasky LA, Berman PW (1986) Variable and conserved neutralization antigens of human immunodeficiency virus. Nature 324: 572–575
Westervelt P, Gendelman HE, Ratner L (1991) Identification of a determinant within the human immunodeficiency virus 1 surface envelope glycoprotein critical for productive infection of primary monocytes. Proc Natl Acad Sci USA 88: 3097–3101
Willey RL, Rutledge RA, Dias S, Folks T, Theodore T, Buckler CE, Martin MA (1986) Identification of conserved and divergent domains within the envelope gene of the acquired immunodeficiency syndrome retrovirus. Proc Natl Acad Sci USA 83: 5038–5042
Wolfs TFW, Zwart G, Bakker M, Valk M, Kuiken CL, Goudsmit J (1991) Naturally occurring mutations within HIV-1 V3 genomic RNA lead to antigenic variation dependent on a single amino acid substitution. Virology 185: 195–205
Wong-Staal F, Shaw GM, Hahn BH, Salahuddin SZ, Popovic M, Markham P, Redfield R, Gallo RC (1985) Genomic diversity of human T-lymphotropic virus type III (HTLV-III). Science 229: 759–762
Zhang J, Temin HM (1993) Rate and mechanism of nonhomologous recombination during a single cycle of retroviral replication. Science 259: 234–238

Towards a Vaccine Against AIDS: Lessons from Simian Immunodeficiency Virus Vaccines

E.J. STOTT

1 Introduction

The discovery in 1983 that HIV-1 is the causative agent of AIDS stimulated a widespread search for an animal model in which prophylactic vaccines and therapeutic approaches could be assessed. A wide range of mammals, including several species of non-human primates, were inoculated with strains of HIV-1 (GIBBS et al. 1986). It soon appeared that only chimpanzees and gibbon apes were susceptible to infection by HIV-1. However, their use in AIDS research was severely limited because these animals are rare and protected species. Furthermore, persistently infected chimpanzees did not appear to develop the characteristic clinical signs of AIDS. Alternative animal models for AIDS and HIV infection in man were sought.

In 1983 simian AIDS was described in a colony of macaque monkeys (LETVIN et al. 1983). Two years later a lentivirus, SIV_{mac}, was isolated from these animals and shown to induce the disease in rhesus macaques (DANIEL et al. 1985; LETVIN et al. 1985). Subsequently other strains of SIV were isolated; SIV_{sm} from sooty mangabeys (*Cercocebus atys*; FULTZ et al. 1986; LOWENSTINE et al. 1986; MURPHY-CORB et al. 1986), SIV_{mne} from pig-tailed

National Institute for Biological Standards and Control, Blanche Lane, South Mimms, Potters Bar, Hertfordshire EN6 3QG, UK

Current Topics in Microbiology and Immunology, Vol. 188

macaques (*Macaca nemestrina*; BENVENISTE et al. 1988), SIV_{agm} from East African green monkeys (*Cercopithicus aethiops*; FUKASAWA et al. 1988) and SIV_{mnd} from mandrills (Mandrillus sphinx; TSUJIMOTO et al. 1988; OHTA et al. 1988). Subsequent comparison of these SIV isolates indicated that strains of SIV_{mac}, SIV_{sm} and HIV-2 are closely related and form a discrete group of primate lentiviruses (KESTLER et al. 1988; DESROSIERS 1990). Isolates of HIV-1 and SIV strains from chimpanzees form a second group. The third group is composed of strains of SIV_{agm}, and the isolate of SIV_{mnd} comprises a fourth group. Strains of SIV_{agm} appear to be non-pathogenic. In contrast, infection of rhesus, cynomolgus or pig-tailed macaques with strains of SIV_{mac}, SIV_{sm}, SIV_{mne} or HIV-2 eventually results in clinical disease in the majority of animals.

Several features of the SIV infection of macaques are shared by HIV-1 infection in man. First, the viruses are both lentiviruses, and during infection their proviral DNA integrates into the host chromosome. Secondly, both viruses have a similar morphology and morphogenesis. Thirdly, both SIV and HIV-1 have a specific tropism for the CD4 receptor on T-helper cells and subsequently cause a significant decline in this population. Fourthly, SIV and HIV-1 have a similar, but not identical, genomic organisation. Fifthly, the viruses are more closely related than other known lentiviruses and share an antigenic relationship at least in the gag protein. Finally, both viruses cause an immunodeficiency syndrome characterised by extensive pathology in the lymphoid tissue, neurological lesions and opportunist infections. A major drawback of the SIV macaque model is that SIV is not HIV-1, and their envelope proteins are quite distinct. As a result, specific HIV-1 vaccines cannot be tested for efficacy in the SIV macaque model. Nevertheless, the model does have the major advantage that sufficient numbers of macaques are available to permit statistical analysis of experimental results.

During the past 5 years a variety of strategies for vaccination against lentiviruses has been tested in the SIV macaque model. There have been successes, setbacks and surprises, as in any scientific endeavour. Nevertheless, the model has proved to be a powerful tool with which to test approaches to AIDS vaccination and to provide a sound scientific basis for decisions about HIV-1 efficacy trials in man. While HIV-1 vaccine manufacturers have concentrated extensively on recombinant proteins of envelope, a much broader range of vaccine strategies has been examined in the SIV macaque model, including traditionally successful methods of protection against virus infection. This review describes experience with inactivated virus vaccines and the role of cellular components in protection. Experiences with subunit vaccines, delivered either as purified proteins or via live or inactive vectors, are considered. Finally, protection by live attenuated virus vaccines is described, and the exciting opportunities now available in the simian model are discussed.

2 Inactivated Virus Vaccines

The initial task of those working with SIV vaccines was to overcome the profound pessimism about the feasibility of vaccination against lentiviruses. Certainly, the obstacles to success are formidable. First, the proviral DNA of the virus is capable of integrating into the host chromosome, implying that a successful vaccine must prevent the initial interaction of virus with host cells. Secondly, both HIV and SIV infect and destroy the T-helper population of lymphocytes, an essential component of the immune system. Thirdly, the virus is capable of rapid antigenic variation. Fourthly, the primary route of infection is via the genital mucosa, a portal of entry notoriously difficult to protect. Finally, the likely vehicle of transmission is not cell-free virus but an infected cell carrying proviral DNA integrated into the chromosome. The cell does not express any viral antigen on the surface and therefore is not recognisable by a host immune system primed against the virus.

In order to demonstrate that vaccination against SIV is feasible, initial experiments were conducted with simple inactivated virus vaccines. Early experiments demonstrated that such vaccines delayed or even prevented the onset of disease when animals were subsequently challenged and a proportion were protected against infection (DESROSIERS et al. 1989; SUTJIPTO et al. 1989). In 1989 MURPHY-CORB and colleagues convincingly demonstrated that most animals immunised with formalin-killed virus are protected against infection with SIV_{sm} (B670 strain). These results were subsequently confirmed by several laboratories using virus or virus-infected cells, inactivated by aldehydes, β-propiolactone, detergent or psoralin and UV light. Several different isolates of SIV, or infectious molecular clones derived from them, were used to prepare the vaccines. A variety of adjuvants including Freund's, aluminium hydroxide, threonyl muramyl dipeptide, Quil-A (a purified saponin) and a Ribi adjuvant composed of lipoprotein A and cell wall skeleton were employed. In every case, vaccinated macaques were protected against infection by an intravenous challenge of between 10 and 50 MID_{50} (50% in monkey infectious doses). The protected animals remained free of virus, even when followed for prolonged periods of over 1 year (CARLSON et al. 1990; STOTT et al. 1990; PUTKONEN et al. 1991a, 1992; LE GRAND et al. 1992; OSTERHAUS et al. 1992; JOHNSON et al. 1992a). Even more surprising was the failure to find proviral DNA in the lymphocytes of protected animals (STOTT et al. 1990; JOHNSON et al. 1992a). Hence it appeared that the first two obstacles to successful vaccination had been overcome.

The next task was to demonstrate protection against heterologous challenge. This was achieved in a series of experiments in which vaccines prepared from SIV_{mac} were challenged with strains of SIV_{sm} and HIV-2 and vice versa (STOTT et al. 1991a; CRANAGE et al. 1992b; GARDNER and MURPHY-CORB 1990; JOHNSON et al. 1992a). These isolates share approximately 80% nucleotide sequence homology in the envelope gene but were

Table 1. Protection by inactivated vaccines against challenge with heterlogous SIV

Vaccine		Challenge			
Virus	Cells	Virus	Cells	Outcome[a]	Reference
SIV_{smB670}	H9	SIV_{mac}	HPBMC	0/4	GARDNER and MURPHY-CORB (1990)
SIV_{mac}	HUT-78	SIV_{smB670}	HPBMC	0/3	GARDNER and MURPHY-CORB (1990)
SIV_{mac}	C8166	SIV_{smB670}	HPBMC	0/4	CRANAGE et al. (1992b)
		SIV_{sm3}	HPBMC	0/4	PUTKONEN et al. (1992)
		HIV-2	SPBMC	4/4	STOTT et al. (1991a)
SIV_{smH4}	CEM × 174	SIV_{mac}	C8166	0/4	JOHNSON et al. (1992a)
HIV-1	C8166	SIV_{mac}	C8166	0/4	CRANAGE et al. (1993)
		SIV_{smB670}	HPBMC	0/4	CRANAGE et al. (1993)

All vaccines protected against challenge with homologous virus grown in human Cells. SIV_{mac} and SIV_{smH4} vaccines failed to protect against homologous virus grown in simian cells. HPBMC, Human peripheral blood mononuclear cells; SPBMC, simian peripheral blood mononuclear cells.
[a] Number infected/number challenged.

shown to be antigenically distinct with a panel of monoclonal antibodies (KENT et al. 1992). The result indicated that inactivated vaccines protect against heterologus strains of SIV, but that the protection does not extend of HIV-2. However, subsequent analysis of the origin of the challenge viruses suggests an alternative explanation for these observations (Table 1, and section on host cell proteins).

Challenge via a mucosal route was next investigated. Up to 1000 times more virus was required to infect macaques intrarectally or intravaginally, suggesting that these are relatively inefficient routes of transmission. Animals vaccinated parenterally with inactivated virus were protected against infection by intrarectal challenge (CRANAGE et al. 1992a); however, local immunisation was required to give partial protection against challenge via intravaginal route (SUTJIPTO et al. 1990; Marx, Personal Communication).

The success of inactivated vaccines given prophylactically in preventing infection encouraged attempts at therapeutic vaccination. However, immunisation of macaques between 1 and 6 months after initial infection with SIV failed to affect the course of the virus infection (GARDNER et al. 1989; STOTT et al. 1990). In a comprehensive and carefully controlled experiment using 30 rhesus monkeys neither the clinical outcome nor time to death were effected by therapeutic vaccination (Murphy-Corb et al. personal communication).

Apart from the failure of therapeutic vaccination, inactivated vaccines had proved remarkably effective in preventing SIV infection. However, immune responses which correlated with this protection were difficult to find. Although neutralising antibodies were detected in vaccinated animals, they were often low, and the levels did not correlate with protection

(Murphy-Corb et al. 1989; Carlson et al. 1990; Stott et al. 1990, 1991b); similarly there was no obvious difference in the level of T-helper cell responses in protected and unprotected animals (Mills et al. 1992, 1993; Stott et al. 1990).

3 The Role of Host Cell Proteins in Inactivated Vaccines

The surprising success of inactivated vaccines in preventing infection with SIV led to a widespread optimism that the essential protective viral component or components would soon be identified. One approach to this task was to use viral subunits derived either from native virus or by recombinant DNA technology (see next section). An alternative approach developed from the successful protective effect of fixed SIV-infected human T-cells, such as the C8166 cell line (Stott et al. 1990). As part of this strategy, groups of macaques were vaccinated with C8166 cells infected with vaccinia recombinant viruses expressing either SIV env or gag proteins. All four animals in both groups resisted challenge with SIV grown in human T-cells (unpublished observations). Simultaneously, during experiments designed to reduce the number of doses of fixed-cell vaccine required to induce immunity, animals were given two doses of SIV-infected C8166 cells, and as a control a second group were vaccinated with uninfected C8166 cells. Following challenge with 10 MID_{50} of SIV grown in human T-cells, two of the four animals vaccinated with uninfected C8166 cells were protected (Stott et al. 1991b). This experiment was repeated with a more prolonged vaccination schedule and gave the same results. Measurement of antibodies against C8166 cells by enzyme-linked immunosorbent assay revealed a highly significant correlation between protection and the titre of anti-cell antibodies in an analysis of 52 vaccinated macaques (Stott et al. 1991b). Although this analysis included some animals which had received inactivated virus vaccines, most had received fixed SIV-infected cell vaccines. Subsequent studies on macaques vaccinated with partially purified SIV have indicated that they all possess high levels of antibody to human cell components (Langlois et al. 1992; Murphy-Corb et al. 1992; Le Grand et al. 1992; Cranage et al. 1992b, 1993). However, analysis of anti-cell antibody in 35 sera from the Delta Primate Research Centre, and over 40 sera from the European Community Concerted Action revealed either no correlation with protection (Murphy-Corb et al. 1992) or a correlation which was no greater than that between antibodies to SIV envelope and protection (Cranage and Norley, personal communication).

Virtually all macaques vaccinated with partially purified SIV or with virus-infected cells were protected against 10–50 MID_{50} of virus administered after the last dose of vaccine. Over 200 macaques have been protected by such inactivated vaccines. In contrast, vaccination with uninfected cells protects only approximately half of the animals against challenge. These observations have given rise to suggestions that at least part of the protection induced by inactivated vaccines is due to viral components. However, when vaccinated animals were challenged with SIV grown in simian cells, there was no evidence of protection against infection (CRANAGE et al. 1992b; LE GRAND et al. 1992; JOHNSON et al. 1992b). There are two exceptions. A proportion of cynomolgus macaques given a prolonged vaccination with inactivated HIV-2 resisted challenge with HIV-2 grown in simian cells (PUTKONEN et al. 1991a). Similarly, four of eight vaccinated animals were protected against a challenge with 10 MID_{50} of SIV-infected simian peripheral blood mononuclear cells (OSTERHAUS et al. 1992). The protected macaques shared an MHC class I allele with the donor of the infected cells used in the challenge and cytotoxic T-cells have been invoked to explain the protection (Heeney, personal communication). Hence there is little evidence that inactivated SIV vaccines induce specific antiviral responses which can protect against cell-free virus grown in simian cells. An alternative explanation for the increased protection which virus-infected cells induce when compared with uninfected cells is that virus infection may upregulate or alter the presentation of a cellular protein which is responsible for protection against virus grown in human cells.

Animals which were vaccinated with uninfected cells and protected against SIV_{mac} which was grown in C8166 cells were given a further vaccination with uninfected C8166 cells and challenged with SIV_{sm3} which had been grown in pooled human peripheral blood mononuclear cells from at least three different donors. The animals remained uninfected, although four of four naive controls became infected. These protected animals were given a further vaccination with uninfected C8166 cells and 2 weeks later challenged a third time with SIV_{mac} grown in simian peripheral blood mononuclear cells. None were protected (unpublished observations). These results are entirely consistent with those obtained from cross-challenge of animals vaccinated with inactivated virus. Protection against heterologus SIV was always observed, provided the challenge virus was grown in human cells (Table 1). This cross-protection extended even to an HIV-1 vaccine protecting against SIVmac (CRANAGE et al. 1993). In contrast, no protection was observed against homologous or heterologous virus which had been grown in simian cells, thus the cross-protection which has been observed with SIV vaccines relates to the cellular origin of the vaccines and challenge virus rather than to any antigenic relatedness between the viruses (Table 1). It may also be relevant that the substrate used to prepare the vaccines and challenge viruses does not need to be identical, and it is unlikely, for example, that C8166 cells carried the same MHC class I and II antigens as a pool of human peripheral blood lymphocytes from three separate donors.

There is much speculation about the identity of the cellular protein or proteins responsible for protection against SIV. The incorporation of MHC antigens into retrovirus particles was described over 15 years ago (BUBBERS and LILLY 1977; AZOCAR and ESSEX 1979). In 1987 there were reports of incorporation of HLA DR into both HIV and SIV (HOXIE et al. 1987; GELDERBLOM et al. 1987; HENDERSON et al. 1987). More recently, CRANAGE et al. (1993) identified the heavy chain of HLA class I and both alpha and beta chains of HLA class II DR in vaccine preparations of HIV-1 and SIV_{mac}. ARTHUR and colleagues (1992) analysed purified immunodeficiency viruses by high-pressure liquid chromotography and suggested that molecules of HLA DR out number gp 120 on the surface of the virion by a ratio 2:1–3:1. Antibodies to human MHC class I have been identified in the sera of protected macaques which were vaccinated with inactivated virus, but antibodies to HLA DR have not so far been identified (CHAN et al. 1992; CRANAGE et al. 1993). In an effort to define the role of MHC class I or II molecules in protection, groups of four macaques were vaccinated either with mouse L-cells or with L-cells transfected with either MHC class I or II. When these animals were subsequently challenged with SIV grown in human T-cells, all became infected except two of the four vaccinated with L-cells transfected with MHC class II. Efforts to identify antibodies to HLA DR in the protected animals have so far failed (unpublished observations).

The mechanisms by which human T-cell antigens protect against challenge with SIV grown in human cells remain to be identified. Passive transfer of antibodies from macaques vaccinated with inactivated virus does confer protection on naive recipients (Gardner and Johnson, personal communication). ARTHUR et al. (1992) reported that antisera to class I, β_2-microglobulin and HLA DR inhibit infection by HIV and SIV and, by implication, suggest that this might be a mechanism for vaccine protection. However, such a hypothesis must explain how neutralising antibodies directed to host cell proteins are protective, whereas those directed against viral glycoproteins (and produced in high titres by vaccination with recombinant envelope proteins) do not protect. Furthermore, sera from macaques vaccinated with uninfected human T-cells and protected against infection by SIV do not neutralise the challenge virus (STOTT et al. 1991b).

An alternative hypothesis is that vaccination with human T-cells generates antibodies to the CD4 molecule, and that it is these antibodies which protect against infection. Immunisation of SIV-infected rhesus macaques with soluble human CD4 does elicit an antiviral response (WATANABE et al. 1991). After immunisation SIV could not be isolated easily from peripheral blood lymphocytes or bone marrow macrophages. The antibodies elicited by soluble CD4 did not neutralise SIV but did inhibit the infection of target cells which had been pre-incubated with the antibody. Antibody to soluble CD4 is unlikely to be the mechanism by which normal human T-cells protect against SIV infection since such antibodies would be expected to prevent infection by action on the host cell rather than the virus and therefore would protect against SIV grown in either human or simian cells. In addition, failure to find antibodies to CD4 in vaccinated protected macaques (CRANAGE et al.

1993) does not support the hypothesis that such antibodies are mediating the observed protection.

The initial recognition that most, if not all, of the protection induced by inactivated vaccines is due to the immune responses to host cell components rather than viral components was initially greeted with dismay and disappointment. The finding did require a re-evaluation of some of the previously obtained results. The cross-protection against heterologus viruses which had been observed was now related to the host origin of the challenge virus rather than to any antigenic characteristics. Furthermore, the failure to influence the course of infection by therapeutic vaccination of an infected animal with vaccines which had been shown to be successful prophylactically was now predictable since the vaccines were effective only against virus grown in human cells, and an infected animal was generating virus in simian cells. Despite these disappointments it is important not to overlook the positive aspects of the finding. The protection induced by inactivated vaccines is potent; neither virus nor proviral DNA is detected in the protected animals over prolonged periods of time. Furthermore, if protection could be induced against HIV by a similar mechanism, the major problem of antigenic variation of the virus would become irrelevant. A number of crucial questions must be answered before anti-cell immunity can be exploited in the development of vaccines against AIDS. The first question is, does anti-cell immunity operate in a allogeneic system as well as the xenogeneic human-simian system, so far observed? Experiments are currently underway to establish whether immunisation with allogeneic simian lymphocytes protect against challenge with SIV grown in autologous and heterologus lymphocytes. Secondly, what is the nature of the T-cell antigens which induce the observed protection? If, as seems likely, an epitope on HLA class I or II molecules is involved, it is not easy to define. Thirdly, is the mechanism relevant to HIV-1 infection in man or merely a characteristic of SIV infection in macaques? Antibodies to MHC class I and II molecules are widespread in the human population. Most mothers have generated antibodies to the alloantigens of their children. Recipients of multiple blood transfusions and some homosexuals carry high levels of antibodies to foreign MHC antigens. Studies of these populations should indicate whether there is a significant correlation between the presence of antibodies and resistance to HIV-1 infection.

4 Subunit Vaccines

The success of inactivated virus vaccines accelerated the development of subunit SIV vaccines based either on purification of native envelope protein from purified SIV or on the production of individual SIV gene products by recombinant DNA technology.

Envelope glycoprotein-enriched and core-enriched SIV subunit vaccines were prepared from detergent-disrupted, gradient purified virions by lentil lectin affinity chromatography. Immunisation with a glycoprotein-enriched preparation prevented SIV infection in two of four monkeys. The core-enriched subunit vaccine failed to prevent infection in all four vaccinates tested (MURPHY-CORB et al. 1991). Immunization with 560 µg Tween-ether disrupted SIV_{mac} in four doses protected four of seven rhesus macaques against infection. At the higher dose of 2 mg four of five macaques were protected (VOSS et al. 1992). However, the role host cell proteins was not excluded from either of these experiments, and in the second experiment when protected animals were challenged with SIV grown in simian cells, all became infected (STAHL-HENNIG et al. 1993).

Recombinant envelope protein of SIV_{mne} induced protection in four of four cynomolgus macaques against challenge with one to nine MID_{50} of a biological clone of SIV_{mne} genetically similar to the infectious molecular clone CL8 used to generate the vaccine (HU et al. 1992a). The animals were immunised at 0 and 12 weeks with a live recombinant vaccinia virus expressing gp160 of SIV. At weeks 62 and 70 the animals received booster immunisations of gp160 produced by insect cells infected with a recombinant baculovirus. Although the challenge virus used was grown in a human T-cell line, anti-cell antibodies were present at only low levels in immunised animals and did not appear to play a role in protection (HU et al. 1992b). The protection induced by this recombinant vaccine probably has a very narrow specificity since only two of the four protected macaques resisted subsequent challenge with an uncloned pool of SIV_{mne} (Hu et al., personal communication).

In contrast, attempts to protect macaques against SIV_{mac} using recombinant proteins have proved uniformally unsuccessful. Macaques immunised with yeast Ty virus-like particles expressing SIV p27 on their surface produced strong T-helper cell responses and high antibody titres to the SIV protein but were not protected against challenge with 10 MID_{50} of SIV_{mac} (MILLS et al. 1991). Similarly, animals immunised with gp160 and gag p55 produced, respectively, by vaccinia recombinant and baculovirus expression systems had strong T-helper responses and high titres of binding and neutralising antibodies but were not protected against infection when challenged with 10 MID_{50} of SIV_{mac} (MILLS et al. 1992, 1993). Recombinant gp130 produced in Chinese hamster ovary (CHO) cells, recombinant gp140 produced in insect cells or virus-like particles carrying gag and env proteins and generated by a recombinant vaccinia virus have all been used as immunogens. Despite the induction of significant immune responses and often high titres of neutralising antibody, none of these preparations protected against challenge with SIV grown in either human or simian cells (STOTT et al. 1991a and unpublished observations). In an effort to repeat the success of HU et al. (1992a), GIAVEDONI et al. (1993) immunised groups of four rhesus macaques with either (a) a live recombinant vaccinia virus expressing gp130 of the 1A11 clone of SIVmac; (b) recombinant gp130 of the 1A11 clone expressed in chinese hamster ovary cells, or (c) a combination of priming with the live recombinant vaccinia virus followed by boosting with the CHO-expressed gp130. Although all the animals developed antibodies and T-cell responses against gp130, none was protected against challenge with SIV_{mac}. A similar experiment based on the J5 infectious molecular clone of RUD et al. (1992) in which vaccine and challenge virus were genetically identical, also failed to show protection (unpublished observation).

Failure to repeat the success of HU et al. (1992a) in a series of experiments using recombinant proteins generated from SIV_{mac} suggests that the protection generated is either restricted to the SIV_{mne} system, or that it requires a close genetic relationship between vaccine and challenge virus. Neither conclusion is reassuring to those who believe that successful AIDS vaccines will be based on recombinant envelope proteins of HIV.

5 Peptide Vaccines

The progress of peptide vaccines has not matched their initial promise. Nevertheless, vaccination with peptides analogous to four conserved domains of SIV and HIV envelope proteins have yielded intriguing results (SHAFFERMAN et al. 1991). The peptides were produced as β-galatosidase fusion proteins. Two of the peptides were analogous to regions in gp120 and two to regions in the transmembrane protein gp41. Three macques immunised with these peptides produced neutralising antibodies. Following challenge with SIV_{mne}, virus was recovered from the blood of two of the three animals. By 36 months after infection the three immunised monkeys remain healthy, while two of three controls have died, and the remaining animal is virus culture positive and has declining CD4 lymphocytes. Proviral DNA was detected in the lymph nodes of all vaccinated animals. However, infection could not be transferred to three naive recipients by inoculation of viable lymph node cells and peripheral blood lymphocytes taken from the three vaccinated animals (SHAFFERMAN et al. 1993). In an extension of this work, inactivated plasma collected from either SIV-infected or peptide-vaccinated macaques were transferred into 17 naive rhesus monkeys. Following transfer these macaques, together with two naive controls, were challenged with SIV_{mne}. Virus was recovered from both controls but not from three of six recipients of SIV peptide vaccine plasma nor from 9 of 11 recipients of SIV-infected monkey plasma. More surprisingly, the 12 protected animals were re-challenged 100 days later and 4 remained uninfected, as judged by virus isolation (LEWIS et al. 1993). Analysis of the antibodies induced by the peptide vaccine indicates that the two peptides originating from gp41 are responsible for inducing neutralising anti-syncytial activity. It remains to be shown whether these antibodies alone are an adequate explanation for the results obtained.

6 Live Attenuated Vaccines

Immunisation with live, attenuated SIV was described by MARTHAS and colleagues (1990). The attenuated virus was an infectious molecular clone (1A11) derived from SIV_{mac}. Rhesus monkeys inoculated intravenously with this virus showed transient viraemia and a persistent antibody response, but no clinical disease. Three macaques were each given two intravenous inoculations of SIV_{mac} 1A11 7 months apart. They were then challenged, together with four naive control animals, with 100–1000 infectious doses of pathogenic SIV_{mac251}. All seven animals became persistently vireamic, and the four unimmunised animals developed severe clinical signs of immunodeficiency. Illness in the immunised monkeys was delayed. Thus, although the live attenuated virus did not protect macaques against infection with the pathogenic virus, it did prevent severe, early disease and death.

In 1990 Putkonen and colleagues described an experiment in which infection of cynomolgus monkeys with HIV-2 protected against the pathogenic consequences of a subsequent SIV_{sm} infection. Three macaques were infected with an avirulent strain of HIV-2 (SBL-K 135). Approximately 5 months later these animals were challenged with 10–100 MID_{50} of SIV_{sm3}. At the time of SIV challenge HIV-2 could no longer be recovered from the blood, and no clinical signs of disease were observed. During the 9 months following SIV challenge virus was recovered from the lymphocytes of the monkeys, but no disease developed. In contrast, four naive control monkeys inoculated with the same dose of SIV developed a decrease in CD4 cells and marked lymphadenopathy. Two of the four animals died. Lymph node biopsies revealed abundant viral antigen in tissues of the control monkeys but absence of SIV or HIV-2 antigens in the biopsies from the three HIV-2 pre-infected and SIV superinfected monkeys.

The most impressive live, attenuated SIV vaccine has recently been described by DANIEL et al. (1992). A deletion in the *nef* gene of the pathogenic infectious molecular clone SIV_{mac239} rendered the virus non-pathogenic. Animals infected with the nef-deleted SIV_{mac239} maintained moderate levels of antibody but very low levels of virus in the circulating lymphocytes. Animals infected with the *nef*-deleted virus have remained healthy for over 3 years. Four animals infected with the *nef*-deleted virus were subsequently challenged with 10 MID_{50} of either the virulent infectious clone SIV_{mac239} or the uncloned SIV_{mac251}. The virulent challenge viruses were recovered repeatedly from four naive macaques, all of which were dead by 36 weeks after challenge. In contrast, virulent virus was never recovered from the vaccinated animals, which all remained healthy. Two of the vaccinated, protected animals were subsequently challenged with 1000 infectious doses of uncloned SIV_{mac251}. There was no evidence of infection by the wild-type challenge virus. This dramatic protection opens up new opportunities for the SIV macaque model. If the protection is immunologically mediated, the immune correlates of protection may be identified. The viral components essential for inducing protection can also be defined by intra- and intergenic exchange within the vaccine and challenge viruses.

7 Conclusions and Future Prospects

The overwhelming majority of macaques which have been protected against infection by SIV were vaccinated with inactivated virus or virus-infected cells. This protection was almost certainly due to immune response to cell components and not to virus antigens. Nevertheless, the significance and potential of this achievement in AIDS vaccine development should not be overlooked. If the relevant cellular epitopes can be identified and used in the immunisation of man, the approach would circumvent the major problem of antigenic variation in HIV-1. Even if this ideal is unattainable, these findings emphasise that all future vaccine trials need to take account of the substrate in which antigens and challenge viruses are produced. The observations have also highlighted the extraordinary intimate relationship between immunodeficiency viruses and the host cells in which they replicate, and will have implications for future studies on protection and pathogenesis.

Vaccines based on recombinant proteins of SIV have in most cases been unsuccessful. The only exception is the work of Hu et al. (1992a), using SIV_{mne} where vaccine and challenge viruses were genetically matched. These results must raise doubts about the likely efficacy of vaccine trials in man based on recombinant envelope proteins of HIV-1. These vaccines do protect chimpanzees against HIV-1, but it remains to be seen whether the infection of chimpanzees with the IIIB strain of HIV-1 is a better predictor of the outcome of human trials than experiments with SIV in macaques.

The most impressive protection against SIV so far described was obtained using a *nef*-deleted attenuated SIV_{mac}. Clearly there are enormous questions of safety which must be answered before this approach could be used for HIV-1 infection in man. Nevertheless, further attenuation of SIV by additional deletions in non-essential genes can be effectively studied in macaques. Already, infectious SIV with a deletion in the integrase domain of the *pol* gene has been reported (Prakash et al. 1992). Clearly, attenuated SIV vaccines are potentially powerful tools with which to study protection against immunodeficiency viruses.

In summary, immune responses specific to virus, and not to cellular, antigens do appear to protect macaques. Passive transfer of serum or plasma from monkeys infected with SIV_{sm} or SIV_{mne} protects naive recipients against homologous virus (Putkonen et al. 1991b; Lewis et al. 1993). In contrast, passive transfer of serum does not protect against SIV_{mac} (Kent et al.1994; Gardner, personal communication). Inactivated vaccines protect a proportion of animals against low doses of SIV_{mac} or HIV-2 grown in simian cells (PUTKONEN et al. 1991a; OSTERHAUS et al. 1992). Vaccination with recombinant envelope protein or peptides protects against SIV_{mne} (HU et al. 1992a, b; SHAFFERMAN et al. 1992, 1993; LEWIS et al. 1993). Finally, and most impressively, live attenuated SIV protects against 1000 infectious doses of virulent virus (DANIEL et al. 1992). Thus, there remain grounds for cautious optimism that effective vaccination against immunodeficiency viruses is feasible.

A major disadvantage of the SIV-macaque model has been that it cannot be used to test the efficacy of HIV-1 vaccines. Even this problem may soon be overcome. AGY et al. (1992) have infected pig-tailed macaques with HIV-1. The infection is not pathogenic so far, and virus was recovered only transiently. More promising is the recent finding that chimaeric viruses of SIV carrying the *tat*, *rev*, *vpu* and *env* genes of HIV-1 (SHIBATA et al. 1991; LI et al. 1992; SAKURAGI et al. 1992) will replicate in cynomolgus macaques. Such viruses offer a model in which HIV-1 vaccines based on recombinant envelope proteins could be compared for efficacy against a challenge in macaques with the appropriate chimaeric virus. Chimaeric viruses expressing envelope proteins from different isolates of HIV-1 could also be used as vaccine and challenge viruses to establish the extent of cross-protection between divergent envelope sequences.

During the past 5 years the SIV macaque model has provided important information about vaccines against immunodeficiency viruses. The new discoveries of live attenuated SIV and chimaeric viruses of SIV with HIV-1 provide exciting opportunities to define the essential components of effective vaccines and the immune correlates of protection against SIV and HIV. This fundamental information is urgently required so that the most effective HIV-1 vaccines are taken forward into phase 3 efficacy trials in man.

References

Agy MB, Frumkin LR, Corey L, Coombs RW, Wolinsky SM, Koehler J, Morton WR, Katze MG (1992) Infection of Macaca nemestrina by human immunodeficiency virus type 1. Science 257: 103–106

Arthur LO, Bess JW, Sowder RC, Benveniste RE, Mann DL, Chermann J-C, Henderson LE (1992) Cellular proteins bound to immunodeficiency viruses: implications for pathogenesis and vaccines. Science 258: 1935–1938

Azocar J, Essex M (1979) Incorporation of HLA antigens into the envelope of RNA tumour viruses grown in human cells. Cancer Res 39: 3388–3391

Benveniste RE, Morton WR, Clark EA, Tsai C-C, Ochs HD, Ward JM, Kuller L, Knott WB, Hill RW, Gale MJ, Thouless ME (1988) Inoculation of baboons and macaques with simian immunodeficiency virus/mne, a primate lentivirus closely related to human immunodeficiency virus type 2. J Virol 62: 2091–2101

Bubbers JR, Lilly F (1977) Selective incorporation of H-2 antigenic determinants into Friend virus particles. Nature 266: 458–459

Carlson JR, McGraw TP, Keddie E, Yee JL, Rosenthal A, Langlois AJ, Dickover R, Donovan R, Luciw PA, Jennings MB, Gardner MB (1990) Vaccine protection of rhesus macaques against simian immunodeficiency virus infection. AIDS Res Hum Retroviruses 6: 1239–1246

Chan WL, Rodgers A, Hancock R, Taffs F, Kitchin P, Farrar G, Liew FY (1992) Protection in SIV-vaccinated monkeys correlates with anti-HLA class I antibody response. J Exp Med 176: 1203–1207

Cranage MP, Baskerville A, Ashworth LAE, Dennis M, Cook N, Sharpe S, Farrar G, Rose J, Kitchin PA, Greenaways PJ (1992a) Intrarectal challenge of macaques vaccinated with formalin-inactivated simian immunodeficiency virus. Lancet 339: 273–274

Cranage MP, Ashworth LAE, Greenaway PJ, Murphy-Corb M, Desrosiers R (1992b) AIDS vaccine development. Nature 355: 685–686

Cranage MP, Polyanskaya N, McBride B, Cook N, Ashworth LAE, Dennis M, Baskerville A, Greenaway PJ, Corcoran T, Kitchin P, Rose J, Murphey-Corb M, Desrosiers RC, Stott EJ, Farrar GH (1993) Studies on the specificity of the vaccine effect elicited by inactivated simian immunodeficiency virus. AIDS Res Hum Retroviruses 9: 13–22

Daniel MD, Letvin NL, King NW Jr, Kannagi M, Sehgal PK, Kanki PJ, Essex M and Desrosiers RC (1985) Isolation of T-cell tropic HTLV-III-like retrovirus from macaques. Science 228: 1201–1204.

Daniel MD, Kirchoff F, Czajak SC, Sehgal PK, Desrosiers RC (1992) Protective efforts of a live attenuated SIV vaccine with a deletion in the nef gene. Science 258: 1938–1941

Desrosiers RC (1990) A finger on the missing link. Nature 345: 288–289

Desrosiers RC, Wyand MS, Kodama T, Ringler DJ, Arthur LO, Sehgal PK, Letvin NL, King NW, Daniel MD (1989) Vaccine protection against simian immunodeficiency virus infection. Proc Natl Acad Sci USA 86: 6353–6357

Fukasawa M, Miura T, Hasegawa A, Morikawa S, Tsujimoto H, Miki K, Kitamura T, Hayami M (1988) Sequence of simian immunodeficiency virus from African green monkey, a new member of the HIV-SIV group. Nature 333: 457–461

Fultz PN, McClure HM, Anderson DC, Swenson RB, Anand R, Srinivasan A (1986) Isolation of a T-lymphotropic retrovirus from naturally infected sooty mangabey monkeys (Cercocebus atys). Proc Natl Acad Sci USA 83: 5286–5290

Gardner MB, Murphey-Corb M (1990) Vaccine protection of rhesus monkeys against homologous and heterologous strains of SIV. In: Grard M, Valette L (eds) Cinquieme Colloque des Cent Gardes. Fondation Marcel Merieux, Lyon pp 221–226

Gardner MB, Jennings M, Carlson JR, Lerche N, McGraw T, Luciw P, Marx P, Pedersen N (1989) Postexposure immunotherapy of simian immunodeficiency virus (SIV) infected rhesus with an SIV immunogen. J Med Primatol 18: 321–328

Gelderblom H, Reupke H, Winkel T, Kunze R, Pauli G (1987) MHC antigens: constituents of the envelopes of human and simian immunodeficiency viruses. Z Naturforsch (C) 42: 1328–1334

Giavedoni LP, Planelles V, Haigwood NL, Ahmed S, Kluge JD, Marthas ML, Gardner MB, Luciw PA, Yilma TD (1993) Immune response of rhesus macaques to recombinant simian immunodeficiency virus gp130 does not protect from challenge infection. J Virol 67: 577–583

Gibbs CJ, Gajdusek DC, Epstein LG, Asher DM, Goudsmit J (1986) In: Salzman LA (ed) Animal models of retrovirus infection and their relationship to AIDS. Academic, London, pp 457–462

Henderson LE, Sowder R, Copeland TD, Oroszlan S, Arthur LO, Robey WG, Fischinger PJ (1987) Direct identification of class II histocompatibility DR proteins in preparations of human T-cell lymphotropic virus type III. J Virol 61: 629–632

Hoxie JA, Fitzharris TP, Youngbar PR, Matthews DM, Rackowski JL, Radka SF (1987) Nonrandom association of cellular antigens with HTLV-III virions. Hum Immunol 18: 39–52

Hu S-L, Abrams K, Barber GN, Moran P, Zarling JM, Langlois AJ, Kuller LR, Morton WR, Benveniste RE, (1992a) Protection of macaques against SIV infection by subunit vaccines of SIV envelope glycoprotein gp160. Science 255: 456–459

Hu S-L, Travis BM, Stallard V, Abraams K, Mischer L, Moran P, Zarling JM, Langlois AL, Kuller L, Morton WR, Benveniste RE (1992b) Immune responses to SIV_{mne} envelope glycoproteins protect against macaques from homologus SIV infection. AIDS Res Hum Retroviruses 8: 1489–1494

Johnson PR, Montefiori DC, Goldstein S, Hamm TE, Zhou J, Kitov S, Haigwood NL, Misher L, London WT, Gerin JL, Allison A, Purcell RH, Chanock RM, Hirsch VM (1992a) Inactivated wholevirus vaccine derived from a proviral DNA clone of simian immunodeficiency virus induces high levels of neutralizing antibodies and confers protection against heterologous challenge. Proc Natl Acad Sci USA 89: 2175–2179

Johnson PR, Montefiori DE, Goldstein S, Hamm TE, Zhon J, Kitov S, Haigwood NL, Misher London WT, Gerin JL, Allison A, Purcell RH, Chanock RM, Hirsch VM (1992b) Inactivated whole SIV vaccine in macaques: evaluation of protective efficacy against challenge with cell-free virus infected cells. AIDS Res Hum Retroviruses 8: 1501–1505

Kent K, Rud E, Corcoran T, Powell C, Thiriart C, Collignon C, Stott EJ (1992) Identification of two neutralizing and 8 non-neutralizing epitopes on simian immunodeficiency virus envelope using monoclonal antibodies. Aids Res Hum Retroviruses 8: 737–741

Kent KA, Kitchin P, Mills KHG, Page M, Taffs LF, Corcoran T, Silvera P, Flanagen B, Powell C, Rose J, Ling C, Aubertin AM, Stott EJ (1994) Passive immunization of cynomolgus macaques with immune sera or a pool of neutralizing monoclonal antibodies failed to protected against challenge with SIV_{mac251}. AIDS Res Hum Retroviruses 10: 189–194

Kestler HW, Li Y, Naidu M, Butler CV, Ochs MF, Jaenel G, King NW, Daniel MD, Desrosiers RC (1988) Comparison of simian immunodeficiency virus isolates. Nature 331: 619–620

Langlois AJ, Weinhold KJ, Matthews TH, Greenberg ML, Bolognesi DP (1992) Detection of antihuman cell antibodies in sera from macaques immunized with whole inactivated virus. AIDS Res Hum Retroviruses 8: 1641–1652

Le Grand R, Vogt G, Vaslin B, Roques P, Theodoro F, Aubertin A-M, Dormont D (1992) Specific and non-specific immunity and protection of macaques against SIV infection. Vaccine 10: 873–879

Letvin NL, Eaton KA, Aldrich WR, Sehgal PK, Blake BJ, Schlossman SF, King NW, Hunt RD (1983) Acquired immunodeficiency syndrome in a colony of macaque monkeys. Proc Natl Acad Sci USA 80: 2718–2722

Letvin NL, Daniel MD, Sehgal PK, Desrosiers RC, Hunt RD, Waldron LM, MacKey JJ, Schmidt DK, Chalifoux LV, King NW (1985) Induction of AIDS-like disease in macaque monkeys with T-cell tropic retrovirus STLV-III. Science 230: 71–73

Lewis MG, Elkins WR, McCutchan EF, Benveniste RE, Lai CY, Montefiori DC, Burke DS, Eddy GA, Schafferman A (1993) Passively transferred antibodies directed against conserved regions of SIV envelope protect macaques from SIV infection. Vaccine 11: 1347–1355

Li J, Lord CI, Haseltine W, Letvin NL, Sodroski (1992) Infection of cynomolgus macaques with a chimearic HIV-1/SIVmac virus that expresses the HIV-1 envelope glycoprotein. J Acquir Immune Defic Syndr 5: 639–646

Lowenstine LJ, Pedersen NC, Higgins J, Pallis KC, Uyeda A, Marx PA, Lerche NW, Munn RJ, Gardner MB (1986) Seroepidemiologic servey of captive Old World primates for antibodies to human and simian retroviruses, and isolation of a lentivirus from sooty mangabeys (Cercocebus atys). Int J Cancer 38: 563–575

Marthas ML, Sutjipto S, Higgins J, Lohman B, Torten J, Luciou PA, Marx PA, Redersen NC (1990) Immunization with a live attenuated simian immunodeficiency virus (SIV) presents early disease, but not infection in rhesus macaques challenged with pathogenic SIV. J Virol 64: 3694–3700

Mills KHG, Barnard AL, Williams M, Page M, Ling C, Stott EJ, Silvera P, Taffs LF, Kingsman AS, Adams SE, Almond N, Kitchin PA, Jones WC (1991) Vaccine induced $CD4^+$ T-cells against the simian immunodeficiency virus gag protein. J Immunol 147: 3560–3567

Mills KHG, Page M, Chan WL, Kitchin P, Stott EJ, Taffs LF, Jones W, Rose J, Ling C, Silvera P, Corcoran T, Flanagan B, Burny A, Bex F, Delchambre M, Van Opstal O, Fabry L, Thiriart C, Delers A, DeWilde M, Bruck C (1992) Protection against SIV infection in macaques by immunization with inactivated virus from the BK28 molecular clone but not with BK28-derived recombinant env and gag proteins. J Med Primatol 21: 50–58

Mills KHG, Page M, Kitchin P, Chan WL, Jones W, Silvera P, Corcoran T, Flanagan B, Ling C, Thiriart C, DeWilde M, Bruck C, Rud E, Clarke B, Stott EJ (1993) Immunization of macaques with SIV env recombinants: specificity of T-cell and antibody responses and evaluation of protective efficacy. J Med Primatol (in press)

Murphey-Corb M, Martin LN, Rangan SRS, Baskin GB, Gormus BJ, Wolf RH, Andes WA, West M (1986) Isolation of an HTLV-III retrovirus from macaques with simian AIDS and its possible origin in asymptomatic mangabeys. Nature 321: 435–437

Murphey-Corb M, Martin LN, Davison-Fairburn B, Montelaro C, Miller M, West M, Ohkawa S, Baskin GB, Zhang J-Y, Putney SD, Allison AC, Eppstein DA (1989) A formalin-inactivated whole SIV vaccine confers protection in macaques. Science 246: 1293–1297

Murphy-Corb M, Montelaro RC, Miller MA, West M, Martin LN, Davison-Fairburn B, Ohkawa S, Baskin, Zlang JY, Miller GB (1991) Efficacy of SIV/dalta B670 glycoprotein enriched and glycoprotein depleted subunit vaccines in protecting against infection and disease in rhesus monkeys. AIDS 5: 655–662
Murphy-Corb M, Ohkawa S, Davison-Fairburn B, Martin LN, Baskin GB, Langlois AJ, McIntee M, Narayan O, Gardner MB (1992) A formalin-fixed whote SIV vaccine induces protective responses that are cross protective and durable. AIDS Res Hum Retroviruses 8: 1475–1478
Ohta Y, Masuda T, Tsujimoto H, Ishikawa K-I, Kodama T, Morikawa S, Nakai M, Honjo S, Hayami M (1988) isolation of simian immunodeficiency virus from African green monkeys and seroepidemiology survey of the virus in various non-human primates. Int J Cancer 41: 115–122
Osterhaus A, de Vries P, Morein B, Akerblom L, Heeney J (1992) Comparison of protection afforded by whole virus ISCOM versus MDP adjuvanted fromalin-inactivated SIV vaccines from i.v. cell-free or cell-associated homologous challenge. AIDS Res Hum Retroviruses 8: 1507–1510
Prakash K, Ranganathan PN, Methus R, Reddy P, Srinivasan A, Plotkin (1992) Generation of deletion mutants of simian immunodeficiency virus incapable of proviral integration. J Virol 66: 167–171
Putkonen P, Thorstensson R, Albert J, Hild K, Norrby E, Biberfeld P, Biberfeld G (1990) infection of cynomolgus monkeys with HIV-2 protects against pathogenic consequences of a subsequent simian immunodeficiency virus infection. AIDS 4: 783–789
Putkonen P, Thorstensson R, Walther L, Albert J, Akerblom L, Granquist O, Wadell G. Norby E, Biberfeld G (1991a) Vaccine protection against HIV-2 infection in cynomolgus monkey. AIDS Res Hum Retroviruses 7: 271–277
Putkonen P, Thorstensson R, Ghavamzadeh L, Albert J, Hild K, Biberfeld G et al. (1991b) Prevention of HIV-2 and SIV infection by passive immunization in cynomolgus monkeys. Nature 352: 436–438
Putkonen P, Thorstensson R, Cranage M, Nilsson C, Ghavamzadeh L, Albert J, Greenaway P, Biberfeld G (1992) A formalin inactivated whole SIVmac vaccine in Ribi adjuvant protects against homologous and heterologous SIV challenge. J Med Primatol 21: 108–112
Rud EW, Yon JR, Larder BA, Cook N, Cranage MP, Clarke BE (1992) Infectious molecular clones of SIVmac 32H: nef deletion controls ability to reisolate virus from rhesus macaques. In: Brown F et al. (eds) Vaccines 92: modern approaches to new vaccines including prevention of AIDS. Cold Sping Harbor Laboratory Press, New York, pp 229–235
Sakuragi S, Schibata R, Mukai R, Komatsu T, Kukasawa M (1992) Infection of macaque monkeys with a chimearic human and simian immunodeficiency virus. J Gen Virol 73: 2983–2987
Shafferman A, Jahrling PB, Benviniste RE, Lewis MG, Phipps TJ, Edin-McCutchan F, Sadoff J, Eddy GA, Burke DS (1991) Protection of macaques with an SIV envelope peptide vaccine based on conserved HIV-1 sequences. Proc Natl Acad Sci USA 88: 7126–7130
Shafferman A, Lewis MG, McCutchan FE, Benveniste RE, Jahrling PB, Burke DS, Eddy GA (1993) Prevention of transmission of simian immunodeficiency virus from vaccinated macaques that developed transient virus infection following challenge. Vaccine (in press)
Shibata R, Kawamura M, Sakai H, Hayami, M, Ishimoto A, Adachi A (1991) Generation of a chimeric human and simian immunodeficiency virus infectious to monkey peripheral blood mononuclear cells. J Virol 65: 3514–3520
Stahl-Hennig C, Voss G, Dittmer V, Coulibaly C, Petry H, Luke W, Hunsmann G (1993) Protection of monkeys by a split vaccine against SIVmac depends on biological properties of the challenge virus. AIDS (in press)
Stott EJ, Chan WL, Mills KHG, Page M, Taffs F, Cranage M, Greenaway P, Kitchin PA (1990) Preliminary report: protection of cynomolgus macaques against simian immunodeficiency virus by a whole fixed cell vaccine. Lancet 336: 1538–1541
Stott EJ, Cranage MP, Kitchin PA, Kent K, Farrar G, Taffs LF (1991a) Vaccination against simian immunodeficiency virus infection of macaques. In: Girard M, Valette L (eds) Sixieme Colloque des Cent Gardes. Fondation Marcel Merieux, Lyon, p. 261–266
Stott EJ, Kitchin PA, Page M, Flanagan B, Taffs LF, Chan WL, Mills KHG, Silvera P, Rodger A (1991b) Anti-cell antibody in macaques. Nature 353: 393
Sutjipto S, Pedersen NC, Miller CJ, Gardner MB, Hanson CV, Gettie A, Jennings M, Higgins J, Marx PA (1990) Inactivated simian immunodeficiency virus vaccine failed to protect

rhesus macaques from intravenous or genital mucosal infection but delayed disease in intravenously exposed animals. J Virol 64: 2290–2297

Tsujimoto H, Cooper RW, Kodama T, Fukasawa M, Miura T, Ohta Y, Ishikawa KI, Nakai M, Frost E, Roelants GE, Roffi J, Hayami M (1988) Isolation and characterisation of simian immunodeficiency virus from mandrills in Africa and its relationship to other human and simian immunodeficiency viruses. J Virol 62: 4044–4050

Voss G, Stahl-Hennig C, Petry H, Coulibaly C, Nick S, Fuschs D, Wachter H, Luke W, Hunsmann G (1992) Immunization of rhesus monkeys with high and low dose Tween-ether disrupted SIVmac. AIDS Res Hum Retroviruses 8: 1397–1400

Watanabe M, Levine CG, Shen CG, Fisher RA, Letvin NL (1991) Immunisation of simian immunodeficiency virus-infected rhesus monkeys with soluble human CD4 elicits an antiviral response. Proc Natl Acad Sci USA 88: 4616–4620

Subject Index

Current Topics in Microbiology and Immunology

Volumes published since 1989 (and still available)

Vol. 145: **Oldstone, Michael B. A. (Ed.):** Molecular Mimicry. Cross-Reactivity between Microbes and Host Proteins as a Cause of Autoimmunity. 1989. 28 figs. VII, 141 pp. ISBN 3-540-50929-1

Vol. 146: **Mestecky, Jiri; McGhee, Jerry (Ed.):** New Strategies for Oral Immunization. International Symposium at the University of Alabama at Birmingham and Molecular Engineering Associates, Inc. Birmingham, AL, USA, March 21–22, 1988. 1989. 22 figs. IX, 237 pp. ISBN 3-540-50841-4

Vol. 147: **Vogt, Peter K. (Ed.):** Oncogenes. Selected Reviews. 1989. 8 figs. VII, 172 pp. ISBN 3-540-51050-8

Vol. 148: **Vogt, Peter K. (Ed.):** Oncogenes and Retroviruses. Selected Reviews. 1989. XII, 134 pp. ISBN 3-540-51051-6

Vol. 149: **Shen-Ong, Grace L. C.; Potter, Michael; Copeland, Neal G. (Ed.):** Mechanisms in Myeloid Tumorigenesis. Workshop at the National Cancer Institute, National Institutes of Health, Bethesda, MD, USA, March 22, 1988. 1989. 42 figs. X, 172 pp. ISBN 3-540-50968-2

Vol. 150: **Jann, Klaus; Jann, Barbara (Ed.):** Bacterial Capsules. 1989. 33 figs. XII, 176 pp. ISBN 3-540-51049-4

Vol. 151: **Jann, Klaus; Jann, Barbara (Ed.):** Bacterial Adhesins. 1990. 23 figs. XII, 192 pp. ISBN 3-540-51052-4

Vol. 152: **Bosma, Melvin J.; Phillips, Robert A.; Schuler, Walter (Ed.):** The Scid Mouse. Characterization and Potential Uses. EMBO Workshop held at the Basel Institute for Immunology, Basel, Switzerland, February 20–22, 1989. 1989. 72 figs. XII, 263 pp. ISBN 3-540-51512-7

Vol. 153: **Lambris, John D. (Ed.):** The Third Component of Complement. Chemistry and Biology. 1989. 38 figs. X, 251 pp. ISBN 3-540-51513-5

Vol. 154: **McDougall, James K. (Ed.):** Cytomegaloviruses. 1990. 58 figs. IX, 286 pp. ISBN 3-540-51514-3

Vol. 155: **Kaufmann, Stefan H. E. (Ed.):** T-Cell Paradigms in Parasitic and Bacterial Infections. 1990. 24 figs. IX, 162 pp. ISBN 3-540-51515-1

Vol. 156: **Dyrberg, Thomas (Ed.):** The Role of Viruses and the Immune System in Diabetes Mellitus. 1990. 15 figs. XI, 142 pp. ISBN 3-540-51918-1

Vol. 157: **Swanstrom, Ronald; Vogt, Peter K. (Ed.):** Retroviruses. Strategies of Replication. 1990. 40 figs. XII, 260 pp. ISBN 3-540-51895-9

Vol. 158: **Muzyczka, Nicholas (Ed.):** Viral Expression Vectors. 1992. 20 figs. IX, 176 pp. ISBN 3-540-52431-2

Vol. 159: **Gray, David; Sprent, Jonathan (Ed.):** Immunological Memory. 1990. 38 figs. XII, 156 pp. ISBN 3-540-51921-1

Vol. 160: **Oldstone, Michael B. A.; Koprowski, Hilary (Eds.):** Retrovirus Infections of the Nervous System. 1990. 16 figs. XII, 176 pp. ISBN 3-540-51939-4

Vol. 161: **Racaniello, Vincent R. (Ed.):** Picornaviruses. 1990. 12 figs. X, 194 pp. ISBN 3-540-52429-0

Vol. 162: **Roy, Polly; Gorman, Barry M. (Eds.):** Bluetongue Viruses. 1990. 37 figs. X, 200 pp. ISBN 3-540-51922-X

Vol. 163: **Turner, Peter C.; Moyer, Richard W. (Eds.):** Poxviruses. 1990. 23 figs. X, 210 pp. ISBN 3-540-52430-4

Vol. 164: **Bækkeskov, Steinnun; Hansen, Bruno (Eds.):** Human Diabetes. 1990. 9 figs. X, 198 pp. ISBN 3-540-52652-8

Vol. 165: **Bothwell, Mark (Ed.):** Neuronal Growth Factors. 1991. 14 figs. IX, 173 pp. ISBN 3-540-52654-4

Vol. 166: **Potter, Michael; Melchers, Fritz (Eds.):** Mechanisms in B-Cell Neoplasia 1990. 143 figs. XIX, 380 pp. ISBN 3-540-52886-5

Vol. 167: **Kaufmann, Stefan H. E. (Ed.):** Heat Shock Proteins and Immune Response. 1991. 18 figs. IX, 214 pp. ISBN 3-540-52857-1

Vol. 168: **Mason, William S.; Seeger, Christoph (Eds.):** Hepadnaviruses. Molecular Biology and Pathogenesis. 1991. 21 figs. X, 206 pp. ISBN 3-540-53060-6

Vol. 169: **Kolakofsky, Daniel (Ed.):** Bunyaviridae. 1991. 34 figs. X, 256 pp. ISBN 3-540-53061-4

Vol. 170: **Compans, Richard W. (Ed.):** Protein Traffic in Eukaryotic Cells. Selected Reviews. 1991. 14 figs. X, 186 pp. ISBN 3-540-53631-0

Vol. 171: **Kung, Hsing-Jien; Vogt, Peter K. (Eds.):** Retroviral Insertion and Oncogene Activation. 1991. 18 figs. X, 179 pp. ISBN 3-540-53857-7

Vol. 172: **Chesebro, Bruce W. (Ed.):** Transmissible Spongiform Encephalopathies. 1991. 48 figs. X, 288 pp. ISBN 3-540-53883-6

Vol. 173: **Pfeffer, Klaus; Heeg, Klaus; Wagner, Hermann; Riethmüller, Gert (Eds.):** Function and Specificity of γ/δ T Cells. 1991. 41 figs. XII, 296 pp. ISBN 3-540-53781-3

Vol. 174: **Fleischer, Bernhard; Sjögren, Hans Olov (Eds.):** Superantigens. 1991. 13 figs. IX, 137 pp. ISBN 3-540-54205-1

Vol. 175: **Aktories, Klaus (Ed.):** ADP-Ribosylating Toxins. 1992. 23 figs. IX, 148 pp. ISBN 3-540-54598-0

Vol. 176: **Holland, John J. (Ed.):** Genetic Diversity of RNA Viruses. 1992. 34 figs. IX, 226 pp. ISBN 3-540-54652-9

Vol. 177: **Müller-Sieburg, Christa; Torok-Storb, Beverly; Visser, Jan; Storb, Rainer (Eds.):** Hematopoietic Stem Cells. 1992. 18 figs. XIII, 143 pp. ISBN 3-540-54531-X

Vol. 178: **Parker, Charles J. (Ed.):** Membrane Defenses Against Attack by Complement and Perforins. 1992. 26 figs. VIII, 188 pp. ISBN 3-540-54653-7

Vol. 179: **Rouse, Barry T. (Ed.):** Herpes Simplex Virus. 1992. 9 figs. X, 180 pp. ISBN 3-540-55066-6

Vol. 180: **Sansonetti, P. J. (Ed.):** Pathogenesis of Shigellosis. 1992. 15 figs. X, 143 pp. ISBN 3-540-55058-5

Vol. 181: **Russell, Stephen W.; Gordon, Siamon (Eds.):** Macrophage Biology and Activation. 1992. 42 figs. IX, 299 pp. ISBN 3-540-55293-6

Vol. 182: **Potter, Michael; Melchers, Fritz (Eds.):** Mechanisms in B-Cell Neoplasia. 1992. 188 figs. XX, 499 pp. ISBN 3-540-55658-3

Vol. 183: **Dimmock, Nigel J.:** Neutralization of Animal Viruses. 1993. 10 figs. VII, 149 pp. ISBN 3-540-56030-0

Vol. 184: **Dunon, Dominique; Mackay, Charles R.; Imhof, Beat A. (Eds.):** Adhesion in Leukocyte Homing and Differentiation. 1993. 37 figs. IX, 260 pp. ISBN 3-540-56756-9

Vol. 185: **Ramig, Robert F. (Ed.):** Rotaviruses. 1994. 37 figs. X, 380 pp. ISBN 3-540-56761-5

Vol. 186: **zur Hausen, Harald (Ed.):** Human Pathogenic Papillomaviruses. 1994. 37 figs. XIII, 274 pp. ISBN 3-540-57193-0

Vol. 187: **Rupprecht, Charles E.; Dietzschold, Bernhard; Koprowski, Hilary (Eds.):** Lyssaviruses. 1994. 50 figs. Approx. 350 pp. ISBN 3-540-57194-9

Springer-Verlag and the Environment

We at Springer-Verlag firmly believe that an international science publisher has a special obligation to the environment, and our corporate policies consistently reflect this conviction.

We also expect our business partners – paper mills, printers, packaging manufacturers, etc. – to commit themselves to using environmentally friendly materials and production processes.

The paper in this book is made from low- or no-chlorine pulp and is acid free, in conformance with international standards for paper permanency.

MIX
Papier aus verantwortungsvollen Quellen
Paper from responsible sources
FSC® C105338

FSC
www.fsc.org

If you have any concerns about our products,
you can contact us on
ProductSafety@springernature.com

In case Publisher is established outside the EU,
the EU authorized representative is:
Springer Nature Customer Service Center GmbH
Europaplatz 3, 69115 Heidelberg, Germany

Printed by Libri Plureos GmbH
in Hamburg, Germany